Histories of Heresy in Early Modern Europe

Histories of Heresy in Early Modern Europe

For, Against, and Beyond Persecution and Toleration

Edited by
John Christian Laursen

palgrave
macmillan

HISTORIES OF HERESY IN EARLY MODERN EUROPE

First published 2002 by PALGRAVE™
175 Fifth Avenue, New York, N.Y.10010 and
Houndmills, Basingstoke, Hampshire RG21 6XS.
Companies and representatives throughout the world

PALGRAVE is the new global publishing imprint of St. Martin's Press LLC Scholarly and Reference Division and Palgrave Publishers Ltd (formerly Macmillan Press Ltd).

ISBN 0–312–29404–2 hardback

Library of Congress Cataloging-in-Publication Data
Histories of heresy in early modern europe : for, against, and beyond persecution and toleration / edited by John Christian Laursen.
p. cm.
Includes bibliographical references and index.
ISBN 0–312–29404–2 (cloth)
1. Heresy—History. 2. Europe—Church history. I. Laursen, John Christian.

BT1313.H49 2002
273'.7'094—dc21

2002022096

A catalogue record for this book is available from the British Library.

Design by Letra Libre, Inc.

First edition: June 2002
10 9 8 7 6 5 4 3 2 1

Printed in the United States of America.

Transferred to digital printing in 2007.

Contents

Contributors

SAMMY BASU is Associate Professor of Political Science at Willamette University in Salem, Oregon. He is the author of numerous articles and the forthcoming book, *A Liberal Theory of Self-Ownership: Richard Overton and the English Revolution.*

PATRICK COLEMAN is Professor French at the University of California, Los Angeles. He is the author of *Rousseau's Political Imagination* (1984), *Reparative Realism: Mourning and Modernity in the French Novel, 1730–1830* (2001), and numerous articles.

KATHLEEN HARDESTY DOIG is Professor of French at Georgia State University in Atlanta. She is the author of *The Supplement to the Encyclopedia* (1978) and many articles and book chapters.

CLORINDA DONATO is Professor French and Italian at California State University, Long Beach. She is the author of *Inventory of De Felice's* Encyclopedie d'Yverdon (2001) in both hard copy and CD-Rom, as well as many articles and book chapters.

MARYANNE CLINE HOROWITZ is Professor of History at Occidental College in Los Angeles. She is the author of *Seeds of Virtue and Knowledge* (1998) and numerous articles and book chapters.

SALLY JENKINSON is Visiting Professor of Political Science at U.C.L.A. and the author of many articles and chapters as well as translator and editor of *Pierre Bayle: Political Writings* (2000).

JOHN CHRISTIAN LAURSEN is Professor of Political Science at the University of California at Riverside. He is the author of *The Politics of Skepticism in the Ancients, Montaigne, Hume, and Kant* (1992) and many articles; editor of several books including *Religious Toleration: "The Variety of Rites" from*

Cyrus to Defoe (1999); and co-translator of Carl Friedrich Bahrdt's *The Edict of Religion and The Story and Diary of My Imprisonment* (2000).

ANTONY MCKENNA is Professor of French Literature at the University of Saint-Etienne. He is the author of *De Pascal á Voltaire* (1990), editor of numerous books and co-editor of the journal, *La Lettre clandestine,* and author of many articles in French and English.

MARTIN MULSOW is Research Scholar at the University of Munich. He is the author or editor of numerous books and articles in English and German and of a recent *Habilitationschrift.*

JOHN G. A. POCOCK is Professor of History *Emeritus* at the Johns Hopkins University in Baltimore. He is the author of numerous books and articles including *The Machiavellian Moment* (1975) and, most recently, *Barbarism and Religion* (2 vols., 1999).

RICHARD POPKIN is Professor *Emeritus* of Philosophy at Washington University in St. Louis and Adjunct Professor of History and Philosophy at U.C.L.A. He is the author or editor of numerous books including *The History of Skepticism from Erasmus to Spinoza* (1979) and *Isaac La Peyrère (1596–1676)* (1987), as well as editor of *The Columbia History of Western Philosophy* (1999).

STACEY SEARL-CHAPIN is Assistant to the Book Review Editor of the *American Political Science Review* and teaches political theory at the University of California, Riverside.

LUISA SIMONUTTI is Research Professor at the Italian National Center for Research in Milan. She is the author of many articles in English, French, and Italian and of *Arminianesimo e tolleranza nel seicento Olandese* (1984); and editor of *Dal necesario al possibile* (2001).

MARTYN P. THOMPSON is Associate Professor of Political Science at Tulane University in New Orleans. He is the author of numerous articles and co-editor of the journal, *Politisches Denken.*

SIMONE ZURBUCHEN is a Research Scholar at Potsdam and the author or editor of several books and numerous articles and book chapters in German, French, and English, including the chapter on Religion and Society in the *Cambridge History of Eighteenth Century Philosophy.*

Acknowledgments

This book is a product of the conviction that we do not hear much when we listen to one hand clapping. We will not understand most important religious and political ideas if we look at only one side, because they are the product of debate, dialectic, and dispute. In this particular case, we will never really understand religious toleration if we do not understand persecution, and we will never understand the latter if we do not understand the reasons persecutors believe they should persecute. Part of their reasoning is based on the theory of heresy, or that which deserves persecution. That is why the study of toleration requires the study of ideas about heresy.

The editor would like to thank all of the authors for their contributions to this project. He would also like to thank the William Andrews Clark Memorial Library of the University of California at Los Angeles, and its director, Peter Reill, for sponsoring the conference in 2000 that shaped these chapters.

It is unusual for an editor to dedicate a book to two writers who have contributed chapters to a book. But without wishing in any way to diminish his appreciation of the other authors, the editor dedicates this book to John G. A. Pocock and Richard H. Popkin, in gratitude for the inspiration of their writings and conversation over many years.

Introduction

Histories of Heresy in the Seventeenth and Eighteenth Centuries

Why Were They Written, and How Were They Read?

John Christian Laursen

Heresy. *Blasphemy. Deviance. Evil.* These words evoke fear and fascination. They have always represented much that is threatening, and nevertheless much that attracts our curiosity. They are also very much the products of and participants in history. What counts as heresy, blasphemy, deviance, and evil, and what should be done about them, have changed over time, and probably will continue to change. That process of change is one of the most interesting things about these terms and the concepts they represent. This volume explores the ways in which the writing of the history of heresy contributed to the understanding of the term and its related concepts in Europe in the seventeenth and eighteenth centuries.

There were many reasons to read, and many reasons to write, a history of heresy in the seventeenth and eighteenth centuries. One context that helps us understand these reasons is the age-old debate between the forces of persecution and the voices for toleration, as we shall see in the chapters that follow.[1] Histories of heresy were a major factor in the intellectual conditions for persecution and toleration in the period. They could be written to justify either of those policies, and they could be read in accord with or at cross-purposes with their authors' intentions.

In the largest sense, the history of heresy was part of nearly every major work of theology, history, and philosophy written in Europe. It was not possible to deal with any of the largest issues in these fields without taking a stand, implicitly or explicitly, on what heresy was, how it had developed over time, who were the heretics, and what should be done about them. But there were also volumes, essays, and chapters of books dedicated specifically to the topic, and some of them are the subjects of this book. By no means have we surveyed the whole field, but we have selected a good number of influential and revealing texts. Introductions to each Part of this volume set them in the contexts of other important histories of heresy and the problematics that all of them addressed.

The Contributions of this Volume

If we may speak of the history of heresy as a literary genre, it was not monolithic. Rather, it was made up of a variety of sub-genres. As we shall see, there were hard-line attacks on heretics in the name of some outraged, self-proclaimed orthodoxy. Sammy Basu's chapter on Thomas Edwards and Stacey Searl-Chapin's chapter on Francis Lee are good examples of those who tarred their opponents with the brush of heresy (see Part One). It is fascinating to see, and perhaps not surprising in view of human psychology, that one of them—Lee—may have been all the more strident because he recognized subconsciously how close his own position was to the heretics he excoriated.

Then there were defenses of heretics and calls for toleration (see Part Two). Some, like Gabriel Naudé, tried to prove that those who had been accused of heresy were harmless victims of envy and malice. Maryanne Horowitz's chapter on Naudé reminds us that writers could not keep control over their texts: later editions of their books could be twisted by translators or editors to say things the original author probably did not intend. Martin Thompson's chapter shows that another defender of heresy, Thomas Hobbes, took a legalistic approach to defending himself. Whatever demerits it might have, he argued, heresy was not illegal in his time and place. Other authors have found a more substantial doctrine of toleration in Hobbes's writings, but Thompson emphasizes the limited nature of Hobbes's defense.[2]

Defenses of heresy could be defenses by offense. Luisa Simonutti's chapter on Phillip van Limborch shows how his *History of the Inquisition* portrays the persecutors as worse than the heretics. Sally Jenkinson's chapter on Louis de Maimbourg, Pierre Bayle, and Jean Le Clerc shows how the first and third of these called each other heretics (as well as "atheists"), but Bayle called for going beyond heresy-hunting. Bayle argued against Maimbourg that one could not rely on disputed events in history to justify persecution; that too many facts on which Maimbourg relied were uncertain; that he did not read his sources with integrity; and that alternative documents that he ignored would justify different conclusions. Against Le Clerc he argued that scholars should distance themselves from party prejudice if they want to convince. Beyond both, he argued for a thoroughgoing toleration of all who might be accused of heresy.

Part Three of this volume explores heretics who went beyond defending themselves and promoting toleration. Some of the authors of the clandestine manuscripts explored by Antony McKenna went on the offensive against the orthodox by spelling out what Bayle had largely left implicit: that all orthodoxy was the result of fantasy, ignorance, superstition, and political power. The hundreds of manuscripts in multiple copies of the underground tradition spread throughout Europe as a sort of disease, eating away at the legitimacy of the orthodox. Martin Mulsow shows how one group, the Socinians, who were accused by everyone else of heresy, could go on the offensive by rewriting history such that the self-proclaimed orthodox were the true descendants of heretics, and those accused of heresy were the truly orthodox.

Richard Popkin's chapter explores two Jewish heresies. Here we have a couple of paradoxes. On the one hand, both heresies represented major assaults on the Jewish establishment. Sabbatai Zevi proclaimed major changes in Jewish ritual and practice and Spinoza called for an end to Jewish ceremonial law. Both were accepted by many Jews and Christians at the time, even as the orthodox of both religions accused them of heresy. And on the other hand, both were influential for centuries, and the ideas of both survive to this day.

Part Four of this volume examines the campaign against enthusiasm that spread throughout Europe. Self-proclaimed moderates and centrists learned to describe their enemies at each end of the spectrum as enthusiasts. In Simone Zurbuchen's account, Heinrich Corrodi wrote a history in which chiliasts or millenarians were the latest fanatics. In his account, it would be ultimately impossible to eradicate some forms of this fanaticism. Edward Gibbon, on the other hand, made fun of both the orthodox and the heretics, calling down a plague on both their houses. In John Pocock's account, Gibbon belongs in a genre including David Hume, Voltaire, and others who thought that both of them were out of their minds.

The century of Enlightenment did not end with a victory for those who wanted to go beyond heresy and orthodoxy, beyond persecution and toleration. Rather, as we see in Part Five, something that might be called "Enlightened Orthodoxy" played a major role. In Patrick Coleman's chapter, the abbé Pluquet wrote histories of heresy that promoted the view that heresy did indeed exist, but without insisting that anything drastic needed to be done about it. Clorinda Donato's chapter on the *Encyclopedia* of Yverdon shows how Protestants could write up their enemies as heretics, without calling for holy war on them. Finally, Kay Doig's chapter on the abbé Bergier shows how he put a similar message, but from the Catholic side, into the *Methodical Encyclopedia,* which sold so widely in the nineteenth century.

A Survey of the Concepts and Terminology

In their most general senses, "heresy" stood against "orthodoxy" as the wrong way, mistaken ideas, sinful behavior. These concepts were immanently subject to debate, and their meanings could change according to time and place. Every party to every debate or controversy about heresy has contested the labeling and its implications. The interpretation of past texts is part of the process of definition of these terms, and thus this volume is part of ongoing efforts to understand them.[3] Perhaps it should be noted that all of the interpretations in this book are subject to debate. Rather than claiming the high ground as the definitive treatment of the historiography of heresy in the seventeenth and eighteenth centuries, the authors of this volume would prefer to provoke thought and stimulate further work in the area.

Heresies could be categorized in several ways. There were the historical heresies based on theological doctrines going back to ancient times. Some were named after a specific heresiarch, or leader. Those from the times of the early church included the Apollinarians, Arians, Donatists, Eutychians, Manichaeans, Montanists, Nestorians, Pelagians, Sabellians, and more. (See the glossary at the end of this book for details about each heresy.) Late medieval and early modern times produced the Hussites, Waldensians, Mennonites, Socinians, Spinozists, and more. Some heresies were named after their doctrines, for or against trinitarianism, salvation by grace or works, or transubstantiation. There were also heresies associated with activities such as prophecy, sorcery, magic, divination, alchemy, and witchcraft.

Almost everyone called their opponents or enemies heretics: it was a multi-purpose term of slander and abuse. Institutionalized religions called their rivals heretics: Protestants called Catholics heretics, and vice-versa; Lutherans called Calvinists heretics, and vice-versa; everyone else called the Quakers heretics; Catholics called the Eastern Church heretics, and vice-versa. We must learn to sort our heretics by names, doctrines, holders of the heresy, sources of the charge, and practical, political, and ideological effects.

Contemporary Applications

The nineteenth and twentieth centuries completed the process of changing our attitudes such that many people know little or nothing about the great traditions of heresy and persecution. When descendants of the Bogomil heresy in Bosnia who converted to Islam centuries ago are referred to as heretics by those who kill them, and when the Afghan Taliban accuse their ethnic rivals of heresy and blow up Buddhist statues, some observers find it hard to understand what they are thinking. This is because at least in what have been referred to as the "lucky, rich, literate democracies,"[4] neighbors of different religions do not often think of each other as heretics. This has certainly contributed a great deal to the creation of relatively tolerant societies. But the ignorance of what others are thinking makes it harder to deal with them.

A major theme in John Rawls's recent *The Law of Peoples* is the question of toleration in the international arena. He concludes that we should tolerate other liberal democracies and also peoples he is willing to call "decent"—which presumably means something like "not-quite-heretical." He is not shy about calling those he will not tolerate "outlaw" peoples or states that are "wrongful, evil, and demonic"; would he recognize that this is his modern terminology for what used to be called "heretical?"[5] Rawls is puzzled by what he calls the "redemptive antisemitism" of the Inquisition and Hitler. Indeed, he asks, but does not answer, the specific question: "How could one believe such fantasies?" (p. 22). It is probably less helpful to characterize other people's ideas as fantasies than it is to get at their roots through careful study of the history of their ideas. The answer to Rawls's question would surely be that one could believe such things if one had inherited a long tradition of ideas that supported them. In this particular case, study of the history of millenarianism would probably be more helpful than name-calling.[6] A comparison to some of the attitudes toward heresy discussed in this volume should help us understand the theoretical and practical implications of Rawls's stance.

Intolerance and treatment of people as if they were heretics is endemic everywhere. Different things might count as heresy in different places. On college campuses in the United States today, various views about race, ethnicity, and gender—including the view that these are not the only important issues and not our only significant identities—are sometimes treated as heresies. In return, in certain business and religious circles, some of the doctrines taught on campuses are treated as heresies.

And so the fights continue over who will control the National Endowment for the Humanities, the National Endowment for the Arts, and so forth. If any of those who are involved in vilifying their opponents on these questions read any of the chapters in this book, they may see analogies to themselves.

This book, naturally, does not cover the intervening two centuries since the seventeenth and eighteenth centuries, but its exploration of histories of heresy from that period provides a rich survey of the development of ideas about the history of heresy in Europe in the early modern period.

Notes

1. A companion volume to the present work is John Christian Laursen, ed., *Religious Toleration: 'The Variety of Rites' from Cyrus to Defoe* (New York: St. Martin's Press, 1999). See also Cary Nederman and Laursen, eds., *Difference and Dissent: Theories of Toleration in Medieval and Early Modern Europe* (Lanham: Rowman and Littlefield, 1996) and Laursen and Nederman, eds., *Beyond the Persecuting Society: Religious Toleration Before the Enlightenment* (Philadelphia: University of Pennsylvania Press, 1998).
2. Alan Ryan, "A More Tolerant Hobbes?" in Susan Mendus, ed., *Justifying Toleration* (Cambridge: Cambridge University Press, 1988); Richard Flathman, *Thomas Hobbes: Skepticism, Individuality and Chastened Politics* (Newbury Park: Sage, 1993); see also Glenn Burgess, "Thomas Hobbes: Religious Tolerance or Religious Indifference?" in Nederman and Laursen, eds., *Difference and Dissent,* pp. 139–161.
3. See, e.g., Melvin Richter, *The History of Political and Social Concepts* (New York: Oxford University Press, 1995).
4. Richard Rorty, *Contingency, Irony, and Solidarity* (Cambridge: Cambridge University Press, 1989), p. xv.
5. John Rawls, *The Law of Peoples* (Cambridge: Harvard University Press, 1999), p. 22.
6. See J. C. Laursen and Richard Popkin, eds., *Continental Millenarians: Catholics, Protestants, Heretics* (Dordrecht: Kluwer, 2001) and David Katz and Richard Popkin, *Messianic Revolution: Radical Religious Politics to the End of the Second Millennium* (New York: Hill and Wang, 1999). A recent special issue of the journal *Millennium* was dedicated to religion and international relations (vol. 29, no. 3, 2000). Pp. 893–924 contain reviews of other books on the mutual influences of religion and international politics.

Part One

Outraged Orthodoxy—The Hard Line

Introduction

John Christian Laursen

The stereotype of intolerance is the often widely held view that heretics are real threats and deserve to be burned at the stake. This was certainly a common position, even in earlier times. It is worth noting that it was often based on a sophisticated intellectual position. As John Pocock has reminded us, orthodoxy has its own history and it should not be assumed that it is backward, unsophisticated, and doomed from the beginning.[1] So-called Whig history, which assumes that all change and all heterodoxy is progressive, is an orthodoxy of its own that deserves appropriate skepticism.

For a baseline, let us briefly look at three histories of heresy from the sixteenth century. Bernard de Luxembourg's *Catalogue of Heretics* of 1522 described the Reformer Martin Luther as a fool, a piece of garbage, and a "toad of the Apocalypse," calling for the use of the secular sword in defense of the Church.[2] Alphonse de Castro, in *Thirteen Books Against the Heretics* (1543), defined heresy as persistence in error, and focused on the doctrinal differences.[3] Gabriel du Préau's *Of the Life, Sect, and Dogma of All Heretics* (1569) was much richer in terms of classifying almost all dissidence of any sort as heresy, and of bringing out the personal evils and socially and politically seditious nature of every heresy.[4] For example, du Préau associated the Turlupins of his own day with the ancient Cynics and Epicureans, also understood as atheists and rebels.[5] The upshot of this sort of work was that heretics did indeed deserve punishment as threats to religion and authority.[6]

Such work continued to appear throughout the following centuries. The proliferation of sects during the English Civil War provoked a flood of indignant pamphlets with titles like *A Swarme of Sectaries, and Schismatiques* and *Discovery of 29 Sects here in London. All of which, except the first are most Divelish and Damnable* (both 1641).[7] Ephraim Pagitt's *Heresiography. Or a Description and History of the Hereticks and Sectaries Sprung Up in These Latter Times* (1645) was reprinted many times, indicating that it was considered useful in a variety of contexts (London, 6th ed. 1662). Religious toleration in the Netherlands led foreigners such as J. B. Stouppe to write books like *The Religion of the Hollanders* (1673), which was really about the perceived heresy of the Netherlanders.[8] In 1692 Paul Stockmann followed in the self-proclaimed orthodox mode with *Elucidation of Heresies, Schisms, and other Opinions and Dogma, Fought with Orthodox Faith.*[9] These writers found heresy everywhere, and it is not too much to label them as heresy-mongers.

One of the most comprehensive examples of this attitude can be found in Sammy Basu's chapter on Thomas Edwards's *Gangraena* (1646). Edwards was worried about the practical results: heresies deprived the Church of tithes, undermined hierarchy, and ignored the university-educated and formally ordained preachers. He was also worried about the doctrines: adult baptism, mortalism (no afterlife), and universal salvation would lead people to sin. Both of these would lead to a failure of social discipline, "revolution," and "strangeness." As Basu interprets him, Edwards campaigned against heresy on behalf of truth. Toleration became his image of the central heresy, which would open the door to all the others. Tolerance was a sin of omission, and restraining heretics expressed the virtue of mercy by saving souls.

The polemics of the orthodox often led to paradoxical results. As contemporaries pointed out, Edwards's compulsion to record and rebuke so many heresies had the unintended consequence of providing a full catalogue of heresies that the reader might otherwise never have seen. The imperatives of historiography ran against a prudent and cautious silence that might have been more effective in discouraging the spread of heresies. Many years later, authorities in Vienna had to ban the sale of the Vatican's *Index of Prohibited Books* because people were buying it as a guide to reading.[10]

Francis Lee was another heresy-monger with paradoxical results. As we shall see in Stacey Searl-Chapin's chapter, Lee belonged to a charismatic religious group organized by Jane Leade, the Philadelphians. He wrote a history of heresy with the apparent intention of undermining the legitimacy of enthusiastic, prophetic sects by showing that they were really Montanists, named after a second-century prophet, Montanus, who was persecuted as a heretic. But the history carried him away. By the time he was finished, Manicheanism, Socinianism, Judaism, Islam, and even atheism were also traced to Montanism as the root of almost all evil. His work was published as part of a campaign against the French Prophets who had appeared in London in the early 1700s.

One irony is that Lee's own affiliation was with an enthusiastic group, and Jane Leade was as close to being a prophet as any of his targets. It is easy to see his attacks on all other prophets as a psychological defense mechanism against admitting that he, too, had been skating very close to heresy. This is a theme that will emerge again and again in the history of heresy literature.

The very fact that Jane Leade and many other religious leaders were women evidently stimulated widespread social-psychological insecurities. In the same period Johann Heinrich Feustking published an attack on their pretensions as *Fanatical Women Heretics, or, History and Description of the False Female Prophets, Quakers, Fanatics . . .* (1704).[11] Jakob Friedrich Reimmann thought that women should be forbidden to read the fanatical, spiritualist, radical Pietists.[12] This attitude goes back at least to Wycliff, who wrongly traced the etymology of "blasphemy," the term for the performative or communicative dimension of heresy, to "blas-femina," foolish or harmful woman.[13]

But not all of the attacks on heretics were couched in the language of indignation, gross exaggeration, or gender paranoia. By the end of the seventeenth century Bishop Bossuet's histories of schisms and Louis de Maimbourg's many volumes of history of heresy could achieve the same purpose with a calmer, more magisterial prose, as we shall see in Sally Jenkinson's chapter.

Notes

1. J. G. A. Pocock, "Within the margins: the definitions of orthodoxy" in R. D. Lund, ed., *The Margins of Orthodoxy: Heterodox Writing and Cultural Response, 1660–1750* (Cambridge: Cambridge University Press, 1995), pp. 33–53. See also Justin Champion, "Introduction" to John Toland, *Nazarenus* (Oxford: Voltaire Foundation, 1999), p. 49: "'criticism' was the product of a battle over rival processes of cultural authentification. 'Criticism' was not a discourse used solely by the heterodox against the status quo."
2. F. Bernardo Lutzenburgo, *Catalogus haereticorum* (Cologne, 1522). See Marie-Madeleine Fragonard, "Le danger des hérésies: Le risque social évalué par les Dictionnaires des Hérésies au XVIème siècle" in Michel

Peronnet, ed., *Protestantisme et Révolution* (Montpellier: Sauramps, 1990), pp. 24.

3. *Adversus Haereses libri XIII* (Paris, 1543). See Fragonard, "Le danger," p. 24.
4. *De vitis, sectis et dogmatibus omnium haereticorum* (Cologne, 1569). See Fragonard, "Le danger," pp. 25–30.
5. See R. Bracht Branham and Marie-Odile Goulet-Cazé, eds., *The Cynics* (Berkeley: University of California Press, 1996), p. 258.
6. See, further, Konrad Schlüsselburg, *Catalogus Haereticorum* (Frankfurt, 1597).
7. John Taylor, *A Swarme of Sectaries, And Schismatiques: Wherein is discovered the strange preaching (or prating) of such as are by their trades Coblers, Tinkers, Pedlers, Weavers, Sowergelders, and Chimney Sweepers* (London, 1641); Anonymous, *Discovery of 29, Sects here in London. All of which, except the first are most Divelish and Damnable* (London, 1641).
8. *La religion des Hollandais* (Cologne, 1673).
9. *Elucidarius Haeresium, Schismatum Aliarumque Opinionum et Dogmatum cum Fide Orthodoxa Pugnantium* (1692).
10. See Leslie Bodi, *Tauwetter in Wien* (Frankfurt: Fischer, 1977), p. 57.
11. *Gynaeceum haeretico fanaticum, oder, Historie und Beschreibung der falschen Prophetinnen, Quackerinnen, Schwärmerinnenn . . .* (Leipzig, 1704).
12. Hanspeter Marti, "Litterärhistorie und Ketzergeschichte" in M. Mulsow and H. Zedelmaier, eds., *Skepsis, Providenz, Polyhistorie: Jakob Friedrich Reimmann (1668–1743)* (Tübingen: Niemeyer, 1998), p. 72n.
13. See David Lawton, *Blasphemy* (Philadelphia: University of Pennsylvania Press, 1993), p. 105; Alain Cabantous, *Histoire du blasphème en occident, fin xvie milieu xixe siècle* (Paris: Albin Michel, 1998).

1

"We are in strange hands, and things are come to a strange passe"

Argument and Rhetoric Against Heresy in Thomas Edwards's *Gangraena* [1646]

Sammy Basu

I am perswaded that the Histories of these times will be so strange to after Ages, that though Truth herselfe be the Compiler, yet there will need the Sanction of some publike Act or Statute, to command incredulous posterity to beleeve them.[1]

The English Civil Wars of the 1640s and the paper battles that preceded and accompanied them involved deeds and words too strange for sensible history. Those who undertook to write such a history of heresy, then, had to do so not merely to make sense to their contemporaries but more so for a posterity that would view their unprecedented age with horror and disbelief. Such was the aim of the London heresiographer, Thomas Edwards (1593–1647), a prominent spokesman for the Presbyterian orthodoxy during the mid-1640s. Edwards was determined for "the benefit both of the present and succeeding generations, to write an Historicall narration of all the Proceedings and wayes of the English Sectaries."[2] The result was *Gangraena: or A Catalogue and Discovery of many of the Errours, Heresies, Blasphemies and pernicious Practices of the Sectaries of this time, vented and acted in England in these last four years* (1646). In what follows I shall parse from Edwards's text some of the major orthodox objections he advanced against heresy, and rely in part upon what struck responsive chords with sympathetic and critical contemporaneous readers. My hope thereby is to provide a robust heuristic schema of mid-century Presbyterian argument and rhetoric against heresy.[3]

"[I]n *England,*" observed Robinson, "it is ordinary with Protestants, to reproach one another with the nickname of Puritan or Separatist, Presbyterian or Independent, . . ."[4] From the onset of the 1640s, rival visions of a New Jerusalem split English Protestants. "Presbyterians" disowned Laudian Episcopacy but wanted a hierarchical national church in control of ordination and excommunication. "Independents" preferred a consultative established church and independent "gathered" congregations. By 1646, after heated and increasingly public debates,[5] the Presbyterian Westminster Assembly of Divines prevailed, leaving the Independents to become increasingly militant.

On the margins of the Independents, but beyond their sympathies, separatist sects dubbed "Sectaries" proliferated. They coalesced locally and idiosyncratically, typically around an inspiring "mechanic" (i.e., manual worker turned prophet). Though their total membership was less than 5 percent of the population, they caused great alarm. Already in 1641, Taylor had published his report of *A Swarme of Sectaries and Schismatiques,* and the Anonymous *Discovery* had identified 29 sects. That same year, the anonymous author of *The dolefull Lamentation* observed that "It is easier to reckon up all the Species and kinds of nature, than to describe all the Sects, Divisions, and opinions in Religion, that now is in *London.*"[6]

Edwards, a Cambridge-trained Divine and noted preacher against both Laud and Independency, was an energetic but not untypical Calvinist Presbyterian.[7] In August 1641, he had published his 60-plus page *Reasons Against the Independant Government of particular Congregations,* and in July 1644, a 300-plus page *Antapologia,* against the Independents' *Apologeticall Narration.* In the latter he predicted a second civil war, this time not between the king and Parliament but among the people themselves over religion.[8] His 650-plus page *Gangraena* was the most vigorous and thorough of Presbyterian efforts to catalogue heretical beliefs, describe heretical practices, and name Independent and Sectarian heretics. It consisted of Edwards's firsthand experiences, those of orthodox informants (whose testimony he assures he corroborated) and, remarkably, extensive (and faithful) reprints of "their [the Independents and Sectaries] own words and phrases syllabically." He evidently thought that properly framed public exposure would illicit horror in his godly readers, and the recantation of some of those named.[9] *Gangraena* was published in three parts: on 26 February 1646, 28 May 1646, and 28 December 1646; with the second responding to several published critics, including John Saltmarsh, William Walwyn, and John Goodwin; and the third to the political implications of the future leaders of the Leveller movement, John Lilburne, Richard Overton, et al.[10]

According to Edwards, the Reformation was giving way to deformation. The rapid proliferation of heresies over the course of 1646—from 176–80 to 199 to 300 by Edwards's count—made the situation all the more urgent. Critics could only rather blandly contend that such numbering was wholly arbitrary. Edwards's aim accordingly was to sway public opinion against toleration, shock Parliament into punitive action, and further solidify a Presbyterian settlement of the Kingdom.

If language can be thought of as a realm of practice, words as acts, and texts as events, this was especially true during the early 1640s, when the collapse of censorship gave newly self-conscious authors and opportunistic peripatetic printers access to a burgeoning literate and politicized public. "In this last and worst of ages," bemoaned an anonymous author, "wee set on foot a worser fiercer warre, a warre of the pen; every man that hath but sufficient ability to spell his owne name right, and to subscribe to an acquittance, summons his wits, and hee will needs bee invective against some one, and divulge his folly in print."[11] Mid-century pamphleteering was indeed genre-bending urban warfare: intellectually mobile and rhetorically rapid-fire. Not only were the fundamental religious, social, cultural, economic, and political institutions called into question, but the very vocabularies used to characterize them were in flux. New kinds of authors and new conceptions of authorizing and authority advanced in tandem with new sorts of readers and expectations.

Edwards, for his part, voiced the frustration that while he labored assiduously to produce his antidote, *Gangraena,* the gangrene of heresy continued to be spread by "the weekly Pampleteers venting some desperate passages, and putting forth strange books upon the nick of things," which were then sold indiscriminately on street corners, so that "some scores of such Books written by Sectaries being in all mens hands." Still, he resolved to "maintain this warfare at my own charges," and produce a discourse that was "short, a manuall that might be for every ones reading."[12]

Even as he argued that Sectarian claims of persecution were grossly exaggerated, Edwards called upon Parliament to "remedy all."[13] Why remedy? Undeniably, much of Edwards's concern was institutional and hence arguably "ideological."[14] The Independents were direct competitors, and the Sectaries advocated a form of church organization—absent tithes, hierarchy, or ordained preachers—that rendered him obsolete.

By the mid-1640s, anti-tithe agitation was widespread among Sectaries. Edwards noted with alarm, and regarded as heretical, the vigorous pamphlets and sermons against tithes, and condemned the refusal to pay tithes and beating of tithe collectors. The Sectaries not only rejected tithes but hierarchical "church" intermediation. Instead, they "gathered" in self-governing meetings. While they might defend their voluntary "agreement" on the grounds of spiritual equality, to Edwards such meetings were rude, raucous, theatrical, and frightening venues. More problematically still, some of the most overt statements for discursive, democratic politics would subsequently issue from them.[15] Finally, to abolish the state-sanctioned ordination on preaching meant that anyone, including uneducated laborers and women, could gather listeners "without all control."[16] "Control" required a settled church-government and orthodox university-trained ministers maintained by parishes.[17]

Having acknowledged Edwards's institutional objections to heretics, I want to concentrate on the intellectual and rhetorical dimensions of *Gangraena*. In 1644, Edwards had accused the Independents of writing a "Popular and Rhetoricall Discourse to take the people with, and to vindicate your selves and way" using invented and insinuating words and phrases.[18] In the subsequent months, London continued to be littered with "popularizing" texts. In *Gangraena,* then, Edwards finally fully mobilized his own rhetorical arsenal, which I shall characterize as three sorts of "slippery-slope" arguments: "error," which necessitated an insistence on "truth" and "doctrines"; "sin," which warranted rigorous "discipline"; "revolution," which required resistance to the "new" and "strange"; and a fourth motivational argument, according to which the imperatives of other-regarding "mercy" and self-preserving "dependency" converged.

Error

According to this first slippery-slope argument, "one absurdity being granted, a thousand follow." Edwards made the same point in *Reasons:* "upon the same grounds . . . if ever the dore of Tolerations should be but a little opened, there would be a great crouding by al sorts to enter in at it." That is, as a logical matter, it is difficult to justify why one tiny incremental error beyond the esteemed norm is so terrible. But if one permits the norm +1, then why not the new norm (norm +1) +1, another marginal difference, and so on?[19] The only way to prevent such slippage was to hold fast to "truth" and "doctrines."

Truth

A central feature of arguments for religious unity (and against toleration) was insistence on the incontrovertible obviousness of some fundamental truths. Members of the Assembly of Divines possessed an infallible sense of assurance. The goal for Presbyterians and Independents alike was unity in truth and against error. For Edwards, "truth is one and uniforme," "the narrow and straight way," and yet "All error now is christned and call'd Conscience. . . . Some say they know no Errors nor Sects at all."[20]

In contrast, those Sectaries who argued for toleration tried to drive a wedge of doubt between truth and purposive action. Goodwin, for example, argued that it was imprudent to act against that "which they know not certainly whether it be from God, or no." Robinson reminded of human partiality: "*Is the same minde in you which is in Christ Jesus?* Unless you have attained to this perfection, according to the measure and fulnesse of Christ Jesus, there is yet a part" that is all too human. Walwyn responded to Edwards that "by heresie you understand all doctrines that are not agreeable unto yours: though you are not infallible."[21] To Edwards, however, the claims of fallibility and uncertainty were "grosse Scepticisme," and tantamount to the overthrow of all religion, to "Atheisme," "as if there were no received known principls and Doctrines of Christian Religion, lay'd down so plainly and clearly in Scriptures."[22]

The Parable of the Tares (Matthew 13.24–30, 36–43) was invoked regularly by "Tolerationists," such as Williams and Overton, and just as readily refuted by critics. The former argued that one must suffer Tares to grow amid Wheat until the 11th Hour or very Day of Harvest. The latter contended that "when a man is sure that he plucks up nothing but tares, or rather thornes," such forbearance is unnecessary.[23]

Some Tolerationists also professed radically individualistic conceptions of the relationship between God and man, while regarding with suspicion the actual assertion of divine transparency. That is, the inspirational interpretation of the divine will—be it in the form of providence, the historical record, or specific scriptural doctrine—is fraught with epistemological obscurities. Consequently, molesting others on one's own theological grounds is unjustifiable. Instead, they stressed experimentation, trying all things, and the salience of being fully persuaded in one's own mind. Edwards also had his doubts, of course, about those Sectaries who "plead miracles, revelations, visions for their way," and invent millenarian futures. Still, he was not above perusing providence himself for assurances large and small (such as a monstrous birth) of the Sectaries' impending demise.[24]

If contemporary events might be read for inklings into the divine mind, Edwards also turned to the ecclesiastical and secular histories to criticize contemporary heretics. According to him, history demonstrated that the Erastian carnal policy of suffering heretics contravenes disastrously both divine purpose and mundane prudence. The Anabaptists' violent, polygamous, communistic chiliastic misrule over Münster in 1534 was almost ritually invoked, and the label extended not only to its nominal adherents, the Particular (Calvinist) Baptists[25] and General (Arminian) Baptists, but to anyone beyond the Lutheran and Calvinist folds.[26]

Tolerationists, for their part, submitted a historical counter-record of the effects of persecution. They also challenged Edwards's sources. Whence, Overton asked: "who writ the Histories of the Anabaptists but their Enemies?" Although Edwards had himself noted the partiality of some historians, e.g., that the *Moderate Intelligencer* is "the great Historian and Chronicler of the Sectaries," he argued that by using such reasoning the Sectaries "will cry all downe for false written of themselves."[27]

To Tolerationists, underlying this specific debate was the larger issue of the epistemological status of history. They contended that historical knowledge is cumulative and progressive but unfinished, and that orthodoxy/heresy are implicated in this contingency: "That those tenets which are now accounted heresies, may be in the countenanced truthes of the next age; as what formerly was accounted errour is now esteemed truth."[28] This was not much of an argument for Edwards. Though he had noted that Calvin "was called a Heretick," he nonetheless regarded such historical shifts as further reason for increased fidelity to the now authoritatively (and correctly) orthodox.[29] Thus some among the Sectaries came to reject history, typically written by the winners, as too ideological and contingent a matter for moral and political foundations.[30]

Doctrines

"Its a sad thing to thinke," lamented one of Edwards's correspondents, "that so much blood hath been spilt, and vast sums of money spent, if in the end we should have a Toleration; Oh what promises have we had of Vniformitie in Religion, both in Doctrine and Discipline!" To Edwards toleration was the central and all-encompassing heresy that "overthrows all relations, both Politicall, Ecclesiasticall, and Oeconomicall." Thus, all the great Ministers of the Reformed Churches: Calvin, Zwingli, Martyr, Melanchthon, de Bèze, Knox, Bullinger, and (the mature) Luther, had been against it; an observation Tolerationists had difficulty refuting.[31]

In fact, few meant by "toleration" actual religious "liberty." That is, the pragmatic willingness to accept less than absolute unanimity in doctrine and rite did not entail the liberty to adhere to any beliefs or subscribe to any practices. Although Edwards, like Prynne, sought to damn the Independents by association with the Sectaries, the Independents were just as insistent on godly discipline. When Independents such as Goodwin or Burroughes avowed toleration, it was for themselves, not everyone; or at best, like Robinson, for all Protestants. Whence, Walwyn could observe that the Independents were "one in doctrine" and "very little differing . . . in Discipline" with the Presbyterians.[32] Rare indeed was the full-fledged inclusionary defense of liberty of conscience, of the sort broached by Walwyn, Williams, and Overton, for all Protestants and Catholics, Muslims, Jews, and Infidels.[33]

The controversy over heresy obviously involved specific doctrines. In addition to the conspicuous and central heresies—repudiating the status of Scripture (by setting the New Testament against the Old, allegorizing, or asserting its multiplicity, opacity and even irrelevance), the Trinity, and Jesus (by characterizing him as human or even fallen)—several specific ones struck Edwards as especially problematic: adult baptism, mortalism, universal salvation, *apokastasis,* and ultimately unbelief.

It was heretical to be against paedobaptism and to refuse to catechize children. Notoriously, some reckless Sectaries profaned the sacrament of baptism by submitting their domesticated cats and horses, and soiling the liturgical accessories with urine. Others, the Baptists, practiced "dipping," i.e., bodily immersion during an exclusively adult rite of baptism. Edwards responded to dipping, as did other Puritans, by imputing ulterior sexual motives, and claiming that those dipped invariably (and appropriately) drowned or subsequently died.[34]

The Mortalist heresy—orally transmitted for centuries but newly in print in works such as Overton's *Mans Mortalitie* (1644)—came in several forms. In all of its variants, however, argued Edwards and other critics, it justified this-worldly, beastly, Epicurean hedonism.[35]

The seemingly ubiquitous heresy of universal salvation and resurrection contradicted the Calvinist doctrines of election and predestination. According to Edwards, to hold that all might be saved or that through good works one might receive grace "does as the Apostle *Peter* speaks, *with fained words make marchandize of mens soules.*"[36]

Apokastasis was the most radical version of the heresy of universal salvation, according to which "all other *Creatures* as well as Man shall be raised and delivered from death at the Resurrection." The orthodox view held that man, by virtue of his constitutive combination of soul and body, spiritual and material, was at the apex of the hierarchical chain of mundane being, below angels but above (and dominant over) the animal world. Hence, the claims that men and beasts are endowed with the same essence, and that before God a saint was no better than a toad, were extremely offensive. The *imago dei* was sullied by association.[37]

Unbelief was heresy at its most dramatic. Edwards saw atheism lurking in the heart and mind of many a Sectary. It is here, if anywhere, that he exaggerated. Even those heretics who were prone to accept all forms of belief excepted unbelief. Nothing spoke to the shared fears of the otherwise disparate godly more pointedly than the threat of articulate irreligion. There could be no arguments for the liberty of conscience of atheists because without God, conscience had no functional reality. Moreover, unbelief undermined the moral order generally. That is, by definition amoral, atheists could not but be loose threads in the fabric of society, and threats to the public trust.[38]

Sin

According to the second slippery-slope argument, "little sins make way for great, and one brings in all." Psychologically, having indulged in a transgressive pleasure, one is tempted to edge downward, mixing and compounding misbehaviors, and seducing others in the process. Indeed, according to Edwards, this was "the true reason and ground of that great growth and increase of the Sectaries among us." Such concern about moral corruption was a recurring motif in the many pamphlets that warned against and listed new sects in a descending order from the mistaken to the malevolent.[39] The only corrective was renewed discipline.

Discipline

Discipline was vital to social stability and spiritual solidarity. Given Adamic origins, man is prone to terrestrial temptations and is inherently rebellious. Indeed, Edwards maintained that "God appointed government both Ecclesiastical and Civil to restrain mens nature and wantonesse, as knowing what men would fall to, that there be no end of his follies and madnes if tolerated." So, much as Parliament in the 1620s had "kept them [Arminian heretics] in awe, and was a rod over them," it must, again, "remonstrate against, punish, suppresse."[40]

Liberty, on the other hand, tends to heresy and sin. Edwards presented a litany of the Sectaries' sins and specific sexual "uncleannesses," all deserving of divine and political sanctions. He also offered an acute rendering of the psychological decay and moral indiscipline that accompanies uncertainty and doctrinal variation:

> They generally walke loosely and at large, over what they did before they turned Sectaries, and in comparison of the godly Presbyterians; they do many things under pretence of Christian liberty, which professours hereto-

> fore were not wont to do, nor do not, neither durst they have done, of which I could give many instances both in persons and things; I do not know, nor hear of a Sectary in *England,* that is so strict and exact in his life, as he was before, and as thousands of the Presbyterians are; and this is not my observation alone, but a generall observation: many of them play at Cards and Tables, are very loose on the Sabbath days, go to Bowls and other sports on dayes of publike Thankesgiving . . . they wear strange long hair, go in such fine fashionable apparell beyond their places as 'tis a shame; they will feast, ride journeys, do servile businesses on the Fast dayes, and give their Parish Churches no Sermons, no Prayers at all on those dayes; they make little conscience of family duties, they will sit and tipple, be joviall and merry together. I could tell true and certain stories of many Sectaries who were exceeding precise and strict before they fell into those wayes, but are abominable loose now; and let but a man turne Sectary now adayes, and within one halfe year he is so metamorphosed in apparell, hair, & c. as a man hardly knows him.[41]

One is struck, in reading *Gangraena,* by the frequency of such objections not only to the doctrines but to the indiscipline of the Sectaries qua heretics vis-à-vis the Sabbath, playing, theater, dress, and even laughter.

Among Puritan rites, the Sabbath and the days of public humiliation or thanksgiving were revered. Puritans had been profoundly offended by James I's *Book of Sports* (1617), and in a Parliamentary Bill of 1621 sought restrictions on Sunday activities. Charles I, unlike his father, endorsed Sabbath legislation in 1625 and 1628, although he also saw to the reissue of the *Book of Sports* in 1633. During the 1640s there were frequent demands, like A. B.'s, for *The Sabbath Truly Sanctified* (1645), and accusations of Sabbatarian lapses such as those of Edwards against Lilburne and Goodwin, were particularly rhetorically damning.[42]

In effect, however, Edwards and others called for Sabbatarianism seven days a week. That is, godly discipline ought to order a daily life purged of frivolous recreations. Gaming, as Philpot explained, was the start of a downward slope through "drunkennesse," "lying," "swearing," "adultery," "beggery," and "theevery," to "eternall damnation."[43]

If playing was objectionable, so were plays. The Elizabethan and Stuart stages were liminal spaces. Set up beyond the city walls, in the suburbs or "Liberties" as they were known, amid trade establishments and brothels, these theaters were also beyond the rules and reach of the City proper, and did indeed take liberties. That is, plays could be staged that challenged the political, social, and sexual parameters of the status quo. Not surprisingly then, given a long-standing Puritan critique of their open prurience, Parliament closed the public theaters in 1641. To Edwards, affection for the theater was symptomatic of Sectary libertinism.[44]

Dress and deportment were also problematic. Erasmus's adage, "Vestis virum facit" (*Adagia,* 731E), was as true here as ever. Clothing and carriage distinguished the gendered, socio-economic, and even political orders of society (e.g., the long-haired and sartorially excessive Royalist "cavaliers" contrasted, in an overdrawn stereotype, against the severely pruned and modestly attired Puritan "roundheads"). To Edwards, socially transgressive dressing was an unsettling manifestation of body- and self-obsession in contravention of divine intentions.[45]

All in all, for Edwards "damnable heresies, strange opinions, fearfull divisions, loosenesse of life and manners" went together. One wants to avoid the caricature of the humorless Puritan, but Edwards's juxtaposition of strictness, exactitude, precision, and circumspection against looseness, levity, irregularity, ease, and pleasure, invites it. Indeed, Edwards was particularly vexed by the "jesting and laughing" of others.[46]

How those accused of indiscipline by Edwards responded is telling. Independents and most Sectaries alike accepted the criteria but denied their applicability to them while accusing the Presbyterians of the same or worse. Early on, the Independent Goodwin assured his (Presbyterian) readers that "the very truth is, that this way, above all her fellows, is so farre from holding any intelligence or correspondence with loosenesse and prophanenesse, that her face is set to advance righteousnesse and true holinesse in all her quarters." Likewise, late in the day, the Particular Baptists Kiffen et al. came to take up Edwards's charges as their own in criticizing Walwyn's dubious lifestyle and Overton's "notorious profanness."[47] On the other hand, such disciplinary micro-management of personal aesthetics also drew complaints. Chidley responded that "I know some indeede that have beene constrained to change their *apparell* for feare of *persecution.*" Others anonymously asked "Shall we kill all that are not of our mindes? Or count them unworthy to live, because they keep not a day as we keep it," and complained "This is an unheard of tyranny, when a man may neither order his estate, nor weare a garment, nor eat, nor speake, nor looke, but at their liking." In a precociously liberal distinction, Overton argued "So that businesse is, not how great a sinner I am, but how faithfull and reall to the Common-wealth; that's the matter concerneth my neighbour, and whereof my neighbour is only in this publick Controversie to take notice; and for my personall sins that are not of Civill cognizance or wrong unto him, to leave them to God, whose judgment is righteous and just. And till persons professing Religion be brought to this sound temper, they fall far short of Christianity."[48]

Revolution

Finally, the familiar trope of the "world turned upside down," a literal "revolution," already paradigmatically rendered by Taylor in 1642, was an extreme form of slippery-slope argument according to which black and white reverse locations. As Edwards put it, "Sectarisme [is] . . . the spring that feeds all our evils, the *primum mobile,* or great wheele that turns all the rest about." That is, what were formerly and rightly regarded as the lowest forms of belief and behavior are now not merely tolerated but held morally obligatory. Such a reversal threatened to transform London into another Amsterdam, Munster, or worse, Sodom, Egypt, or Babylon, or into something "new" and "stranger" still.[49]

Innovation

Significantly, many heresies were new and unfamiliar. In this fervently conservative age, innovation smacked of presumption, and invited the suspicion that intellectual machinations were afoot. King and Parliament, Episcopacy and Presbytery, Presbytery and Independency ritually traded such accusations. The radically novel was, then, extremely difficult to articulate. New words, new ideas, and those articulating them suffered censure.

Neologisms were not to be introduced lightly. It was widely acknowledged that the English language was heterogeneously constituted. Still, many nonetheless argued for control over vocabulary and the repudiation of both affected archaisms and insolent neologisms.

New words threatened to bespeak strange new ideas. Brinsley cautioned (women especially) "Let not *Affectation* of *Novelty,* of *New wayes,* inordinatly transport any of us, least thereby unawares we be brought to reject and renounce *Old Truth,* as some of late times have done." For Wortley, a true English Protestant was someone "not daring to broach new opinions." Edwards agreed with his peers that all the Independents' and Sectaries' talk of New Light and New Truth is either "an old Error long ago condemned . . . or else a New Error lately invented."[50]

Not only were the words and ideas new, but their adherents dangerously unrepentant. The Independents and Sectaries were "wanton witted men who are conscious to themselves of singularities and novelties." Indeed, already by 1641, they were a character-type: "Novellists."[51] Knowingly novel then, Sectaries had to beg their readers' indulgence. Robinson opened with the request: "Let not the seeming noveltie of opinions deter thee from searching out the Truth."[52] If it was a challenge to say something new, Sectaries from the lower orders faced a second constraint. In a society of peers, persons, and parsons, they were disallowed from speaking in their own voice. Those who spoke radically for the lower and middling orders concentrated on inconsistencies in the dominant theological and political discourse, and availed themselves of the defense that their criticism of the status quo "is no more then their owne Doctrine hath taught us, and therefore no Blasphemy, Treason or Heresie."[53]

Strangeness

As an evaluative matter Edwards saw limitations in those beyond his ken. He ignored Tolerationists' arguments that to condemn those who are novel or other is overactive Procrusteanism. Instead, he insinuated "strangeness," imbuing the unfamilar with disquieting connotations. Sectary beliefs and practices were of "the strangest manner that ever was in any age." In *Reasons,* Edwards observed that "whereas we have now too many new and strange doctrines, we shall have nothing but errours and novelties broached, and so greater contentions and breaches amongst us."[54] The "strange" figured at least 25 times in *Antapologia,* and in *Gangraena,* it was a virtual refrain (occuring over 75 times). Nothing the Independents and Sectaries espoused or did escaped the adjective: clauses, conceits, designs, doctrines, errors, expressions, expositions of Scripture, gods, letters, long hair, meetings, opinions, passages, petitions, positions, practices, sermons, speeches, ways, words, and worships. Viewed historically, "we are in strange hands, and things are come to a strange passe," or rather, paradoxically, "now things are grown to a strange passe, (though nothing is now strange)." That is, heresies were so abundant and metamorphizing so rapidly that they had become a numbing norm, leaving many oblivious to their presence. Edwards sought to remind, to the contrary, that toleration was itself "a strange unheard of suffering and bearing with."[55]

Again, Edwards's idiom was no idiosyncracy. Rehearsing Erasmus's adage "Semper Africa novi aliquid apportat" (*Adagia,* 886A), Presbyterians and many Independents drew a ready association between novelty, strangeness, and monstrousness.[56] Prynne regarded the separation of church and state as "a most strange and senselesse whimsie," and noted that the Independents "invent many strange Apocryphalll new shifts, Interpretations, Distinctions, to evade direct Scriptures and fundamentall Truths." Gatford took "a strange raving giddinesse or light-headnesse" to be "a notorious symptome" of the infection of the mind by heresy. "R.H." wrote *The True Guide* because "never did private opinion or strange Heterodoxie more prevail." The anonymous author of *A Spirit Moving In the Women-Preachers* spoke of an "affronted, brazen-faced, strange, new *Feminine Brood.*" Clarke, amid repeated references to Edwards's works, took up the charge: "*Beware* of strange, & new, and unfound opinions, doctrines, positions."[57] In sum, the Sectaries were alien, incomprehensible, and intolerable. Their beliefs and practices could be explained away, mundanely, as functions of error, vanity, and ideology, or metaphysically, as the machinations of the devil, but they were too strange to be understood on their own terms.

Strikingly, Edwards's radical Independent and Sectary targets occasionally responded in kind. Chidley, for example, found Edwards's arguments to be, in turn, "strange evasions," and volunteered what "is much more strange to me." She also defended the Independents' alleged behavior, noting "it is no marvell though their spirits grow narrow, towards such an Adversarie as your selfe, and great cause they have to be *strange* towards you, and *reserved* and *subtill* also." More to the point, Walwyn simply contended that "how strange soever this [liberty of conscience] may seem to you, unto me it seemeth most equal."[58]

Motivating Orthodoxy

Although Edwards's rhetorical objections to "error," "sin," and "revolution" were already quite evocative, he supplied two further motivations to immediate action on behalf of orthodoxy and against heresy: "mercy" and "dependency."

Mercy

Edwards's theological ethics turned on fearing and honoring God, but it also appealed to saving souls out of kindness:

> I would aske . . . whether are they more mercifull, tender, that will not let little children, sick persons, mad men do what they will, go where they please, eate what they list; or they who use all good means to hinder them, and provide wholsome food and medicines for them; yea, I am of the mind, before a yeer comes about the Magistrates and Kingdom will say Master *Edwards* and those men who either by writing or preaching spoke against the Errors, Heresies, Schisms of the times and a Toleration, and stirred us up to hinder and suppresse their growth, were most mercifull both to the souls and bodies of men, and to the kingdoms, and far from bloudinesse, bitternesse, or inveterateness.

In effect, by their heresies and indiscipline the Sectaries evinced their "rage and madnesse" in a desperate cry for help that should not go unheeded.[59]

The merciful impetus to save souls was reiterated by many of Edwards's peers. According to one, "connivence and indulgence to Hereticks is a *cruell mercy:* correction is a *mercifull severity,* and a *wholesome medicine,* as well to themselves as to the Church." Not only the body can be sick and diseased, but the soul, the mind, the belief system. Hence, the Magistrate can and should apply strong medicine: "the terror of civill power, which may and doth (when blessed of God) prove a preparation to free obedience." Walwyn's argument that "compulsion and enforcement may make a confused masse of dissembling hypocrites, not a Congregation of beleevers," was found unconvincing.[60]

In some Sectary arguments one can detect a novel conceptualization of ethics. The implications of epistemological uncertainty and presumptive universal salvation lead some to affirm a minimal and post-theological ethic beyond good/evil *qua* orthodox/heterodox. What made an act moral and hence politically salient, be it divinely ordained or humanly posited, was its consistency with natural creaturely reasonableness. Thus, for example, Burroughes, an Independent with radical leanings, questioned the legal punishment of all "evil" because of the "infinite variety of men's apprehensions about what is good or evill," while affirming it for those few acts which "is fit in reason to be subjected to." For some, God could even be erased from the moral equation. Whereas someone like Williams justified toleration out of a millenial piety, others such as Overton and Walwyn argued from creaturely sympathy that though we might differ about the nature of our soul and summum bonum, given minimal assumptions about the universality of self-preservation we ought to agree reasonably that physical suffering unto death is our summum malum. Affective horror toward "cruelty" figured prominently in such arguments.[61]

Edwards, though, could also redirect the charge of cruelty against the Sectaries. Indeed, "Machiavell and the Jesuits are but punies and fresh men to them." Thus, in towns where Sectaries predominated they persecuted the orthodox.[62] Moreover, their infiltration of the New Model Army (NMA) shows that they "do in a sort deifie this Army" and intend a "Domination." To Edwards, the NMA was more dangerous than the King's late Papist Army because while both killed bodies the former also spread heresies "without controle" and thereby endangered Protestant souls. That the NMA would go on forcibly to shift the balance of power from the Presbyterians to the Independents in both the City and Parliament vindicated Edwards.[63]

Dependency

Gangraena, in its very title, affirmed an idiomatic line of argumentation in Early Modern English political discourse, namely that of the organic body politic. In its absolutist form, to which James I had frequent recourse, it positioned the King as head of the body politic (and also as its proper physician). More generally, this language was systematically functionalist and conservative. Political unrest was invested with correspondences to the human body diseased and schizophrenic and the cosmos unhinged and chaotic.[64]

Echoing 2 Timothy II.17, *Gangraena* conveyed the visceral image of the body politic in jeopardy while also justifying medicinal imperatives. "I have played the part of a Physitian," Edwards said of his authorial role, and Parliament must supply the "Antidote" he has prescribed. The radical sects were socio-political rot (variously poison, infection, plague, canker, leprosy) in need of dramatic excision (cutting, scarification, washing) before they infected others.[65]

Those singled out by Edwards sometimes responded with mock diagnostic rhetoric of their own. According to Burroughes, Edwards was suffering from distemper, a superfluity of choler and malice, and "excrementious filthinesse at the mouth." To Drapes, *Gangraena* was an "Vlcerous Treatise." Maddock and Pinnell sought to apply *A Plaister,* Lanseter a *Lance.*[66]

In any event, Edwards linked individual fates to such a degree that one felt compelled out of healthy self-interest to take an interest in the indiscipline of others. That is, bolstered by abundant scriptural corroboration, and like Prynne and Baylie, Edwards countered the Independents by affirming mutual moral "dependency." Patience, indifference, and agnosticism were malign "lukewarmenesse." Indeed, inaction, in the face of Sectarian heresy, was a sin of omission.[67] This was Edwards's final trumping argument: "I do as a Minister pronounce that the plague of God will fall upon the heads of those" guilty by inaction. Put positively, each man had to do his part. The presence of Sectaries was evidence "that Magistrates, Ministers, and other Christians Masters of families, Parents, & c. have been a sleep, and too carelesse."[68] Instead, Magistrates should command a period of fasting and humiliation, renew the National Covenant, burn heretical books, perform exemplary punishment, and settle Church-government. They should pass Ordinances against unordained preachers, unlicensed texts, and especially heresy (such as *An Ordinance . . . for the Preventing of the Growing and Spreading of Heresies,* 1646).[69] Ministers should redouble their efforts to protect doctrine and discipline. They must be more vigilant over their spiritual children, organize, and pray. Finally, private Christian fathers, for their part, should secure their households, shun Sectaries, pray, and petition for protection, lest they "never have peace in their families more, or ever after have command of wives, children, servants."[70]

In sum, then, in contrast to Catholic seasonal shriving, the godly Puritan had to be perpetually publicly self-examining, and rely upon private confessional devices to intensify his self-fashioning. He had to acknowledge truths, abide by doctrines, be disciplined, eschew innovation, and avoid the strange, if he was to save others' souls and ultimately his own.

Edwards died in December 1647, having fled to the United Provinces in the aftermath of the failed Presbyterian counter-revolution in August of that year. Had he lived he would have been aghast at the demise of the

Assembly, triumph of the Independents, and Regicide, but might have taken some small comfort in the fact that the most incendiary of the heretics, the Levellers, were extinguished, and that he had played a role in their demise. The English 'Revolution' did not turn the world upside down and the Leveller movement succeeded in moving little level. The latter disintegrated, during 1649, as its Sectary supporters in the gathered congregations and amid the rank and file of the NMA (appeased by the payment of arrears) came to decry the Levellers' precociously liberal-democratic assumptions and proposals as strange and untoward.

In closing, given the malignity and infectiousness of heresy, one might wonder if Edwards's "history of heresy," with all of its animadversions and narrative abundance, was the most appropriate orthodox response. *Gangraena* was a giant version of, and alluded to, well-established voyeuristic pamphlet genres: "the true relation," the "looking glass," and the "discovery." Still, should Edwards have been describing the strange and despoiling ideas and practices "in a methodicall way for memories sake?" Where he had excerpted, should he have referred "the Reader for further satisfaction to many Pamphlets and Books daily printed and openly sold?" It is ironic, to say the least, that modern historians turn to *Gangraena* as a source of information on mid-century radicalism. Edwards's performative contradiction was not lost on his contemporaneous readers, either. Even a critic could wonder aloud why, if Edwards was right, he nonetheless "spread poyson; infect[ed] many souls who by this shall come to the knowledge of such things as they never heard before?"[71]

Notes

1. John Bond, *Occasus Occidentalis* (London, 1645 [Thomason Tracts British Library No. E25/22; hereafter cited as 'E25/22']), 44.
2. Thomas Edwards, *Gangraena* (Exeter: Rota, 1977), Book I, 42 [hereafter 'I:42']. There are pagination problems with this modern composite edition from which all citations in this paper are drawn: I:76, then 49–76 recur, then 77-on resumes: I have listed the second set of numbers with an ⋆. An Appendix to the second part, II, runs from 117–123; which I have identified with an A. In III, there are several pages missing pagination inserted between 16 and 17, which I have numbered 16a–16q; likewise between 240 and 241, which I have numbered 240a–240I.
3. J. G. A. Pocock, "Within the margins: the definitions of orthodoxy," in *The Margins of Orthodoxy*, ed. R. D. Lund (New York: Cambridge University Press, 1995), 33–53, asks us to write history so that orthodoxy appears on its own terms, which is as sensible as it is difficult.
4. Henry Robinson, *Liberty of Conscience* in *Tracts on Liberty in the Puritan Revolution, 1638–1647,* ed. W. Haller (NY: Columbia University Press, 1934 [1644]), 41. All italics appear in original unless otherwise stated.

5. Independents Thomas Goodwin, et al., *The Independants Declaration* (London, 1644 [E399/1]), appealed to Parliament and the people to resist the Assembly, even while cautioning against further sectarianism. It drew acrimonious Presbyterian responses, such as William Prynne, *Independency Examined, Unmasked, Refuted* (London, 1644 [E257/3]), and John Bastwick, *Independency not Gods Ordinance* (London, 1645 [E285/2]). See W. K. Jordan, *The Development of Religious Toleration in England* (Cambridge: Harvard University Press, 1938); Keith Lindley, *Popular Politics and Religon in Civil War London* (Aldershot: Scolar, 1997).
6. John Taylor, *A Swarme Of Sectaries, And Schismatiques* (London?, 1641 [E158/1]); Anonymous, *Discovery Of 29, Sects here in London* (London, 1641 [E168/7]); Anonymous, *The dolefull Lamentation of Cheap-side Crosse* (London, 1641 [E135/9]), 3. See Murray Tolmie, *The Triumph of the Saints* (Cambridge: Cambridge University Press, 1977).
7. For William Haller, Commentary, in *Tracts on Liberty in the Puritan Revolution 1638–1647,* Vol. I (New York: Columbia University Press, 1934), 81, "Though he [Edwards] writes . . . in a kind of vertigo, he nevertheless sees and expresses more vividly than do less hysterical writers the actual color of the time." See also Jordan, *The Development of Religious Toleration,* 281–7; William Haller, *Liberty and Reformation in the Puritan Revolution* (NY: Columbia University Press, 1955), 223–9; Maurice Goldsmith and Ivan Roots, Introduction, in *Gangraena* (Exeter: Rota, 1977), 1–15; Tolmie, *Triumph of the Saints,* 130–8. Ann Hughes, 'Popular' Presbyterianism in the 1640s and 1650s: the cases of Thomas Edwards and Thomas Hall, in *England's Long Reformation 1500–1800,* ed. N. Tyacke (London: UCL Press, 1998), 235–59, is more cautious.
8. Thomas Edwards, *Reasons against the Independant Government of Particular Congregations: As also against the Toleration of such Churches to be erected in this Kingdome. Together with an Answer to such Reasons as are commonly alledged for such a Toleration* (London, 1641 [E167/16]); and *Antapologia: Or, A Full Answer to the Apologeticall Narration* (London, 1644 [E1/1]), 293.
9. Edwards, *Gangraena,* I:4. There is a definite shift in rhetorical strategy between the relatively brief *Reasons,* with Edwards's arguments against Independency, the substantial *Antapologia,* which excerpts extensively from the Independents' *Apologetical Narration* and then submits critical arguments, and *Gangraena,* in which a multitude of brief excerpts and narrative accounts occupy the bulk of the work. Much of the criticism of *Gangraena* was taken up with the refutation of specific attributions. Sympathetic James Cranford, *Plain English* (London, 1646 [E350/11]), 17, found it "to[o] tedious to particularise the persons and actions" of the Sectaries, but applauded Edwards's efforts "to unmaske them" and "give a caveat to well-meaning people."
10. John Saltmarsh, *Groanes For Liberty* (London, 1646 [E327/20]); William Walwyn, *A Whisper In The Eare of Mr. Thomas Edwards Minister* (London, 1646 [E328/2]), and *A Word More To Mr. Thomas Edwards Minister* (London, 1646 [E328/20]); and [John Goodwin], *Cretensis* (London, 1646

[E328/22]). I ignore Edwards's objections to political innovations and leveling. A Fourth Part to *Gangraena,* a fuller reply to his critics, was repeatedly promised but never delivered. In 1647, he published *The Casting Down of the last and strongest hold of Satan, or, A Treatise against Toleration And Pretended Liberty of Conscience* (London).

11. Anonymous, *Match me these two* (London, 1647 [E400/9]), 3. See Nigel Smith, *Literature and Revolution in England, 1640–1660* (New Haven: Yale University Press, 1994); Dagmar Freist, *Governed by Opinion* (NY: Tauris Academic Studies, 1997); and Kevin Sharpe, *Reading Revolutions* (New Haven: Yale University Press, 2000), 3–62.
12. Edwards, *Gangraena,* II:190, III: Pref 4, I: Pref 8, 8. The first installment of *Gangraena* sold well, going through a third printing within 2 months despite its length (and presumably high price) (II:48).
13. Edwards, *Gangraena,* III:281.
14. Several of Edwards's critics, such as Katherine Chidley, *The Ivstification of the Independent Chvrches of Christ* (London, 1641 [E174/7]), 60, and Jeremiah Burroughes, *A Vindication Of Mr Burroughes* (London, 1646 [E345/14]), 13ff, suggested that Edwards's animosity was due to the vigor of his competition. See also Anonymous, *Certain Aditionall Reasons* (London, 1646 [E316/10]), which parodied Anonymous, *A Letter Of The Ministers of the City of London . . . Against Toleration* (London, 1646 [E314/8]), by attributing economic motivations to the Ministers. Much of Edwards's anti-popery was shared with the Independents and Sectaries. See Peter Lake, "Anti-popery: the structure of a prejudice," in *The English Civil War,* eds. R. Cust and A. Hughes (New York: Arnold, 1997), 181–210.
15. Richard Overton, *The Araignement of Mr. Persecution,* in *Tracts on Liberty in the Puritan Revolution 1638–1647,* Vol. III, ed. W. Haller (New York: Columbia University Press, 1934 [1645]), 32, a member of Thomas Lambe's General Baptist congregation, admitted to "knowing their Meetings"; as did William Walwyn, *The Compassionate Samaritane* in *Tracts on Liberty in the Puritan Revolution, 1638–1647,* Volume III, ed. W. Haller (New York: Columbia University Press, 1934 [1646]), 3, 10. Lilburne was attached to Edmund Rosier's separatist congregation and the Particular Baptist William Kiffen. See Lindley, *Popular Politics and Religion.*
16. Edwards, *Gangraena,* I:A120. The idiom, "without controle," was shared by others, such as Anonymous, *Hinc Illae Lachrymae.* (London, 1648 [E540/16]), 6. See also Anonymous, *Tub-preachers overturn'd* (London, 1647 [E384/7]). On female prophets, see Edwards, *Gangraena,* 30–1, 84–88, 106–108, II:10, III:242, 278. See also Anonymous, *A Discoverie of Six women preachers* (London?, 1641 [E166/1])
17. In 1641, Chidley, *The Ivstification,* 22–23, defended "Taylors, Felt-makers, Button-makers, Tent-makers, Shepherds, or Ploughmen, or what honest Trade soever, if they are well-meaning Christians"; and argued that they "are able to divide the Word of God aright." See also Walwyn, *The Compassionate Samaritane,* 31–2; and William Hartley, *The Priests Patent Cancelled* (London, 1649 [E542/17]).

18. Edwards, *Antapologia,* 246.
19. Edwards, *Gangraena,* I:175; *Reasons,* 33.
20. Edwards, *Antapologia,* 297, *Gangraena,* I:18, 175, III:235. The titles of numerous pamphlets illustrate this spirit: William Prynne, *Trvth Trivmphing over Falshood, Antiqvity over Novelty* (London, 1645 [E259/1]); Robert Baylie, *A Dissuasive From The Errours Of the Time* (London, 1645 [E317/5]); and Samuel Bolton, *The Arraignment of Errovr* (London, 1646 [E318/1]).
21. John Goodwin, in *Tracts on Liberty in the Puritan Revolution, 1638–1647.* Volume III, ed. W. Haller (New York: Columbia University Press, 1934 [1644]), 7; Robinson, *Liberty of Conscience,* 50; Walwyn, *A Word More,* 6.
22. Edwards, *Gangraena,* I:47, III:103, 115, 117.
23. Roger Williams, *The Bloudy Tenent, of Persecution* (London, 1644 [E1/2]), 1, 40–49; Overton, *The Araignement;* cf. Anonymous, *Wholsome Severity reconciled with Christian* (London, 1645 [E24/5]), 16.
24. Edwards, *Gangraena,* I:66.
25. See William Kiffen, *A Briefe Remonstrance Of The Reasons and Grounds of those People commonly called Anabaptists, for their Seperation* (London, 1645 [E293/31]); and Anonymous, *A Confession of Faith Of seven Congregations or Churches of Christ in London, which are commonly (but unjustly) called Anabaptists* (London, 1646 [E319/13]).
26. See also William Prynne, *A Fresh Discovery of Some Prodigious New Wandring-Blasing-Stars, & Firebrands, Stiling themselves New-Lights* (London, 1645 [E261/5]); and Frederick Spanhemius, *Englands Warning By Germanies Woe* (London, 1646 [E362/28]).
27. Overton, *The Araignement,* 20; Edwards, *Gangraena,* I:11, III: Pref 6.
28. Anonymous, *A Demvrre To The Bill For Preventing The Growth And Spreading Of Heresie* (London, 1646 [E356/5]), n.p. See also Robinson, *Liberty of Conscience,* 9; Overton, *The Araignement;* and Saltmarsh, *Groanes,* 11.
29. Edwards, *Gangraena,* I: Pref 6.
30. Although scholars have attributed the shift from historical argument to rationalist political philosophy among mid-century radicals to the legacy of the Norman Yoke, such a shift was also necessitated by this religious epistemological controversy.
31. Edwards, *Gangraena,* I:101, 122.
32. Goodwin et al., *An Apologeticall Narration;* Jeremiah Burroughes, *Irenicum, To The Lovers Of Truth and Peace* (London, 1646 [E306/9]), 23–4; Robinson, *Liberty of Conscience;* Walwyn, *The Compassionate Samaritane,* 2.
33. Walwyn, *The Compassionate Samaritane,* 1; Williams, *The Bloudy Tenent,* 86; Overton, *The Araignment.* On toleration, see Haller, *Liberty and Reformation,* 125–34; Conrad Russell, "Arguments for Religious Unity in England, 1530–1650," *Journal of Ecclesiastical History,* 18 (1967), 201–26; Blair Worden, "Toleration and the Cromwellian Protectorate" in *Persecution and Toleration,* ed. W. Sheils (Oxford: Basil Blackwell, 1984), 199–233; Avihu Zakai, "Religious Toleration and Its Enemies: The Independent Divines and the Issue of Toleration during the English Civil War," *Albion,* 18 (1989), 617–35.

34. Edward Drapes, *A Plain And Faithfull Discovery Of A Beame In Master Edwards his Eye* (London, 1646 [E350/22]), 13, challenged the veracity of such reports. See Christopher Durston, *The Family in the English Revolution* (Oxford: Basil Blackwell, 1989), 116–21; and David Cressy, *Travesties and Transgressions in Tudor and Stuart England* (New York: Oxford University Press, 2000), 171–185.
35. Richard Overton, *Man's Mortalitie,* ed. H. Fisch (Liverpool: Liverpool University Press, 1968 [1644]). On mortalism, see Norman T. Burns, *Christian Mortalism from Tynedale to Burns* (Cambridge: Harvard University Press, 1972).
36. Edwards, *Gangraena,* III:161.
37. Overton, *Mans Mortalitie,* 1. See also Anonymous *Divine Light, manifesting the Love of God unto the whole World.* Both were cited by Edwards, *Gangraena,* I: 20–1, 27, 36, III:11. On *apokastasis* see C. A. Patrides, *Premises and Motifs in Renaissance Thought and Literature* (Princeton: Princeton University Press, 1982), 200–17.
38. On unbelief, see Nigel Smith, "The Charge of Atheism and the Language of Radical Speculation, 1640–1660" in *Atheism from the Reformation to the Enlightenment,* eds. M. Hunter and D. Wootton (Oxford: Clarendon, 1992), 131–158.
39. Edwards, *Gangraena,* I:175, 135.
40. Edwards, *Gangraena,* I:121.
41. Edwards, *Gangraena,* I:72–3. See also Cranford, *Plain English.*
42. A. B., *The Sabbath Truly Sanctified* (London, 1645 [E26/15]); Edwards, *Gangraena,* I:96, II:29–30, 76, 103, 112; Goodwin, *Cretensis,* 28–35. See also Ronald Hutton, *The Rise and Fall of Merry England* (New York: Oxford University Press, 1994).
43. John Philpot, *A Prospective-Glasse for Gamesters* (London, 1643 [669f7/64]); Thomas Adams, *A Whip for a Drunkard, and a Curbe for Prophanesse* (London, 1646 [669f10/14]); Thomas Shepard, *Theses Sabbaticæ* (London, 1649 [E559/19–21]), 45ff.
44. William Prynne, *Historio-Mastix* (London, 1633) marks an early high point in this critique. See also Edwards, *Gangraena,* II: 77, 178, 187. Anonymous, *The Actors Remonstrance or Complaint* (London, 1643 [E86/8]), promised to make amends. See Martin Butler, *Theatre and Crisis* (Cambridge: Cambridge University Press, 1984).
45. Edwards, *Gangraena,* I:136, II:145, III: 191, 212. See also R. W., *A Looking-Glasse for Women* (London, 1644 [E2/18]).
46. Edwards, *Gangraena,* I:125, III:191. See also Lionell Gatford, *Hyperphysicall Directions in Time of Plague* (Oxford, 1644 [E27/11]), John Whincop, *Gods Call to Weeping and Mourning* (London, 1645 [E26/21]), Bond, *Occasus Occidentalis.* See Sammy Basu, "'Woe unto you that laugh now!': Humor and Toleration in Overton and Shaftesbury" in *Religious Toleration,* ed. John Christian Laursen (New York: St. Martin's Press, 1999), 147–172.
47. Goodwin, 37; William Kiffen, John Price et al., *Walwins Wiles* (London, 1649 [E554/24]), 2.

48. Chidley, *The Ivstification,* 64; Anonymous, *Marsh his Mickle Monument* (London, 1645 [E301/11], n.p.; Anonymous, *The Trojan Horse of the Presbyteriall Government Unbowelled* (London?, 1646 [E353/1]), 19; Richard Overton, *The Picture of the Councel of State,* in *The Leveller Tracts,* eds. W. Haller and G. Davies (New York: Columbia University Press, 1944 [1649]), 231. Such arguments also figured early in Royalist critiques of Presbyterianism, such as Thomas Aston, *A Remonstrance Against Presbitery* (London, 1641 [E163/1]).
49. Edwards, *Gangraena,* III:276; see also *Antapologia,* 280. See also John Taylor, *Mad Fashions, Od Fashions, All out of Fashions* (London, 1642 [E138/3]).
50. John Brinsley, *A Looking-Glasse For Good Women* (London, 1645 [E305/23]), 14; Francis Wortley, *Characters and Elegies* (London, 1646 [E344/21]), 11; Edwards, *Gangraena,* I:134. See also Prynne, *Trvth Trivmphing,* and *A Fresh Discovery;* and Baylie, *A Dissuasive.*
51. Anonymous, *The Brownists Conventicle* (London, 1641 [E164/13]), 2. See also Edwards, *Antapologia,* 174; *Gangraena,* I:125, II:171–2.
52. Robinson, *Liberty of Conscience,* Pref. 1, 49. See also Goodwin, 47.
53. Richard Overton, *An Appeale from the Degenerate Representative Body,* in *Leveller Manifestoes of the Puritan Revolution,* ed. D. Wolfe (New York: Humanities, 1967 [1647]), 174.
54. Edwards, *Gangraena,* 218; *Reasons,* 27.
55. Edwards, *Gangraena,* III:266, I:Ep Ded 3, I:Ep Ded 8.
56. Edwards, *Antapologia,* 262; *Gangraena,* I:121, II:22.
57. William Prynne, *Twelve Considerable Serious Questions touching Chvrch Government* (London: 1644 [E257/1]), 7; and *Trvth Trivmphing,* Ep Ded 8; Lionell Gatford, *Hyperphysicall Directions in Time of Plague* (Oxford, 1644 [E27/11]), 19; R. H. *The True Guide* (London, 1646 [E344/8]), Pref. 1; Anonymous, *A Spirit Moving In the Women-Preachers* (London, 1646 [E324/10]), title; and John Clarke, *Leaven, Corrupting the Childrens Bread* (London, 1646 [E354/15]), 58.
58. Chidley, *The Ivstification,* 35, 64; Walwyn, *A Word More,* 5.
59. Edwards, *Gangraena,* II:65, 127; see also *Reasons,* 50; *Antapologia,* 168ff.
60. Anon. *Wholsome Severity,* 33, 33; Edwards, *Gangraena,* III:125, 263; Walwyn, *A Whisper,* 5. See also Edward Terry, *Lawless Liberty* (London, 1646 [E356/11]).
61. Burroughes, *Irenicum,* 49; Williams, *A Bloudy Tenent;* Overton, *The Araignement;* Walwyn, *A Whisper, A Word More;* see also Anonymous, *A Parable* (London, 1646 [E359/8]). See also John Coffey, "Puritanism and Liberty Revisited: The Case for Toleration in the English Revolution," *Historical Journal,* 41 (1998), 961–985 and J. C. Davis, "Religion and the Struggle for Freedom in the English Revolution," *The Historical Journal,* 35 (1992), 507–530.
62. See Edwards, *Gangraena,* II:182.
63. Edwards, *Gangraena,* III:133, 270, 265.
64. See James Daly, *Cosmic Harmony and Political Thinking in Early Stuart England* (Philadelphia: American Philosophical Society, 1979), and David

Sacks, "Parliament, Liberty, and the Commonweal," in *Parliament and Liberty,* ed. J. H. Hexter (Stanford: Stanford University Press, 1992), 87–91.

65. Edwards, *Gangraena,* II:131, III: Pref 1.
66. Burroughes, *A Vindication,* 3; Drapes, *A Plain And Faithfull Discovery,* 13; John Maddock and Henry Pinnell, *Gangraenachrestum, Or, A Plaister to alay the tumor, and prevent the spreading of a pernitious Vlcery* (Oxford, 1646 [E353/12]); John Lanseter, *Lanseters Lance, for Edwards'es Gangrene* (London, 1646 [E354/17]). Anonymous, *A Parable,* volunteered a consultation with Doctors Love, Justice, Patience, and Truth to cure Edwards.
67. Edwards *Gangraena,* Ep ded 7; *Reasons,* 10; *Antapologia,* 282. See also Prynne, *Independency Examined,* 3; Baylie, *A Dissuasive.*
68. Edwards, *Gangraena,* III:218, II:198. He vigorously reiterated the need for vigilance in *Casting Down.*
69. Anonymous, *An Ordinance presented to the Honorable House of Commons . . . For the preventing of the growing and spreading of Heresies* (London, 1646 [E669f9/50]), would have made punishable by death the failure to subscribe to essential doctrines. It was explicitly endorsed by Edwards, *Gangraena,* III:151, 211, and elicited intense defenses, such as Anonymous, *A Vindication Of A Printed Paper, entitled An Ordinance* (London, 1646 [E362/25]), and critiques such as Anonymous, *A Demvrre;* and Richard Overton, *An Arrow Against All Tyrants* (Exeter: The Rota Press, 1976 [1646]). The latter, to Edwards, were the most "Atheistical peece[s] written by any man since the reformation," *Gangraena* III:117.
70. Edwards, *Gangraena,* I:156; see also *Reasons,* 26; *Antapologia,* 199; and Edward Harris, *A True Relation of a Company of Brownists, Separatists, and Non-Conformists* (London?, 1641 [E172/31]); Ricraft, *A Looking Glasse;* and Brinsley, *A Looking-Glasse.*
71. Edwards, *Gangraena,* I:4, III:185; Saltmarsh, *Groanes,* 27. Anonymous, *A Relation of Severall Heresies* (London, 1646 [E358/2]), 20, although sympathetic to Edwards, chided him that "this is not the way to decrease Errors by a violent furious repetition of them."

2

Francis Lee and The French Prophets

The History of Montanism [1709]

Stacey Searl-Chapin

Francis Lee's *The History of Montanism* of 1709 played a role in the reception of the French Prophets in early eighteenth-century England, in the medicalization of enthusiasm between Burton and Kant, and in the use of history for the demonization of heresies. Its author may seem an unlikely candidate for such roles, since he was one of the founding members of the Philadelphians, a charismatic and enthusiastic movement of the late seventeenth century. We learn from the history of histories of heresy that authors' intentions and authors' results are not always straightforward.

Lee's book appeared anonymously, "By a Lay-Gentleman," as pages 73–352 in a volume entitled *The Spirit of Enthusiasm Exorcised*. The first 72 pages of the volume was an essay under that title by George Hickes, an Anglican clergyman who refused to sign the oath of allegiance to William and Mary in 1689. It was followed by *The New Pretenders to Prophecy Examin'd, and their Pretences Shewn to be Groundless and False* by Nathaniel Spinckes, M.A., also a nonjuror. According to the *Concise Dictionary of National Biography*,[1] Hickes's chief works were *The Case of Infant Baptism* (1683) and *Records of the New Consecrations* (no year given); *The Spirit of Enthusiasm* is not listed.[2] Both Hickes and Spinckes refer explicitly to the French Prophets, whereas Lee's book never mentions them. It is obvious, however, from the context of publication between the other two books, and from the content of the work, that *The History of Montanism* was published as a critique of the Prophets.

Francis Lee was a physician, so it is not surprising that he wrote in the tradition of the medicalization of religious enthusiasm. His book came out between Robert Burton's *Anatomy of Melancholy* of 1621 and Kant's "Essay on Diseases of the Mind" of 1764 and refutation of Swedenborg in *Dreams of a Spirit-Seer* of 1765. Some in this tradition stigmatized religious enthusiasm as a disease partly because they recognized how close it was to their own position and wanted to distance themselves from it. As we shall see, this may have been the case with Lee.

Perhaps the most impressive thing about Lee's book is that after a slow start covering the details of the rise and progress of Montanism, he manages to increase the heat by claims that different branches of Montanism grew into various Jewish heresies, Manicheanism, Socinianism, atheism, and Islam! Starting from a relatively insignificant heresy of the second century A.D. we suddenly have the root of almost all evil. This is demonization and conspiracy theory on a grand scale, and when the reader realizes that the implicit message is that the French Prophets are Montanists, it is now clear just how dangerous they are.

Francis Lee (1661–1719) has been described as a "miscellaneous writer."[3] He was an Orientalist at St. John's College, Oxford, in 1679 and studied medicine at Leiden beginning in 1692. Most important, he became a disciple of Jane Leade and in 1697 was one of the founders, with her, of the Philadelphian Society. The entry for him in the *Lutheran Cyclopedia*[4] lists his works as *Horologium Christianum* (1689), *The Labouring Person's Remembrancer* (1690), and *Considerations Concerning Oaths* (no year given); it does not mention his *History of Montanism*.

Montanism

Montanism was a movement in Christianity in the second century A.D. Sometimes called Phrygians (after the region in Asia Minor where the sect developed), Montanists took their name from Montanus, who started the movement in about 156 A.D. Lee begins with the emergence of Montanus and Montanism, noting that:

> . . . about the Middle of the Second Century, or soon afterward, he made himself known, not only in his own Country, but in the Churches also thereabouts: For he had hardly embrac'd the Christian Faith, but his great Austerities of Life made him to be particularly taken notice of, and to be had in esteem among many of the most zealous (but indiscreet) Christians (Lee, 1709: 74).

Lee argues that at the beginning, Montanus did not have "evil" intentions, and just wanted to be of service to the Church. This quickly changed, however, when Montanus became possessed by the "strange spirit," the "Serpent," rather than the true Spirit of God.

Montanists claimed that a new, or at least renewed, presentation of the Holy Spirit had arrived. The Holy Spirit supposedly spoke in tongues through Montanus at his baptism. This notion revived the charismatic emotionalism of the Church, which to some was lacking in the increased formalism of Christianity.[5] According to Lee, Montanism originally sought to serve the Church: "by bringing back those that should be under his Charge to their *first Love,* and strengthening the things that were *ready to die.*"[6]

Furthermore, according to Lee, the doctrine that Montanus preached was not contrary to the teachings of the Church. Lee writes:

> None of all the Ancients was more express than he, in the Mystery of the Incarnation, or seem'd more to honour the Person of *Christ* and extol his Merits; concerning whom he taught, That the everlasting Son of God . . . took upon him a Body of the Virgin *Mary,* wherein he preached the Gospel of the Kingdom, and at last dy'd for the Salvation of the World. All the Ancient Hereticks generally err'd as to this, from the Truth; only he was most clear in this Point, preaching up Repentance and Faith in the Name of the Lord Jesus only, as the *one Mediator* betwixt God and Man: Of whom he held, that *he* did not only *appear* in our Nature, but that he was *united with* it after a most real manner; thereby condemning most, or all the Gnostics and Enthusiasts.[7]

Montanus's views on the Trinity were also orthodox: he believed and taught that the Father, Son, and Holy Ghost were but "*one* God in Essence and Substance."[8] He was also orthodox in believing in the Church as a Christian fellowship, modeled on the pattern delivered by Christ himself, "and knit together by the Bond of his Spirit, under Pastors and Officers, of several Orders, having a clear and certain Mission from Him whom they represent."[9]

Reacting to the laxity of the Church, Montanus encouraged extreme asceticism. Celibacy was demanded of all Christians, marriage was discountenanced, and second marriages were forbidden. Montanists practiced rigorous fasting and their pre-Easter fast was two weeks instead of the orthodox one week. Montanists also practiced severe penance. Lee notes: "The Discipline of the *Montanists* was in all the Points of it extremely rigorous: They refus'd the Absolution to almost all Sins; that is, to all those which they called *Deadly* and *Capital,* except committed before Baptism; thereby taking even from the Church the power of Opening and Shutting, or receiving Penitents to her Communion."[10]

Martyrdom was also encouraged, and the sect condemned Christians who avoided it. One goal of the Montanists was to be a party/sect of martyrs.

The Montanists preached certain themes that were commonly found in Puritanical and charismatic sects in later centuries. They were millenarians, believing that Christ would return in the immediate future and reign for one thousand years in the New Jerusalem. Following would be the end of history and the final accomplishment of God's will. For example:

> the *Montanists* went up to celebrate certain Mysteries in a Desart place of *Phrygia*, where had formerly stood the town of *Papuza*, but was now quite destroy'd. This place they honour'd exceedingly, fancying, declaring, That the Heavenly *Jerusalem* was there descended. But 'tis not altogether so certain, whether they understood, or no, by this *Jerusalem* said to be amongst them, the Form of a City, which is reported to have apear'd in the Air about the year CXCVIII, and which *Tertullian* took as a Mark of this *Jerusalem* from above.[11]

Montanism caused little stir at first from its beginnings circa 156 A.D. From about 171 A.D. on, however, it was opposed by Apollinarius, Bishop of Hierapolis. The sect was primarily found in North Africa at Carthage, in Rome, and in Gaul. In 177 Irenaeus, who later wrote against the sect in *Against Heresies*[12] and was then a priest at Lyons, accompanied the confessors who discussed Montanism with Eleutherius, Bishop of Rome. Montanus was condemned by the first synod of bishops held in Asia Minor, his followers expelled from the Church and debarred from Communion. For approximately the first six decades of the movement, Montanists did not refer to themselves as such. They referred to their own movement as "The Prophecy," or "The New Prophecy," and called themselves "The Prophets."[13]

Neither the opposition of other Christians nor persecution under Emperor Severus after 193 stopped the spread of the movement. Probably the most famous Montanist was Tertullian, converted to Montanism about 207. He was one of the foremost Christian thinkers of his age, and one of the first proponents of the notion that the Holy Spirit was God equally with the other two persons of the Trinity, the Father and the Son. Though his adherence to Montanism helped strengthen the movement, its success was in large part due to its organization. From Phrygia it sent out preachers or missionaries, supported by the "gifts of the faithful." The Montanist movement continued into the fourth century, when Justinian sought to suppress it, and, according to a recent scholar, "survived into the fifth century when it died from natural causes."[14]

The French Prophets

The French Prophets of the seventeenth and early eighteenth centuries were a millenarian sect with origins among the Camisards, the French Protestants of the Cevennes. Under ruthless persecution the Huguenot resistance movement became militant and developed into a charismatic movement "complete with visionaries, ecstasies, and mass hysteria which made it a perfect embodiment of what the age called 'enthusiasm.'"[15] The Revocation of the Edict of Nantes in 1685 prompted them to hear angelic choirs in the sky, see prophets weeping tears of blood, and witness their children predicting the fall of Babylon.[16]

In response to increased persecution following the Revocation, the faithful turned to the millenarian speculations of Pierre du Moulin, a pastor of the early part of the century, and his grandson, the Pastor Pierre Jurieu, exiled in Rotterdam. Both predicted that the resurrection of the two witnesses of Revelation 11 would occur in 1689.[17] Many Huguenots in Languedoc and Dauphine continued their religious worship despite threats of imprisonment or enslavement in the galleys. Like the Montanists of the second century A.D., they viewed martyrdom as part of a possible millennial scheme.

When the first prophets appeared among them in 1688, the Huguenots of this spiritual "Desert" (as they called it) rejoiced, for it seemed that the last days were upon them and the persecution could not last much longer. Within six months, younger and older prophets appeared in Languedoc and Dauphine, fulfilling the prophecy of Acts 2:17: "In the latter days your young people shall prophesy, and your old men shall dream dreams."[18] The new prophets had physical symptoms such as being passive yet agitated, asleep but awake, unconscious but perceptive, exhausted yet refreshed.

As visions increased, as Judgment Day approached, the uninspired too fell to the ground in "Desert" congregations. This was a public act of repentance and a visible demonstration of faith. Some Catholics maintained that the convulsions and fainting were the result not of inspiration, but of "emotional contagion." The faithful did not reject this metaphor, but rather welcomed the divine epidemic. They sought the disease by fasting, and did not risk the humiliation of immunity.

In the winter and early spring of 1689, royal officials hunted down and imprisoned all prophets they could find, sent the young to convents or prison hospitals, and condemned others to death. This, too, the prophets had foreseen: the short yet cruel persecution that would bring final deliverance.

Prophecy continued to appear in the "Desert" in the subsequent years. Late in 1700, a young man named Daniel Raoux discovered that God had "transfused into him" the spirit of the Old Testament prophet Daniel. The prophetic movement grew so widely and quickly that by the autumn of 1701 there were hundreds of prophets. By November 1701, as more prophets arose, over 350 *fanatiques* were in prison, and some 40 prophets had been condemned to slavery in the galleys. Then, on 22 July 1702, Abraham Mazel, 24, pronounced the divine order to take up arms without delay and rescue Huguenots imprisoned at Pont-de-Montvert. On July 24, some 40 to 60 Huguenots began a guerrilla war against the authorities.[19] Their numbers increased to 500 by September, and by the year 1704, the Camisards had a force of perhaps 2,000 men and women.

The war lasted more than two years, from 1702 through much of 1704. Twenty thousand royal troops could not prevail against the Huguenots, who knew the mountains. The soldiers were battling men and women who considered themselves to be warned, divinely, of traitors, guided to their camps at night by celestial lights, and protected from ambush by guardian angels. Many of the Camisard leaders were themselves prophets.

In order to put an end to this long revolt, especially since Savoy, the English, and the Dutch had attempted to provide the Camisards with supplies, Louis XIV sent the duc de Vilars to direct the royal troops. He offered amnesty to all rebels who surrendered and jailed the parents and relatives of the remaining Camisards. This strategy proved effective, and band after band of Camisards surrendered between May and November of 1704.

Elie Marion[20] and other prophets sought the advice of the Spirit, not wanting to abandon a divinely inspired movement without a divinely inspired command. Most took the silence of the Spirit as a command to surrender. Unlike most, Marion continued to fight, but achieved little. So, Elie Marion, Durand Fage, and Jean Cavalier made their way first to Switzerland, then to London.

The French Prophets in London

England in 1706 was no stranger to prophesy or to millennial speculation. Encouraged by the Reformation identification of the Antichrist with the Papacy, serious Protestant scholars of the seventeenth century looked for other identities between Biblical symbols and historical persons or events.[21] Though some in England might have accepted the possibility of miracle and inspiration, and though some found meaning in the Camisards' movement, religious behavior that drew its validity from war and despair seemed inappropriate to many. London was an increasingly stable and powerful financial center, with an increasingly urban economy, and a population of half a million people. Whigs still feared that the Glorious Revolution of 1688, which replaced the Catholic James II with the Protestant monarchs William and Mary, would be overturned by Jacobites upon the death of Queen Anne.

The differences between London and the Cevennes were not necessarily detrimental to the spread of prophecy. People could move around quite easily, as well as make contacts with diverse Christian groups. People could print and distribute inspirations with fair assurance that one's audience could read. Finally, as a Huguenot, one could locate oneself in a society of forty or fifty thousand other refugees in the capital.[22]

Despite high expectations of enthusiastic reception by London Huguenots, the French Prophets had only 24 men and women members at the end of March 1707, following ten months of prophesy and informal advertisement. Swayed by the associates of the most powerful French Protestant churches, the London Huguenots turned against the Camisard refugees. The hostility may have been due in large part to the feelings of guilt felt by refugee ministers for "leaving their flocks" behind in France to face persecution. In addition, they were put off by the Prophets who had stayed behind and substituted for the absent ministers, performing baptisms and marriage ceremonies without ordination or recognized rank in the official churches. The hostility was also due to the uncertain political position of the Huguenots in late Stuart England. Because they depended on the Whigs for toleration and financial support, the Huguenot clergy could not risk tolerating prophets who were likely to predict or encourage violence. The Huguenots were already disliked by London laborers, accused of stealing the livelihoods of weavers in particular, so Huguenot leaders were eager to prevent incidents that would reflect poorly on their community. On January 5, 1706/7, the Savoy consistory, followed by other consistories, published a condemnation of the three prophets:

> the Agitations of these pretended Prophets are only the Effect of a voluntary Habit, of which they are entirely Masters, though in their Fits they seem to be agitated by a Superior Cause. . . . But the Way in which they make the Spirit speak, is still more unworthy of him, which is by perpetual Hesitations, Childish Repetitions, unintelligible Stuff, gross Contradictions, manifest Lies, Conjectures turned into predictions, already convicted of Falsehood by the Event.[23]

This "Acte Noir," as the French Prophets termed it, stimulated the Huguenot population, who began insulting the Prophets publicly, splattering them with mud and filth and assaulting them. This new persecution allowed the French Prophets to begin a tradition that was similar to, yet independent of, the Camisard movement. Excommunication of the Prophets by the consistories allowed the group to define itself.

The French Prophets and the Philadelphians

The Prophets began to attract a wider English following. Among the new followers were 15 to 20 Philadelphians, men and women influenced by the mystical principles of Jacob Boehme as interpreted by Jane Leade (1623–1704) and by Richard Roach, an Anglican clergyman interested in alchemy and spiritual transformation. Inspired by John Mason,[24] the Philadelphians were primarily a sect devoted to an inward search for revelation and the true peace of Philadelphia, but they soon identified the Camisard prophets and their English followers as another valid Christian dispensation.

Gradually, their quietist influence was felt among the Prophets, accounting for changes in that group in its final years. At the end of 1708 there were 400 followers of the French Prophets in the London area, with twice as many English followers as Huguenot, and 22 English Prophets compared to 15 French.

Jane Leade claimed to have prophetic visions and organized the Philadelphian Society for the Advancement of Piety and Divine Philosophy with Dr. John Pordage (1607–81), a Berkshire rector, and Francis Lee.[25] Lee, who married her daughter, helped spread her doctrines in Europe.[26]

At age sixteen, while dancing at a Christmas gathering, Leade was overcome with a sense of sadness and perceived a voice instructing her to withdraw from this way of life, for it was vanity. In 1663 she became a member of a private nonconformist group gathered around Dr. John Pordage in London. Following her husband's death in 1670, she had several visions. While walking in the country she envisioned "an overshadowing bright Cloud and in the midst of it the Figure of a Woman." Three days later a luminous reappearance commanded, "Behold me as thy Mother." Six days later, in London, came the promise "I shall now cease to appear in a Visible Figure unto thee, but I will not fail to transfigure my self in thy mind; and there open the Spring of Wisdom and Understanding."[27]

In 1674 Leade joined Pordage's household. A former Anglican priest who may also have been an alchemist and astrologer, Pordage lost his parish 20 years earlier through church censure of his own ecstatic practices and Ranter inclinations.[28] Pordage and Leade, who viewed their relationship as a spiritual union, led their congregation, experienced visions, studied, and wrote extended commentary on the work of the German inner-light theologian and mystic philosopher Jacob Boehme (1575–1625). One major element of Boehme's philosophy influenced Leade and Pordage: namely, his adaptation of the mystical tradition of Wisdom, or Sophia. Derived mainly from gnostic heresies, the Kabala, Wisdom literature, and apocalyptic language of Scripture, Boehme's theory of creation held that Wisdom is the primary, feminine ground of being out of which the masculine will-to-manifestation forms existence.[29] Boehme's theology was a re-synthesized blend of the fading systems of Christian gnosticism, Jewish and Christian mysticism, Neoplatonism, alchemy, and astrology, and this hybrid was termed theosophy. Boehme's belief in a transcendent spirit and millennial harmony contributed to the Protestant theology of the "Inner Light."[30]

Lee and Leade's relationship began when, in 1694, one of Leade's books reached Germany. This prompted correspondence among sympathizers, who published translations that prompted Continental interest in Leade's work. Lee, who was training as a physician in Leiden, found her work in the Netherlands and returned to England to meet her. Lee, too, had refused to sign the oath of allegiance to King William. He was drawn to mysticism as a religious person, interested in the literature of the Apocrypha as a scholar, and curious about trances as a physician.[31] About a year after he joined Leade (who had become blind at age 71), Lee married her widowed daughter Barbara and began acting as Leade's secretary and editor, until her death at 80 in 1704. Lee, Leade, and Pordage reorganized their congregation as the Philadelphian Society, advocating individual transformation as the means of bringing about the millennium, but rejecting the anarchism of the Ranters and the social reforms of the Diggers and Quakers. The group took its name from the church of Philadelphia, the remainder of believers at the beginning of the Apocalypse, and was philosophically based on Boehme's ideas.[32]

Originally sympathetic to the French Prophets, the Philadelphians began to move apart from them by the time Lee wrote his book. Desiree Hirst mentions the rift when discussing the meetings of the Philadelphian Society: "Unfortunately these meetings were invaded by groups of what were called the 'French Prophets' or 'Camisards', émigrés from the Cevennes war, who specialized in violent, ecstatic manifestations."[33] The Philadelphians began holding their meetings at a separate location.

Though the quietist manners of the Philadelphians had some influence on the French Prophets, the agitations and warnings of the Prophets were unsettling to the Philadelphians. Critics of the French Prophets addressed themselves to two central questions: one, Is outwardly religious behavior explicable in the same manner as all other human behavior? And two, How can the social appeal of excessive behavior based upon inspirations "known" (to the critics) to be false be explained? Answers to these questions were couched in medical terms as often as they were in theological ones.[34]

Francis Lee's History of Montanism

In *The History of Montanism* (1709), Lee refers to Montanus as a false Prophet: "Some few there were indeed that took *Montanus* for that which he was, or that at least which he afterward prov'd himself to be; that is a False Prophet."[35] He attempts to explain both spiritually and medically what may have caused Montanus's visions, speculating that perhaps he suffered from "melancholy." A physician and fellow of the Royal Society, Lee studied "Fevers of the Mind."[36] He found that Montanists and other enthusiasts had been victims of such fevers, which might "put forth themselves in as great a variety of Symptoms as the Hysterical and Hypochondriacal, if not in greater." Lee delivered a medical diagnosis similar to that of many critics of the French Prophets: "But there is no Enthusiasm of any sort without a Fever in the Soul, though it may be complicated indeed with other Distempers besides. This seems to be seated chiefly in the Animal Spirits, as that of the Body in the Mass of the Blood."[37]

Lee's book is a lengthy, detailed, yet critical examination of Montanism, its development, break with the Church, and its influence upon other sects. His chief historical sources are Church Fathers such as Eusebius, Augustine, and Theodorus. His *History* is more than just that, however, for he infuses his historical account of the sect with his opinion of the "False Prophets," both medical and theological.

According to Lee, where Montanus erred, and hence became heretical, was in his application of doctrine: "But it is notorious, that the Mistake was in the *Application* of the same, by pretending to limit the Catholick Church to those of his own Society, and consequently breaking the Unity of Christ's Mystical Body; which alone was sufficient to answer the Ends and Designs of the grand Enemy, whose constant maxim is *Divide and Reign*."[38]

Thus, the doctrine alone of the Montanists was not heretical. Rather, their notion that they were the "true" church, as well as their violent fits of prophecy, were what, according to Lee, accorded the sect the label of "Heretics."

Also important to Lee is the role of reason. Lee uses this criterion to distinguish Montanus (and implicitly the later French Prophets) from "true prophets." For example, explaining Montanus's visions, Lee reports: "he was possessed by a *strange Spirit,* instead of the True Spirit of God. Which Spirit seizing him of a sudden was wont to Agitate him, just as if he were a distracted person: And he continued all the while that the Spirit thus acted in him without the use of his Reason; and had, they say, no command over his own Members and Organs."[39]

Lee appeals to medicine, as well as theology, to discredit Montanus and his followers: "For the Enthusiasm of *Montanus* appeared to be very much like a Fit of Frenzy, or distemper'd Melancholy, accompanied with violent Agitations of the Body; and which took away from him the Liberty of his Reason, and the Free Exercise of his Faculties. Which was never so in any of the true Prophets of God, either in the Old or the New Testament."[40] Many French Prophets claimed not to have remembered their visions once the fit was over. This was taken, by Lee and other critics, as an indication of the loss of reason, and used to differentiate the French Prophets from true prophets.

Montanism, Lee, and "The Weaker Sex"

Soon after Montanus appeared, "several of the weaker Sex [were] excited by his high Pretensions."[41] Women figure prominently from the outset of Montanism. After Montanus began receiving his visions, Lee notes, he was joined by *Priscilla* (Prisca) and *Maximilla,* who formed with him a prophetical society: "This was a Society, not according to the manner of the Schools of the ancient Prophets of God, but rather according to the Custom of the Heathen Prophets and Prophetesses."[42]

Prophecy was a legitimate ministry for "appropriately gifted" women, as Paul had allowed, in Asia Minor in particular.[43] According to Lee, Priscilla and Maximilla were both married, yet came to favor abstinence once they came into contact with Montanus: "And therefore no sooner were they touch'd by the Power in *Montanus,* and fill'd thence with this Spirit of Prophecy and Divination, but immediately they left their Husbands."[44]

According to Lee, Montanists had a notion of what was an "Archetypal Virgin." Women were thought to be "Divine Nature" (p. 94), organs and vessels of the Holy Spirit, and those that joined Montanus were thought to be "Paradisiacal Virgins" who should be thought of as angels on Earth. The women believed themselves to be heavenly virgins, the wives of Christ, and often wore all white and "the Virginal Veil."[45]

Note that Lee does not seem to object to the prominent role of women in prophecy, but rather that they followed Montanus. In fact, according to Lee, the Church did not condemn the women of Montanism for their prophesying (though Maximilla and Priscilla, having been married, should not have been considered "virgins"):

> it was also an undoubted Custom of the Primitive Christians, conformable to a plain Apostolical Constitution, to have their Inspired Women when they did Prophecy to be veil'd; and since likewise none of the Catholicks ever objected against them a Defect of this kind, which they must certainly have known, and as certainly would have censur'd. For taking notice of their

outward adorning, they condemn'd 'em for their wearing of Gold and Silver; and for their putting on of fine apparel.[46]

According to Lee's *History,* the women of Montanism, and in particular Priscilla and Maximilla, were charged by Apollonius, as quoted in Eusebius, with bribery; namely, performing and providing prophecies in exchange for money and gifts. St. Heirom argued that the women bribed others in order to attract followers. Lee seems to favor the latter argument, "for this Father [Heirom] has acquainted us, that those two, *viz. Priscilla* and *Maximilla* were Ladies of Quality, and exceedingly Rich; so that they could have no need of acting the Prophetesses for hire."[47] Their motivation for "false" prophesying, then, according to Lee, was "the Ambition of being great in the expected Kingdom, and the Confidence of receiving again what was so laid out, with abundant Interest."[48] Thus, the women were led with a "false faith" and mistook "the Imaginations of their own Hearts," as well as suggestions from Montanus, to be the pure inspirations of the Divine Spirit.[49]

Thus, it does not seem that Lee was against the prominent role of women in the Church per se, but rather that these particular women were swayed by false prophecy. This makes sense, given the prominence of women, and Lee's association with Jane Leade, in the Philadelphian society. One difference between the Philadelphians and the French Prophets, which may account for their break, was their attitude toward women. The Philadelphians had a tradition of female leadership, while the French Prophets associated women with false prophecy.[50] It was not that Lee was too respectful to touch Leade's published work: he and Roach often "revised" her poems, and on occasion even replaced some of her work with that of Roach. According to Smith, "These university trained men, one a scholar and critic of esoteric literature and the other a cleric-poet, perhaps felt ambivalence in their role as Boswells to little-educated women who wrote prolifically, sometimes even automatically."[51]

It is not clear when Lee turned against the French Prophets in England, but it is clear that his history of Montanism draws parallels between the French Prophets and "The False Prophets" of the early Church.

The Spread of Montanism

The most remarkable thing about Lee's book is his claim in the second half that Montanism led to nearly every other important heresy, both within and outside of Christianity. According to Lee, Montanism not only branched out into other sects, some of which lost the name of their patriarch, it also, " . . . intermingled with most or all of the ancient Heresies, the

Arians excepted."[52] Of main concern for Lee are the Manicheans, Judaism, Socinians, atheists, and Islam.

Around the latter end of the fourth, and in the fifth century more especially, there was a decline in interest in Montanism. Some members, according to Lee, were convinced that Montanus was deluded, and they returned to the Church. A great many others, however, were "swallowed up" by the "Manichees." Lee argues that both Montanus and Manes set themselves up to be Paraclets, the Spirit and Power of God in the flesh, and expected their disciples to honor them as such. Lee then goes on to list many similarities between the Manicheans and the Montanists.

According to Lee, the "Manichees" and the Montanists continued as two distinct bodies for a period of time, but it is difficult to determine "whether they be originally from one or from the other," given their similarities, as well as their "reciprocal Communication of their Inspirations and Traditions."[53] What is clear to Lee is that Manicheanism, as a heresy, could not have taken hold had Montanism not supported it.

Lee also claims that the Montanists (via the Manicheans) led to the Jewish heresies of the Talmudists, and later the Rabbins. According to Lee: " . . . after having blasphem'd against the Law of *Moses,* as proceeding from a Root of the dark Magick, it [Manicheanism] join't with the Zealots for that Law against Christianity, and united it self with the worst and most degenerate *Judaism* that sprung up in the Dregs of that State after their last captivity. Hence it is observable, that the *Talmudists* and latter *Rabbins* have generally dres'd up just such another God as *Manes* had done."[54] The main error of Manicheanism, a lie "on which all the rest do hang," concerns the consubstantiability of the Spirit with Matter. The Manicheans err in believing that the spirit and matter are one, thus making God material. This heresy, according to Lee, can be found in Jewish texts.[55] There are also several rabbinical fables or riddles that favor Manicheanism, which Lee would elaborate on or demonstrate, "were it worth the while."

According to Lee, one does not hear any more of the Montanists as a distinct sect after the seventh century. At that point, after surviving 500 years and giving birth to other heretical sects like the Manicheans and the Talmudists, Montanism gave way to another pretended Paraclet "who first appear'd out of the Wilderness of *Arabia,* and set up beyond all that had ever been before him for Revelations and Miracles. This was no other than *Mahomet* (of cursed Memory) who was the successor of *Manes* as *Manes* was of *Montanus.*"[56] Lee argues that though accounts of Mahomet are often contradictory, it is clear from all of them (though he does not cite any) that his pretensions were very similar to those of Montanus. Both men were great visionaries, both set out to reform corruption existing in the established church, and both "pretended" to build both their missions and their doctrine upon the foundation of Christ and his Apostles. Both Montanus and Mahomet convinced their followers that they were not setting out to destroy established law or Gospel, but to fulfill them.

Mahomet was subject to agitations similar to those of Montanus, which affected him so violently that he appeared, sometimes, to lay dead. According to Lee, Mahomet, like Montanus before him, claimed that his fits were the result of the majesty of the divine presence overpowering his nature, and that he had an intimate relation with the heavenly objects, hearing God, the angels, and the saints frequently. Following his fits, he was able to give detailed accounts of Heaven and Hell, of Paradise, and of Araf. In this he was unlike the Montanists and the later French Prophets, who did not remember what had happened to them during their fits. Lee manages to claim, however, that Mahomet, like the others, lost the use of his reason. For Lee points out on several occasions that Mahomet was a self-proclaimed simple man, "without letters," who, though he was able to recount his visions, was not really able to understand them. This was often Mahomet's own defense, according to Lee, against those who claimed he was an impostor. In addition to being a man of no letters, Lee also claims that Mahomet suffered from a disease of the brain.[57] Thus, Mahomet, like Montanus, was an "Epileptical Enthusiast"[58] who was able to sway those who did not use their reason.

Lee also linked the Mahometans and the Socinians, implying that the latter are also linked to the Montanists. According to Lee, the Socinians (and other Protestant sects, even Luther himself) undermined Christianity by claiming that Mahometism is more orthodox than Catholicism.[59]

Montanism was also the root of the ultimate evil, atheism. Lee asserts that "it [Montanism] rested not until it came at length to settle either in down right *Anti-Christianism* or in *Atheism*."[60] Thus, according to Lee, every conceivable enemy to Christianity developed from this small second-century sect. Conspiracy theorists tend to believe that all of their enemies share a common source. Lee's book is an example of conspiracy theory on a grand scale.

Conclusion

Francis Lee was a physician, so it is not surprising that he wrote in the tradition of the medicalization of religious enthusiasm. Perhaps Lee set out to stigmatize religious enthusiasm as a disease partly because he saw how close it was to his own position. The most interesting thing about Lee's book is that he claims that different branches of one second-century sect grew into Jewish heresies, Manicheanism, Socinianism, atheism, and Islam. This is conspiracy theory at its most ambitious. From a relatively insignificant heresy, we suddenly have the root of all evil. Since Lee's work appeared sandwiched between two anti-French Prophet tracts, his implicit message is that the French Prophets are Montanists, and thus very dangerous. It seems very unlikely that Lee's *History* would generate any tolerance for them.

Notes

1. *The Dictionary of National Biography: The Concise Dictionary* (Oxford: Oxford University Press, 1953), p. 617.
2. Hickes's *The Spirit of Enthusiasm Exorcised* was originally delivered to the University of Oxford in 1680 as a sermon attacking the Quakers. This corresponds to the time Lee was working on his B.A. there (1679–1683). It is possible that Lee was present when the sermon was delivered, and thus began the link between Hickes and Lee. The volume of *Spirit* in which Lee's and Spinckes's works appear is the fourth edition. Also not mentioned is Hickes's *The True Notion of Persecution Stated in a Sermon Preached at the Time of the Late Contribution for the French Protestants* (London: 1681). In this sermon, Hickes argues that the French Protestants were truly persecuted (they suffered in the name of God and not merely for themselves) by the Catholic French government, and that it was the duty of English Protestants to help them.
3. *The Dictionary of National Biography,* p. 759.
4. Erwin L. Lueker, ed., *Lutheran Cyclopedia* (St. Louis: Concordia Publishing House, 1975), p. 577.
5. Francis Lee, *The History of Montanism,* printed in George Hickes, *The Spirit of Enthusiasm Exorcised* (London: Share, 1709), and David Christie-Murray, *A History of Heresy* (New York: Oxford University Press, 1976). See also John DeSoyres, *Montanism and the Primitive Church: A Study in the Ecclesiastical History of the Second Century* (Lexington: American Theological Library Association, 1965).
6. Lee, *History of Montanism,* p. 75.
7. Ibid., p. 85.
8. Ibid., p. 86.
9. Ibid.
10. Ibid., pp. 145–6.
11. Ibid., pp. 165–6.
12. Saint Irenaeus, Bishop of Lyon, *The Scandal of the Incarnation: Irenaeus Against the Heresies* (San Francisco: Ignatius Press, 1990).
13. Christine Trevett, *Montanism: Gender, Authority, and the New Prophecy* (New York: Cambridge University Press, 1996) p. 2.
14. Christie-Murray, *A History of Heresy,* p. 35. Christie-Murray briefly mentions the French Camisards, but not the French Prophets in England or Francis Lee.
15. Gordon Rupp, *Religion in England: 1688–1791* (Oxford: Clarendon Press, 1986), p. 216.

16. Hillel Schwartz, *Knaves, Fools, Madmen, and that Subtile Effluvium: A Study of the Opposition to the French Prophets in England, 1706–1710* (Gainesville: University Presses of Florida, 1978).
17. Pierre du Moulin, *Accomplissment des propheties* (La Rochelle: 1612), pp. 239–41; [Pierre Jurieu], *Apologie pour l'accomplissement des propheties* (Rotterdam: 1687), pp. 8–9. Noted in Schwartz, *Knaves, Fools,* p. 6.
18. Cited in Schwartz, *Knaves, Fools,* p. 7.
19. The term "Camisard" is derived either from the word for "night attack" (camisade) or from the patois for a white shirt (camisole). See Schwartz, *Knaves, Fools,* pp. 11, 15.
20. For a recent discussion of Elie Marion, see J.P. Charbol, *Elie Marion, le vagabond de Dieu (1678–1713). Prophetisme et millenarisme protestants en Europe a l' aube des Lumieres* (Aix-en-Provence: Edisud, 1999).
21. See Hillel Schwartz, *The French Prophets: The History of a Millenarian Group in Eighteenth-Century England* (Berkeley: University of California Press, 1980), pp. 37–45.
22. Schwartz, *Knaves, Fools,* p. 14.
23. Ibid., p. 17. The original announcement can be found, translated, in N.N., *An Account of the Lives and Behavior of the Three French Prophets* (London, 1708), pp. 9–10.
24. See Christopher Hill, "John Mason and the End of the World," in *Puritanism and Revolution* (London: Schocken Books, 1969), pp. 311–23.
25. *Lutheran Cyclopedia,* p. 577.
26. See Catherine F. Smith, "Jane Lead: The Feminist Mind and Art of a Seventeenth-Century Protestant Mystic," in *Women of Spirit: Female Leadership in the Jewish and Christian Traditions,* ed. Rosemary Ruether and Eleanor McLaughlin (New York: Simon and Shuster, 1979); "Jane Lead: Mysticism and the Woman Clothed with the Sun," in *Shakespeare's Sisters: Feminist Essays on Women Poets,* ed. Sandra M. Gilbert and Susan Gubar (Bloomington: Indiana University Press, 1979); and "Jane Lead's Wisdom: Women and Prophecy in Seventeenth Century England," in *Poetic Prophecy in Western Literature,* ed. Jan Wojcik and Raymond-Jean Fontain (Cranbury: Associated University Press, 1984).
27. Smith, "Jane Lead: Mysticism," p. 57, quoting Leade's *A Fountain of Gardens Watered by the River of Divine Pleasure, and Springing up in all Variety of Spiritual Plants* (London, 1697–1701), 1: 69–71.
28. Ibid., p. 60.
29. Smith, "Jane Lead: Mysticism," p. 8.
30. Smith, "Jane Lead: Mysticism," and "Jane Lead's Wisdom."
31. Smith, "Jane Lead: Mysticism," p. 7. On Lee's interest in prophecy, and specifically in Rosamunde Juliana von der Asseberg, see *Alsted and Leibniz: On God, the Magistrate and the Millennium,* ed. Maria Rosa Antognazza and Howard Hotson (Wiesbaden: Harrassowitz, 1999), pp. 170–187, 192. Lee translated the work of Johann Wilhelm Petersen, who published the prophecies of Rosamunde (and was read by Leibniz).

32. Desiree Hirst, *Hidden Riches: Traditional Symbolism from the Renaissance to Blake* (New York: Barnes and Noble, 1964), p. 103.
33. Ibid., p. 104.
34. For a good discussion of the medicalization of enthusiasm, see the special issue on "Enthusiasm and Enlightenment in Europe: 1650–1850," *Huntington Library Quarterly,* 60, 1998.
35. Lee, *History of Montanism*, p. 79.
36. Schwartz, *Knaves, Fools,* p. 45.
37. Lee, *History of Montanism,* p. 345.
38. Ibid., p. 86.
39. Ibid., p. 75.
40. Ibid., p. 80.
41. Ibid., p. 92.
42. Ibid., p. 93.
43. Trevett, *Montanism,* p. 152.
44. Lee, *History of Montanism,* p. 93.
45. Ibid., p. 94.
46. Ibid., pp. 94–5.
47. Ibid., p. 95
48. Ibid.
49. Ibid.
50. Schwartz, *The French Prophets,* p. 142.
51. Smith, "Jane Lead: Mysticism," p. 12.
52. Lee, *History of Montanism,* p. 317.
53. Ibid., p. 308–9.
54. Ibid., p. 312.
55. Ibid., p. 313.
56. Ibid., p. 320.
57. Ibid., p. 325–6.
58. Ibid., p. 326.
59. Ibid., p. 336–7.
60. Ibid., p. 339.

Part Two

Heretics in Self-Defense and Calls for Toleration

Introduction

John Christian Laursen

Right from the beginning of charges of heresy there were defenses of heresy. But the variety of such defenses was immense. They can be categorized in a number of ways. For example, some were written with the clearly pious intention of showing that alleged heretics were not really heretics, and that there was hope for bringing them back into the fold of orthodoxy. Some were written with the also pious intention of showing that the heretics held the truly orthodox position, and that the supposed orthodox were in the wrong. Some were written with obviously impious intentions, designed to show that neither the heretics nor the orthodox made any sense. Then, they could be read with subversive or nonsubversive intentions, with or against the author's intentions.

It should not be assumed that no one in the sixteenth century ever came to the defense of heretics. Erasmus of Rotterdam was one of the great voices for tolerance and understanding. He has recently been called "The Paradigm of Humanist Toleration" because of his predisposition to persuasion and his aversion to force.[1] Sebastian Castellion's defense of Michael Servetus relied in part on the history of heresy to vindicate that victim of Jean Calvin's heresy-hunting. The companion volume to this book, *Religious Toleration: "The Variety of Rites" from Cyrus to Defoe,* contains an annotated bibliography of sixteenth and seventeenth century works that argued for toleration, some of which explored parts of the history of heresy in defense of heretics.[2]

In the seventeenth and eighteenth centuries we can distinguish two types of works that had special influence in attacking heresy-mongering. One is works by "great names" such as Gabriel Naudé, Thomas Hobbes, Phillip van Limborch, Pierre Bayle, and Gottfried Arnold, which shall be examined in this part. The other is the flood of sometimes anonymous manuscripts, printed pamphlets, and other ephemera put together by relatively insignificant writers, such as the "Tract on the Three Impostors"; the clandestine manuscripts discussed by Antony McKenna; and the Socinian counter-histories discussed by Martin Mulsow, which shall be reviewed in the following part of this volume.

Many later figures reported that it was a reading of the work analyzed in the fourth chapter of this volume that had a breakthrough effect. Christian Thomasius claimed that a careful reading of Gabriel Naudé's *Apology for All the Great Men who have been Falsely Suspected of Magic* (1625) caused "the blinders to fall from my eyes." He encouraged his student Johannes Reiche to translate it into German in 1704.[3] Thomasius himself used Naudé's ideas when he wrote against torture and trials for witchcraft and magic, and succeeded in ending such practices in Prussia.[4] Something of what Naudé and Thomasius were up against is indicated by the fact that Jean Bodin, a tolerationist in several of his writings, was also the author of an influential defense of the burning of witches and presumably failed to see any inconsistency in his positions.[5]

Maryanne Horowitz's chapter in this volume traces some of the changes added to later editions of Naudé's book after the first edition in 1625. Refutations of Naudé's book could be timely contributions to contemporary debates, as in d'Autun's text of 1671 in defense of the Parlement of Normandy's imprisonment of alleged magicians. In turn, such affirmations of witchcraft and magic provoked reprintings of Naudé's text.

Naudé is known in the scholarship as an erudite libertine (*libertin érudit),* which meant that he belonged to a small group of Parisian intellectuals (hence the "erudite") who were skeptical of punishment after death and other orthodox dogmas and concentrated on this life and on their humanistic studies (hence the "libertine").[6] Naudé's book showed that most accusations of magic were the products of ignorance, jealousy, or malice. Written in Catholic France, some points of orthodoxy are scored by clearing popes of charges of heresy made by Protestants. But Naudé's general position of hostility to orthodoxy and authority when they interfere with scholarly humanism was obvious. So the question arose, Was he sincerely pious or really a disguised atheist?

Although some contemporaries and later scholars accused Naudé of atheism, he was employed as librarian for four cardinals (Bagni, Barberini, Richelieu, and Mazarin), so he obviously managed to arrive at an understanding with at least some officials of the Church.[7] Then, of course, even if Naudé was sincere, his later editors could have intended to twist his message in impious directions. As Professor Horowitz points out, notes in the later editions could attract readers' attention to subversive works that had appeared after Naudé's text, such as Richard Simon's biblical scholarship and Bayle's *Dictionary.* The 1712 edition seems to have been part of the same commercial venture that was planning to publish the impious "Tract on the Three Impostors," largely to make a profit. In turn, passages from Naudé's book were summarized in that pamphlet when it came out in print in 1719, without attribution of the source.[8]

Pierre Bayle is a similarly contested case. Many of the entries in his *Historical and Critical Dictionary* deal with famous figures from the history of heresy.[9] Did he write in the impartial pursuit of scholarship that he extolled in many places, or did he intend to subvert all orthodoxies? On the one hand, his materials spread knowledge and tolerance of the ideas of heretics from Manes to Spinoza, and on the other hand he did his best to slander the memory of any heretic that he thought might have descended into fanaticism.[10] On the one hand, many recent scholars have argued that he wrote with genuine Calvinist faith, and on the other hand others have insisted that he was really promoting atheism.[11] Similarly, readers in the eighteenth century drew opposing conclusions from Bayle's texts. Antony McKenna's chapter in this volume demonstrates how many of the subversive clandestine manuscripts of the eighteenth century drew on Bayle's texts for anti-religious purposes.

As Sally Jenkinson's chapter reminds us, Bayle wrote in self-conscious emulation of high standards of historical technique and responsibility. This allowed him to critique his opponents on both the left and the right on technical grounds. He could claim the high ground professionally, but his claims alone do not settle the question of the honesty and integrity of his own position. He used the history of charges and counter-charges of heresy to construct the most far-reaching doctrine of tolerance of the age while still purporting to uphold Calvinism. As Jenkinson shows us, Jean Le Clerc, at any rate, thought he was equivocating.

The bibliophile Prosper Marchand's entry on "The Three Impostors" reprises several of the questions of sincerity that we have broached above. In 27 folio pages of his posthumous *Historical Dictionary* (1758), he reviewed every case he could find of allegations that Moses, Jesus, and Mohammed were impostors. Anyone who read it would have the point driven home that this had been widely suspected. Did he do this for motives of erudition or subversion? His biographer has insisted that he was a believing Calvinist. Chances are that erudition drove his work and prudence dictated publication after his death, but without further evidence concerning his motives, we will probably never know for sure.[12]

A telling case for the interpretation of heterodox histories of heresy is John Toland's *Nazarenus* of 1718. It is a fairly clear case of insincerity, since it was a deliberate literary fraud, but the question is, fraud on behalf of what? Toland "invariably used the great patristic heresiographical sources to provide himself with testimonies in favor of theological deviance rather than to condemn it," so this could look like he favored the heretics.[13] Nevertheless, at least one-eighteenth century reader took Toland at his word that he was a spokesman for orthodoxy.[14] Some contemporary authors have argued for Toland's sincerity as a sort of Christian, and others have argued that he was an atheist. This led his recent editor to argue for placing the text along a different dimension. In the latter interpretation, Toland is best understood as a public intellectual taking positions that irritate all theologians for the strategic purpose of promoting religious pluralism and toleration.[15]

André Boureau-Deslandes is yet another test of our ability to detect insincerity in scholarship. We know from his previous works that Boureau-Deslandes was a freethinker.[16] He considerably exaggerated the number of heretical sects descended from the ancient Greek Cynics in the first French-language survey of the history of philosophy, his *Histoire critique de la philosophie [Critical History of Philosophy]* (1737). In his account, the Ebionites, Manicheans, Adamites, Beghards, Turlupins, Waldenses, Flagellants, Humiliati, Cathars, Anabaptists, Mennonites, Quakers, and the French Prophets were the equivalent of "the history of Christian Cynics or the history of fanaticism in Christianity."[17] Was this, like Francis Lee, a case of paranoia, seeing Cynics everywhere, or was it a spoof on the likes of Lee? Why is a freethinker paralleling the argument of a sincere spiritualist? It could be mockery, but part of the answer must be that freethinkers and the pious could join together in condemning the enthusiasts, who scared them both. We will return to the campaign against enthusiasm below.

History in the Service of Self-defense

More than one writer in the seventeenth and eighteenth centuries wrote with the specific purpose of defending himself against charges of heresy. Thomas Hobbes is a good example. He knew Naudé and his circle in Paris in the 1630s, and shared a number of principles with him. But unlike Naudé, his explorations of the history of heresy, as recovered in Martyn

Thompson's chapter in this volume, were inspired chiefly by the need to defend himself against charges of heresy.

Hobbes's *Leviathan* of 1648 unleashed a flood of critics. One of them reprised the "Three Impostors" theme in an orthodox vein, accusing Hobbes, Spinoza, and Herbert of Cherbury of heresy and impostorship.[18] Others called for serious punishment of Hobbes. Virtually admitting that he was at least a semi-heretic, Hobbes realized that the danger was real. So he ransacked the history of heresy and of the common law of England in order to prove that he should not be punished. The implicit structure of at least part of his argument was that even if someone may be a heretic, that was not against the law.

Hobbes's position was always that the monarch, not the theologians, should determine the law of heresy. This was known as Erastianism, or state control of the church. As Thompson argues, this was not an argument for toleration.[19] It may be true that the authorities cannot affect one's conscience as long as it is kept private, but on Hobbes's account the individual has no right to express his or her conscience. Thus, the history of heresy that Hobbes retailed was designed only to protect himself from prosecution, not to promote tolerance of the expression of dissenting religious views.

The Hobbesian dynamic of writing in self-defense was common. But not everyone defended himself by showing that there was no law against heresy. As we have already seen, Francis Lee was a putative semi-heretic who wrote against other heretics in order to distance himself from them. Another contemporary of Naudé and Hobbes, Johann Heinrich Alsted, tracked down numerous heresies, ancient and medieval, for his influential *Encyclopedia* (1630). He did so not so much because he perceived them as a threat, but in order to demonstrate the essential unanimity of contemporary Protestantism by contrasting it to the errors that all of them avoided.[20] A recent biographer suggests an underlying psychological purpose of focusing on heresy far away and far out: "the further his polemical opponents were beyond the pale, the less he would have to struggle with his own inner contradictions"; "what better way to reassure himself of his basic orthodoxy than in a protracted refutation of Socinianism?"[21] In contrast, Hobbes, one imagines, had little need to reassure himself of anything, but he certainly felt he had to reassure others of his lawfulness.

One of the many paradoxes of the histories of heresies of the seventeenth and eighteenth centuries is that they could be written from a Hobbesian point of departure, but on behalf of orthodoxy. As Sally Jenkinson shows us in her chapter, Louis de Maimbourg's numerous histories of heresy of the 1670s and 1680s were written from the point of view of a Hobbesian *politique*. They drove home the point that the civil authorities should suppress heresy as a matter of political stability and self-interest.

Spinoza, too, could write history of heresy from Hobbesian principles with intentions opposite to those of Maimbourg. His *Tractatus Politico-Theologicus* was designed to give control of the definition of heresy to the political authorities, at the same time advising them not to use it much.[22] Spinozism became one of the most notorious heresies of the eighteenth century because of its materialism, pantheism, and fatalism.[23] La Mettrie, d'Holbach, and others followed in his tradition.

The History of Persecution as Part of the History of Heresy

Part of the history of heresy is the history of responses to heretics. One sub-genre of the history of responses to heresy is the history of the Inquisition, which was the most thorough, most institutionalized, and most notorious response. Some influential examples are discussed in Luisa Simonutti's chapter. Charles Dellon's *History of the Inquisition in Goa* (1687) was naturally popular among Protestants because it made the Catholic Church look bloodthirsty. Also naturally, it was placed on the papal *Index of Prohibited Books* in 1690. But the complex politics of the history of the Inquisition comes into relief when we observe that the author could make a few changes and republish it in 1709 with impunity, under the protection of Bishop Bossuet and the Gallican Church. Gallicanism stood for the relative independence of the French national church from papal control, and the Gallicans did not object to the publication of books that made Rome look bad.

Jacques Marsollier's *History of the Inquisition and its Origin* of 1693, also condemned to the *Index,* similarly appealed to the civil authorities by its stress on the need to subject the Inquisition to civil authority and its discussion of the benefits of toleration. Gregorio Leti's text, *The Inquisition on Trial* of 1681, prefigured Dostoyevski's Grand Inquisitor, using humor and sarcasm to undermine the legitimacy of the Inquisition.

These histories of the punishment of heresy prepared the way for Philipp van Limborch, whose publication of actual records of the Inquisition made several moves. One was to reject the use of the word "heretic" as anything more than faction-speech. Simonutti follows up with an analysis of Limborch's notes for a second edition of his Latin text, which was never published. The notes were published in translation by the English translator of Limborch's introduction, a historical essay on the Inquisition. They included such things as Limborch's account of the torturing of the "heretical" Jewish physician Orobio de Castro and of the trial of the "heretical" Galileo. The reader was encouraged to fear all heresy-hunting and persecutory ideology, and the translator's purpose was surely some combination of promotion of tolerance and discrediting of Catholics.

Another history of persecution that prospered was the study of burned books. A. Westphal brought out a *Dissertation on Books Burned by the Public Authorities* in 1713.[24] For many years following 1712, Johann Lorenz Mosheim planned a *Vulcan's Library,* which would list and comment on all the books that had been burned. Other authors such as Johann Georg Schelhorn beat him into print and he eventually backed away from publishing anything with that title, probably because he realized it could be put to less-than-pious uses. (We shall return to Mosheim in parts three and four.) Just past the end of the century, G. Peignot brought out a *Critical, Literary, and Bibliographical Dictionary of the Principle Books Condemned to the Fire, Suppressed, or Censored* in two volumes (1806).[25]

Hobbesian Erastianism, or the subjection of religion to the political authorities, was not the only way to defend individuals or communities. Isaac de Beausobre's *Critical History of Manes and of Manicheanism* (1734–39) was an indirect defense of Beausobre's Huguenot community by defense of other alleged heretics.[26] His strategy was also based on the techniques of professional scholarship. He showed that the Greek and Latin historians had been hostile and mendacious. In contrast, alternative Persian, Syrian, and Arab sources painted a more reliable and better picture of the Manicheans, in Beausobre's account.

Notes

1. See Gary Remer, *Humanism and the Rhetoric of Toleration* (University Park: Pennsylvania State University Press, 1996), ch. 1.
2. J. C. Laursen, ed., *Religious Toleration: "The Variety of Rites" from Cyrus to Defoe* (New York: St. Martin's Press, 1999), pp. 231–245.
3. Martin Mulsow, "Appunti sulla fortuna di Gabriel Naudé nella Germania del primo illuminismo," *Studi filosofici* 14–15, 1991–92, pp. 147–8
4. See Werner Schneiders, ed., *Christian Thomasius 1655–1728* (Hamburg: Felix Meiner, 1989).
5. See M. L. Kuntz, "The Concept of Toleration in the *Colloquium Heptaplomeres*" in J. C. Laursen and C. J. Nederman, eds., *Beyond the Persecuting Society: Religious Toleration before the Enlightenment* (Philadelphia: University of Pennsylvania Press, 1998), pp. 125–144.
6. René Pintard, *Le libertinage érudit dans la première moitié du XVIIe siècle* (Paris, 1943).
7. Popkin, *The History of Scepticism from Descartes to Spinoza* (Berkeley: University of California Press, 1979), p. 90.

8. See S. Berti, F. Charles-Daubert, and R. Popkin, eds., *Heterodoxy, Spinozism, and Free Thought in Early Eighteenth-Century Europe: Studies on the* Traité des trois imposteurs (Dordrecht: Kluwer, 1996).
9. Accessible translations of some of them can be found in Pierre Bayle, *Historical and Critical Dictionary: Selections,* ed. R. Popkin (Indianapolis: Hackett, 1991) and Pierre Bayle, *Political Writings,* ed. S. Jenkinson (Cambridge: Cambridge University Press, 2000).
10. See J. C. Laursen, "Pierre Bayle's Antimillenarianism" in Laursen and R. H. Popkin, eds., *Continental Millenarians: Protestants, Catholics, Heretics* (Dordrecht: Kluwer, 2001).
11. Elisabeth Labrousse and Richard Popkin read Bayle as a genuinely religious thinker. Gianluca Mori's recent *Bayle philosophe* (Paris: Champion, 1999) makes the case for Bayle's philosophical atheism.
12. See J. C. Laursen, "The politics of a publishing event: the Marchand milieu and *The Life and Spirit of Spinoza* of 1719" in Berti, et al., *Heterodoxy, Spinozism, and Free Thought,* pp. 273–296.
13. Justin Champion, "Introduction" to John Toland, *Nazarenus* (Oxford: Voltaire Foundation, 1995), p. 36
14. Champion, "Introduction" to Toland, *Nazarenus,* pp. 102–3.
15. Ibid., pp. 76–78, 97, 101–102.
16. See J. C. Laursen, "Abel Boyer, the translation of libertines, and the politics of French politeness in Britain" in Antony McKenna, ed., *La Vie intellectuelle aux refuges protestants, II: Les traducteurs huguenots* (Paris: Champion, 2001).
17. See R. B. Branham and M.-O. Goulet-Cazé, eds., *The Cynics* (Berkeley: University of California Press, 1996), pp. 258–9.
18. Christian Kortholt, *De tribus impostoribus magnis* (Hamburg, 1700).
19. Thompson cites arguments by Richard Flathman, Glenn Burgess, and Alan Ryan to the effect that Hobbes can indeed be interpreted as a tolerationist on other grounds.
20. Johann Heinrich Alsted, *Theologia polemica* (1620); *Compendium theologicum* (1624); *Thesaurus chronologia* (1624, 1628); *Turris Babel destructa* (Herborn, 1639). See Howard Hotson, *Johann Heinrich Alsted 1588–1638: Between Renaissance, Reformation, and Universal Reform* (Oxford: Oxford University Press, 2000), pp. 124, 135–6. See also Howard Hotson, *Paradise Postponed: Johann Heinrich Alsted and the Birth of Calvinist Millenarianism* (Dordrecht: Kluwer, 2000).
21. Hotson, *Alsted,* p. 136.
22. See J. C. Laursen, "Spinoza on Tolerance: Arming the State and Reining in the Magistrate" in C. Nederman and Laursen, eds., *Difference and Dissent,* pp. 185–204.
23. See J. C. Laursen, "Spinoza in Denmark and the Fall of Struensee," *Journal of the History of Ideas,* vol. 61, 2000, pp. 189–202.
24. *Dissertatio de libris publica authoritate combustis* (Greifswald, 1713).

25. *Dictionnaire critique, littéraire et bibliographique des principiaux livres condamnés au feu, supprimés ou censurés* (Paris, 1806). See Martin Mulsow, "Bibliotheca Vulcani: Das Projekt einer *Geschichte der verbrannten Bücher* bei Johann Lorenz Mosheim und Johann Heinrich Heubel," *Das Achtzehnte Jahrhundert,* vol. 18, 1994, pp. 56–71.
26. *Histoire critique de Manichée et du manichéisme.* See B. Schwarzbach, "Politics and Ethics in the Huguenot Diaspora" in J. C. Laursen, ed., *New Essays on the Political Thought of the Huguenots of the Refuge* (Leiden: Brill, 1995), pp. 109–130, and J. C. Laursen, "Temporizing After Bayle: Isaac de Beausobre and the Manicheans," in Sandra Pott, Martin Mulsow, and Lutz Danneberg, eds., *The Berlin Refuge 1680–1780* (Leiden: Brill, 2002).

3

Gabriel Naudé's *Apology for Great Men Suspected of Magic*

Variations in Editions from 1625 to 1715

Maryanne Cline Horowitz

In 1625, 25-year-old Gabriel Naudé published *Apologie pour tous les grands personages qui ont esté faussement soupçonnez de Magie [Apology for All the Great Persons who have been Falsely Suspected of Magic]* in Paris.[1] Demonic magic was well-established as a heresy in the seventeenth century. Naudé contended against so-called historians and demonologists from a wide range of religious viewpoints who were sure that demonic magicians had been and continued to be very active. Naudé's amusing book on how to read and evaluate books contributed to questioning such allegations among the lawyers and judges, theologians and academics, and doctors and professionals who frequented libraries open to gentlemen.[2]

Naudé dedicated both the 1625 *Apologie* and the 1627 *Advis pour dresser une bibliothèque [Advice on Establishing a Library]* to "Monseigneur de Mesmes, Conseiller du Roy & President en sa Cour de Parlement de Paris." Henri de Mesmes II, president of the Parlement of Paris, was his patron and employer in the 1620s, entrusting Naudé as librarian for the Bibliotheca Memmiana, an inherited collection of 8,000 printed volumes available to "worthy men."[3] Concurrently, Naudé studied medicine at the

University of Paris, continuing in Padua from September 1626 to May 1627 and then returning to Paris.[4]

Before writing the *Apologie,* Naudé rapidly wrote two other works: *La Marfore, ou discours contre les libelles,* a pamphlet defending King Louis XIII's favorite, the Duke of Luynes, against rabble rousers who would defame him, and *Instruction à la France sur la Verité de la Roze Croix,* a tract ridiculing the Rosicrucians as a fraudulent new concoction created out of false prophecies and heresies,[5] as well as poking fun at popular hysteria concerning their new manifestos.[6] The *Apologie* continued his style of ridiculing popular currents including claims that the activities of recent philosophers and historic figures displayed demonic magic. Naudé's *Advis pour dresser* created a new classification system for the diverse books Naudé deemed appropriate for a library.[7] Naudé's first four works displayed a hierarchy between popular libel and ideas worthy of consideration; a willingness and commitment to read a diversity of works to untangle truth from falsehood; an application of critical judgement and current scientific method taught to medical doctors to evaluating evidence and comparing claims in a wide variety of fields; and an enjoyment in collecting and organizing Latin quotations, odd ideas, and rooms of books and pamphlets.

Naudé's *Apologie* was published in multiple French editions, such as The Hague of 1653 (the year of Naudé's death) and Paris of 1669. Davies's translation into English appeared in 1657. In 1671 a capuchin, Jacques d'Autun (Jacques Chevannes), published a polemic against the *Apologie,* calling his own book "an apology for the belief of scholars who have justly accused great persons of magic."[8] The controversy and notoriety may explain additional publications of the *Apologie* in The Hague in 1679 and two in Amsterdam in 1712. Increased interest in Naudé may also have come from the publication in Paris in 1701 of *Naudaeana et Patiniana,* which accentuated the reputation of their libertine circle. A German translation of the *Apologie* appeared in 1704 and 1787.[9]

Jacques d'Autun spoke with the authority of his presence in Burgundy in 1644–45 during a popular witchcraze epidemic and having authored in 1649 a theology book dedicated to Louis de Bourbon.[10] He agreed with a report from Colbert's inquiry into the magistrates of Dijon in 1648–49 that they preferred their books more than their trade. He quoted a magistrate who said "one must not attribute the vexation of the possessed to the malice of evil spirits."[11] His three-part book concluded that there are magicians and sorcerers as declared in divine and human law; there are a variety of ways of ascertaining their demonic behavior; and there is an obligation to punish them.[12] He then turned to Naudé on the ground that readers of the *Apologie* scoff at the Parlement of Normandy's imprisonment of magicians and sorcerers in 1670. The public mistakenly attributed to criminals the innocence Naudé claimed for some famous authors. d'Autun's next 163 pages reexamined the sources Naudé cited and found evidence of demonic magic among ancient and modern authors.[13]

The Amsterdam edition published by Pierre Humbert has notes that define terms (diverse rituals involved in divination, ten attributes of God in the Jewish Kabbalah) and provide alternative sources and historical accounts of matters such as the King's Touch and medicine by magical herb or incantation. Notes also contend with De l'Ancre's *Traité de la Mescreance du sortilege* and apply interpretations from Pierre Bayle's *Dictionaire,* Huet's *Demonstratio Evangelio,* Lightfoot's *Hora Hebraica,* and Richard Simon's *Biblical Criticism.*[14]

The first word of Naudé's title, *Apologie,* would remind the educated of at least three books: Plato's *Apology,* Montaigne's skeptical essay "Apologie de Raimond Sebond," and the current cause célèbre, Tommaso Campanella's *Apologia pro Galileo* of 1622.[15] The works resonate to each other not only by title. Campanella, jailed by the Church in Rome since 1599 but serving nevertheless as a theologian invited to assess Galileo Galilei's views for the trial of 1616, managed to have his *Apologia pro Galileo* published in Frankfurt in 1622. In it, Campanella defended Galileo in the name of the benefits of freedom of inquiry for the Church, a principle Naudé affirmed. Naudé shared with Socrates, Montaigne, and Campanella an interest in paths to knowledge. Later, Naudé edited several of Campanella's books for the French public, and recorded and published Campanella's autobiographical *Syntagma.*[16]

Naudé discredited accusations of demonic magic as the products of ignorance or malice. He lumped together Socrates's "daemon" and the genii of Giralamo Cardano or Julius Caesar Scaliger as claims of theurgic magic (aid by an angel), and dismissed them as either exaggerated humility about their own natural powers or boasting.[17] Naudé thus puts on the appearance of a "sensible" man, borrowing the techniques of his favorite French authors, Montaigne and Pierre Charron, for ridiculing books reporting demonic or theurgic magic.[18] Cumulatively, readers come to "see" the patterns of absurd allegations, and in amusement become less prone to the absurd litigation of their own day.

The 1625 edition of Naudé's *Apologie* had some special features. After the preface, which mentions that some readers are superstitiously overly impressed by a multi-lingual author's borrowed texts, the reader comes upon an assortment of dedicatory poems, possibly from a "cabal" of associates: a poem in distinctive Hebrew letters; a formal Latin statement by I. C. Frey, doctor and philosopher, that might be interpreted by a passerby as an association of Naudé with magicians; as well as a commendation from physicians Guy Patin and Jacques Jovvin.

Likewise, the original edition has a valuable index that lists an assortment of claims of demonical magic and names of those accused of such practices. The alphabetical list juxtaposes names of eminent authors with a potpourri of popular superstitious ideas and fantasies, serving as a hilarious collection of wild accusations bouncing back on those who fabricated them.[19] Accusers' names are not worthy of mention in the index; abbreviations of the names of accusers and of their books are in the margins of the various editions.

In the index, under "l'heresie" and "heretiques," one learns that heresy is a reason that many have been accused of magic: heretics are enemies of Gregory VII; there is no merit to claims that certain authors are heretics; and many heretics have been magicians.[20] Chapter 9 focuses on Popes Sylvester II and Gregory VII, and clears them of accusations by Jean Wier, Calvinists, and Lutherans. In an effective image, Naudé suggests that the overall conundrum is like a Tower of Babel founded on envy and built with the imitative stonework of historians and demonographers.[21]

While many references in the index to "demons" lead to fictional accusations, a few such references lead to the phenomenen of "la Magie Goetique" or "cette Geotie."[22] His evidence is mainly biblical. Demonic action is charged against holy figures: the magicians of Pharoah, according to Moses; Jammes, Mambres, and Elymas, according to St. Paul; Simon Magus, according to St. Peter; Cynops, according to St. John the Evangelist. Nevertheless, citing demonic magicians Dr. Faustus and the Jew Zedechias and judgements during the reign of Emperor Charles V, Naudé upholds the condemnation of magicians in the Code, *Lege 7, Cod. de malef. & Mathem.*[23]

In chapter 7, on heresy, malice, ignorance, credulity, and want of judgement, Naudé ridicules the accusation of magic against Jan Hus and the Lollards, and the accusation by the English of witchcraft against Joan of Arc. Likewise he views the Pantagruelism of Rabelais as a romance and ridicules the University of Paris for judging Rabelais harshly. Naudé left out of his book the conspicuous and controversial topic of women accused of witchcraft or magic.[24] His book focused on clearing the reputations of famous historic men of influence.

Naudé followed the style of Michel de Montaigne. I have shown that in his essay "On Cripples," Montaigne cited Augustine's skepticism about pagan boastful claims to spoof Bodin's gullibility about Apuleius's tale of a man metamorphized into an ass carrying a burden.[25] Naudé cited Nider to cast doubt on the *Malleus* and, like Montaigne, he extrapolated to the currently influential story-telling of Jean Bodin's *Démonomanie*. Naudé's amusing explanation of how Platonists call the upper part of the soul a "daemon" as if it were a god would very easily undermine Bodin's *The-*

atrum naturae, section 14, "On the active and passive intellect," where Bodin attests to the active intellect as separate from the person, as in a good angel (daemon of Socrates cited) or in a demon of a witch.[26] Naudé touched on the topic again in discussing the influence of Avicenna's idea of active intellect on false prophets Savonarola and Nostradamus.[27] Having struggled to untangle Bodin's sources for his theory, an important link between his science and his theory of both prophecy and witchcraft, I appreciate Naudé's lighthearted dismissal. As for Bodin, whose *Methodus* on critical historical method Naudé is visibly employing, and whose *République* he praised in his *Advis pour dresser* of 1627,[28] Naudé conceded that the *Démonomanie* proves he was just a man, confused by reading Rabbis and Talmudists,[29] but also by reading Platonists and Arabs.

The *Apologie* distinguishes four kinds of magic and only condemns one form, the "geotick" by a demon.[30] Divine magic as in the miracles of Jesus Christ[31] or theurgic magic by an angel[32] is very rare but acceptable. Natural magic by one's own ability, as among Persian wise men, Greek philosophers, and Jewish Kabbalah, are also acceptable; his examples make it broader than what will soon be the new "natural philosophy."[33] Naudé's Preface tells of how in reading *Nouveau Jugement*'s attack on Virgil as a conjurer and magician, an accusation borrowed from Pierre De Lancre, he decided that such false calumny often befalls the great, and it is his task to restore such maligned authors to their proper reputation.[34]

Naudé's first chapter is on how to read and judge historians and demonographers and find one's way through their "labyrinth." The second is on kinds of magic in which the main point is to distinguish demonic magic from natural philosophy, called natural magic. Then follow five chapters explaining the roots of the accusations of demonic magic. These first 7 chapters provide the critical acumen for evaluating the next 14 chapters on those falsely accused. Of these, chapters 8–14 focus mainly on ancient and medieval figures and chapters 15–21 mainly on modern figures, although he does not circle back to Virgil until chapter 21. Numerologists in his readership would certainly see Naudé's magic of 3 times 7, and note how he evens the score in the concluding chapter 22, explaining how false views are maintained and mock-prophecizing what will occur if errors are erased. Ironically, what follows in the 1625 edition is an alphabetical hodgepodge of all the errors people have maintained. As if in fulfillment of the mock-prophecy, the posthumous editions do erase them.

Naudé demonstrated that people are accused of magic on account of their extraordinary erudition or mathematics.[35] A 1611 sudden death may have been a reference point for Naudé when he devoted a whole chapter to accusations against mathematicians, omitting mention of posthumous allegations of demonic activity against le Sieur De Vatan, a mathematician who died suddenly after printing an edition of Euclid.[36] People view those who can do more than they can as magicians; when people learn to do things, they no longer call it "magic."[37] People of the New World at first considered the Spanish—with their magical gunpowder—to be devils out to destroy them, and their ships as products of their magic. Exposing this logic, he points out that likewise people mistake the mechanical inventions of natural philosophers as magic, and suggests that books explaining mechanics would do wonders for demystification.[38] Naudé's comments apply to courtiers as well as commoners, for seventeenth-century palaces and gardens displayed curiosities including automata, fountains, and other amusing moving toys, explained by Salomon de Caus in his *Les Raisons des forces mouvantes* of 1615.[39] On this popular topos of mechanical devices, a decade after Naudé's *Apologie,* Descartes drew his own inference in *Discours de la méthode:* animal and human bodies are automata.[40]

Naudé gave many examples of great individuals initiating the charge of magician. In succession Naudé discusses Greek, contemporary, biblical, and theatrical claims to communicate with a daemon: Socrates, Apollonius, Chicus, Chardan, Scaliger, Campanella, Adam, Abraham, Jacob, St. Paul, and Dr. Faustus.[41] Mixing together a scholar in prison (Campanella) with holy figures who fend off demons (Abraham, St. Paul) and a true demonic magician (Dr. Faustus), Naudé argues that the claim of communicating with God or with an angel often leads to an accusation of demonic magic.

Of particular importance for the spread of Machiavellianism[42] and for increased rumors about a book *De tribus impostoribus* are Naudé's examples of famous founders of religions and of states. He cited Titus Livy, Book 4, Dec. 1, showing that to gain obedience lawgivers claim their laws were given by the gods and feign apparitions and magic to lay a future empire. For example, Draco and Solon founded their states with the magic of Mercury. Naudé applies this paradigm of impostors or fakes to Mohammed's son.[43] He was aware of the similar accusations made against Moses and Jesus, for later in defending Arnaulde de Villeneuve (Michael Servetus) from libel charges, he argued that the latter did not write "*De Tribus impostoribus* par Postel."[44] The editor of the 1712 edition, citing *De Tribus impostoribus* again at the beginning and end of a long note, suggested alternative authors for this book "everyone talks about but no one has seen."[45] Might this be an ad for a book about to be printed in the Hague in 1719?[46]

Naudé's *Apologie* devoted its entire chapter 11 to Numa Pompilius, legendary founder of both Roman religion and law. The attention to this one figure is striking, as is the choice of chapter 11—a chapter number that Agrippa Von Nettesheim's *De occulta philosophia* of 1531 does not provide because it transgresses ten, the number of the Decalogue.[47] Defending

Plutarch's, Tertullian's, and Lactantius's view of him as the historic second founder of Rome, civilizing the city with laws and religious rituals, Naudé quoted the French edition of Plutarch's life of Numa to claim that Lycurgus, Numa, and other great founders "wisely feigned communication with the Gods, anticipating that this fiction would be useful and salutary to those they knew would believe it."[48] Naudé proclaimed it the "maxim of Politicians . . . that the troubles and dissensions of the State and government always follow those which arrive in Religion" to justify the Romans of idolatrous later times burning the rediscovered books of Numa's principles of religion and government.[49] Naudé thus historically explained an attribution of demonic magic to the founder of an earlier and more pure form of laws and religion as a device for preserving a state. He referred to "the monarchy of the Romans" although Livy (40.29) attributed the discovery of the chest of books to the period of the Republic. The implications a seventeenth-century Protestant or Jewish reader might have drawn from this about either the French monarchy or the Roman papacy are far-reaching, possibly beyond Naudé's intent.

Naudé's chapter on Numa appears to be the source of the two pages on Numa in the 1719 composite book containing the earliest extant *Traité des trois imposteurs.*[50] In particular, Naudé's quote from Plutarch that Numa's law code dictated by a nymph was a political strategy appears succinctly as a fact in 1719; there is possible direct borrowing from Naudé here.[51]

The 1712 Naudé edition may be of use in untangling the origins of the 1719 pamphlet, which cited Naudé directly on the topic of Mohammed and *coups d'états.*[52] A small note with an asterix at mention of Numa's law forbidding idol worship states: "M. Huet and others believe that Numa Pompilius has derived from Moses the dogmas and rituals which the author here discusses."[53] The 1777 edition of *Traité des trois imposteurs* cited Guillaume Postel in 1543 as the first to say that the book existed and interpreted Naudé as attributing the book to Arnauld de Villeneuve. It also cited a discussion in the last article of *Naudeana* about the sources of *De Tribus impostoribus.*[54]

Naudé followed King James I in dismissing accusations that Moses learned magic from the Egyptians as absurd, although attesting to the presence of demonic magicians in Egypt. Likewise, he dismissed claims that Abraham or Jacob were astrologers, Joseph a diviner, and Solomon an enchanter.[55] On the subject of Jesus, after dismissing any parallel with Apollonius's *Golden Ass,* Naudé was adamant that there is no evidence of Jesus practicing magic; events of his life involve divine miracles. For the apology for Jesus he cites St. Jerome and St. Augustine.[56] The 1712 edition cites Machiavelli by name in the following, although the English translation follows the 1625 edition: "Polite Italian in his discourse upon

Livy: 'The people of Florence are no fools, yet Brother Savonarola persuaded them that he had conferences with God.'"[57] Directly following is Tertullian's description of the second founding of Rome by Numa and the succession of ancient leaders claiming to be sons of gods (interpreted by some as devils) and thus maintaining authority.[58]

Very wisely, in chapter 6 Naudé distrusts lists of books attributed to authors known for great books, pointing out that booksellers profit from selling such falsely listed works to people building a library of great authors. When we think of all the Aristotle texts now considered pseudo-Aristotle, including the popular eighteenth-century pamphlet *Aristotle's Masterpiece,*[59] we appreciate his concern for false attribution. He also warns against assuming than an author's viewpoint can be deduced from a title: for example, Alexander Aphrodiseus's *Magicall Arts,* Aquinas's *of Judiciary Astrologie,* and Roger Bacon's *Necromancie.*[60] He brings up the problematics of discussing a forbidden science: people think you are invoking it if you can discuss it. "After all, it would be grave simplification to believe that they have entered the circle, have practiced the invocations and exercised the magic, who refer to it in books . . ."[61]

Naudé's *Apologie* communicated a new appreciation of what authors attempt in books and what readers might attempt in reading and evaluating evidence. In the first chapter, he observed that the printing press multiplies the problem by spreading misleading information. To judge, one needs good authors and logic; in practice he is suggesting a scientific approach to history. The good authors are Seneca, Quintillian, Plutarch, Charron, Montaigne, and Vives and the historians Thucydides, Tacitus, Guicciardini, Comines, and Sleidan. In chapter 7, on heresy, he quipped that people use the poor books (like the pamphlets he condemns in his first published works) and leave the good ones in the libraries.[62]

On demonographers, Naudé applied the requisite Catholic belief in the Devil as an explanation for the devilish accusations against the innocent: "it commonly happens that they prove mere *calumnies,* weakly grounded suspicions, and indeed but vaine, light and inconsiderate words; which the Devil doth insensibly impose on the good names of the innocent, to the end they may one day prove occasions that men shall not be able to discern and punish the guilt."[63]

In his *Advis,* discussing selecting books, Naudé separates his categories carefully: Copernicus, Kepler, and Galileo belong together for changing astronomy. Likewise, he groups critics of old sciences such as Sextus Empiricus, Agrippa, and Pico refuting astrologers, and Scaliger opposing Cardano. The category of those writing on "cabbala, mneumonic devices, the Lullian art, the philosopher's stone, divinations, and the like matter" he characterizes as unprofitable except for the purpose of refutation, yet they do require space in the library.[64] In the *Apologie* severe criticism befalls those who think demonic magic is a part of medicine, yet he refuses to believe that Paracelsus practiced demonic magic, although he views Paracelsus as a heretic in philosophy, medicine, and religion.[65] Likewise, he ridicules diviners such as Nostradamus and their followers who find predictions in trivial events.[66] There is a rationalist consistency in Naudé's rejection of belief in occult powers such as magic and divination.

One way Naudé demystifies is by exposing symbols as powerless images. He classes philosophy and science as natural activities by explaining iconography by description (there are no visuals until the 1712 frontispiece). His critical historical method, which took on momentum in the early seventeenth century, contended with another movement also gaining momentum, the Baroque sensibility to manifestations of the transcendent in daily life. Naudé approved of the quadrivium—arithmetic, geometry, music, and astronomy—but condemned astrology, not as a magic, but as an understanding of destiny inconsistent with religion and free will.[67] Naudé lists the tools of magicians among "artificial machines": the physics and clocks of Archimedes.[68] Concerning Pythagoras, he dismissed the myth of a golden thigh as a hieroglyph of the philosopher's stone[69] and amused the reader with the tale of Pythagoras writing with blood on a hollow glass, with the blood made from beans. That topic leads him into a French peasant farce concerning flatulence, from which Naudé emerges as a believer in the Pythagorean prohibition on eating beans.[70]

Naudé saves Virgil for his last example. Virgil was his original reason for writing a defense of maligned great men, and Naudé attributes his so-called magic to his poetic inspiration. Tourist sites play a role in his modern reputation: a cabinet of the Duke of Florence has a large looking glass as physical remains of Virgil exorcising the demon Catoptramancy.[71] Of recent magic, the public associates Agrippa with a bell, book, and candle, as if a magician; and even worse is the image of the devil in the form of one of his dogs.[72]

Naudé undercut neo-Platonic claims of correspondences between stars and human life, evident in astrological decoration of the "grande salle" of the Paduan palace, as well as in numerous popes' tastes, by pointing out that it is not magic but simply a matter of using symbols when Abraham Aben-Ezra in commentary on the Talmud makes a parallel of ten emanations of the Divine, ten celestial spheres, and ten commandments of the law.[73] Yet despite his technique of tracing false allegations to non-Christian sources, he thinks it essential that a library have the Talmud, Kabbalists, and the Koran, as well as "the most learned and famous heretics" such as Luther and Calvin.[74] Naudé dismisses superficial similarities, as in parallels of pagan events to later events in the life of Jesus Christ.[75] Central to his dismissal of all false imagery is a Latin quote from Cicero; a common topos of the juxtaposing of multiple viewpoints on a subject in Bodin's works on historical method and comparative government; Montaigne's *Essais;* and Charron's *De la Sagesse:* "Nothing can be said to be so absurd that it has not been said by some Philosopher."[76]

Naudé applied the popular vogue of vegetative symbolism to point to truths growing within the forests of misleading texts, likened to tall trees devoid of spreading branches of knowledge. "Tedious and fruitlesse discourses are like Forrests of Cypresse trees, fair and flourishing to the eye, but bearing no fruit suitable thereof."[77] He scolded those who "quit the bark of truth."[78] He expressed concern that people glean what they want from printed books without weighing the evidence supplied by the author.[79] He spoofed those so taken with Renaissance printed book graphics and oratory that they interpret as "magicians those who represent some printer's roses and flowers during the worst season of winter: also all these great men who have appeared as stars in the middle of a cloudy night and have produced such admirable effects by their knowledge during a wintry season of letters."[80] While he expressed jaundiced weariness with the sixteenth-century tropes of new light and rebirth—great individuals described as stars of sparkling genius or as flowers growing out of wintry classical ruins—he took very seriously the seventeenth-century tropes for re-classifying the forest of knowledge.[81]

In 1622, three years before Naudé's *Apologie,* Leonard Gaultier, the designer of the wreathed Wisdom frontispiece of Naudé's favorite philosophy book, Charron's *De la Sagesse,* produced enlarged engravings of his 1616 print of the *Laurus metaphysica,* a laurel tree reaching high toward God whose roots are the three subdivisions of the ancient Stoics: physics, ethics, and logic.[82] Naudé cited Cardinal Bessarion (d. 1472) in opposition to the neo-Platonic attribution of Christian truths to Plato, let alone Aristotle.[83] He picked a most appropriate role model, the collector of the Greek manuscripts housed in the Marciana library in Venice. According to Bessarion, "Everyone knows that Plato was not a Christian and that, consequently, no one who professes to be a Christian should follow in his precise footsteps. To be sure some signs and traces, as it were, of our faith can be found in Plato's writings. Their presence is due to the natural lucidity of an enlightened mind." Dismissing both popular pamphlets fabricated from rumor and erudite tombs built on false foundations, Naudé declared that his task was not to pick off false branches of the tree of knowledge, but to pull out the roots: "But since it were vainly to squander time away, to lop off the branches instead of pulling up the roots, we must with that begin the ruine of all these fabulous narrations, and shew, that whatever the Platonicks have advanc'd, either of Daemons or Magick, can neither be prov'd by reason nor experience."[84]

Notes

1. Gabriel Naudé, *Apologie pour tous les grands personages qui ont esté faussement soupçonnez de Magie* (Paris: 1625). The reprint (Westmead: Gregg, 1972), has the same error of the last pages of chapter 22, pp. 641–649, numbered pp. 607–615. Davies translated the book as *The History of magick, by way of apology, for all the wise men who have unjustly been reputed magicians, from the creation to the present age* (London: 1657). See Gabriel Naudé, *Apologie pour tous les grands personages qui ont esté faussement soupçonnez de Magie, Dernière Edition où l'on a ajoûté quelques remarques* (Amsterdam: Pierre Humbert, 1712). I regularly cite the Amsterdam: Pierre Humbert, 1712 edition from Occidental College Special Collections by chapter number; for page references, I indicate city of publication before page number: Paris 1625, London 1657, Amsterdam, 1712.
2. See Robert Mandrou, *Magistrat et sorciers en France au XVIIe siècle: Une Analyse de psychologie historique* (Paris: Éditions du Seuil, 1980) and Elizabeth Wirth Marwick, *Louis XIII: the making of a king* (New Haven: Yale University Press, 1986). For Naudé's role in libertinage, see Antony McKenna and Pierre-François Moreau, *Libertinage et philosophie au XVIIe siècle. 2: La Mothe Le Vayer, Naudé* (Saint-Etienne: Publications de l'université de Saint-Etienne, 1997).
3. L. Jacob, *Traicté des plus belles bibliothèques publiques et particulieres . . .* (Paris: 1644), pp. 548–550.
4. Jack A. Clarke, *Gabriel Naudé 1600–1653* (Hamden: Archon, 1970), pp. 7–17. Lorenzo Bianchi, *Rinascimento e libertinismo: Studi su Gabriel Naudé* (Naples: Bibliopolis, 1996), pp. 207–211.
5. Clarke, *Gabriel Naudé,* pp. 9–10.
6. See Frances Yates, *The Rosicrucian Enlightenment* (London: Routledge & Kegan, 1972), pp. 106–109.
7. Gabriel Naudé, *Advis pour dresser une bibliothèque* (1627) (Leipzig: VEB, 1963). See Robert Damien, *Bibliotèque d'État: Naissance d'une raison politque dans la France du XVIIe siècle* (Paris: Presse Universitaire, 1995), especially pp. 29–33, 57, 75, 82–83, 100.
8. Jacques d'Autun, *L'Incredulité sçavante, et la crédulité ignorante: au sujet des magicians et les sorciers. Avecque la Reponse à un livre de Mr Naudé intitulé Apologie pour tous les Grands Personnages, qui ont esté faussement soupçonnez de Magie* (Lyon: 1671), pp. 935–1108. See Bianchi, *Rinascimento e libertinismo,* p. 36 n. 21.
9. Kristeller, "Between the Italian Renaissance and the French Enlightenment: Gabriel Naudé as an Editor," *Renaissance Quarterly* 39 (1969), 44 n. 11.
10. Mandrou, *Magistrat et Sorciers,* pp. 372–73, 385–86, 549.
11. D'Autun, *L'Incredulité,* pp. 470–80.
12. Ibid., p. 934.
13. Ibid., pp. 935–1108.
14. *Apologie,* Amsterdam: Pierre Humbert, 1712.

15. Thome Campanellae, *Apologia pro Galileo, mathematico florentino* (Frankfurt: 1622); Tommaso Campanella, *A Defense of Galileo, the Mathematician from Florence,* trans. R. J. Blackwell (Notre Dame: 1994). See John Headley, *Tommaso Campanella and the Transformation of the World* (Princeton: Princeton University Press, 1997), pp. 98, 115.
16. Kristeller, "Between the Italian Renaissance and the French Enlightenment," pp. 41–72.
17. *Apologie,* ch. 13, London, pp. 143–4, 147, 163; Amsterdam, pp. 241–246. See Ian Maclean, "The Interpretation of Natural Signs: Cardano's *De subtilitate* Versus Scaliger's *Exercitationes,*" in *Occult and Scientific Mentalities in the Renaissance,* ed. Brian Vickers (Cambridge: Cambridge University Press, 1984), pp. 231–252.
18. See Maryanne Cline Horowitz, *Seeds of Virtue and Knowledge* (Princeton: Princeton University, Press, 1998), on Montaigne and Charron. On Naudé, Marie de Gournay, La Mothe le Vayer, see also M. C. Horowitz, "French Free-Thinkers in the First Decades of the Edict of Nantes," *Early Modern Skepticism and the Origins of Toleration,* ed. Alan Levine (Lanham: Lexington Books, 1999), pp. 77–102.
19. The index to La Mothe Le Vayers's multi-volume *Oeuvres* also has a potpourri of ideas, but there is a minimum of references to heretics or magicians (Paris: 1669).
20. *Apologie,* Paris, citing pp. 100, 109, 551, 552, 577, 578.
21. *Apologie,* ch. 9, London, p. 265.
22. *Apologie,* ch. 8, pp. 135–136.
23. *Apologie,* ch. 2, pp. 56–57.
24. See Maryanne Horowitz, "Doubts about Witchcraft and Magic in Reginald Scot and Gabriel Naudé," *At the Frontiers of the Reformation,* ed. Lee Palmer Wandell (Kirksville: The Sixteenth Century Journal Publishers, forthcoming).
25. Maryanne Horowitz, "Drogue médicinale ou vieux conte: l'histoire et la justice chez Montaigne, Bodin, et saint Augustin," *Montaigne et l'histoire,* ed. Claude-Gilbert Dubois (Bordeaux: Presse de l'Université de Bordeaux, 1991), pp. 173–181.
26. Horowitz, *Seeds of Virtue and Knowledge,* pp. 198–199, p. 322n. 51, and p. 88n. 88. *Universae naturae Theatrum* (Lyon: Roussin, 1596), section 15, pp. 527–28.
27. *Apologie,* ch. 16. See *Jean Bodin: Selected Writings on Philosophy, Religion and Politics,* ed. Paul L. Rose (Geneva: 1980).
28. Horowitz, "French Free-Thinkers," pp. 90–91.
29. *Apologie,* ch. 7, Amsterdam, p. 83.
30. *Apologie,* ch. 2, Amsterdam, pp, 17–18.
31. *Apologie,* ch. 2, Amsterdam, pp. 18–19; ch. 12, Amsterdam, pp. 209–211.
32. *Apologie,* ch.2, Amsterdam, ch. 2, pp. 20–21.
33. *Apologie,* ch. 2, Amsterdam, pp. 27–30.
34. *Apologie,* Preface, Amsterdam, 4r–5r.

35. *Apologie,* ch. 4, Amsterdam, pp. 33–48 and ch. 5 on mathematicians.
36. *Apologie,* Amsterdam, p. 413 note, citing La Mothe le Vayer.
37. *Apologie,* ch. 4, pp. 43–44.
38. *Apologie,* ch. 4, Amsterdam, p. 47.
39. John Dixon Hunt, *Garden and Grove: The Italian Renaissance Garden in the English Imagination, 1600–1750* (Philadelphia: University of Pennsylvania Press, 1986), pp. 77–80.
40. Lorraine Daston and Katharine Park, *Wonders and the Order of Nature 1150–1750* (New York: Zone Books, 1998), pp. 88–100, 281–84 with illustrations, 292–293.
41. *Apologie,* ch. 3, Amsterdam, pp. 35–37.
42. See "Machiavelli and Religion: A Reappraisal," *Journal of the History of Ideas* 60, 4 (Oct. 1999), pp. 579–681.
43. *Apologie,* ch. 3, Amsterdam, p. 33.
44. *Apologie,* Ch. 14, Paris, p. 378, Amsterdam, pp. 268–269. *La Vie et L'Esprit de Spinosa,* 1719, the earliest discovered extant "three impostors," is a patched-together pamphlet. On pp. 123 and 126 it cites passages from Pierre Charron on religion and Naudé on coups d'états. See Silvia Berti, "The First Edition of Traité des trois imposteurs, and its Debt to Spinoza's *Ethics,*" *Atheism from the Reformation to the Enlightenment,* ed. Michael Hunter and David Wootton (Oxford: Clarendon, 1992), p. 193.
45. *Apologie,* Ch. 14, Amsterdam, pp. 268–269 note. The editor suggests that the book was written by Frederick Barbarossa against the papacy.
46. *La Vie et L'esprit de Spinosa* (The Hague: Charles Levier, 1719), available as *Trattato dei tre impostori: la vita e lo spirito del signore Benedetto de Spinoza,* ed. and trans. Silvia Berti (Turin: G. Einaudi, 1994).
47. Agrippa von Nettesheim, *La Philosophie occulte ou la magie* (Paris: Bibliothèque Chacornac, 1910), cited in Maryanne Horowitz, "Montaigne's Doubts," p. 84.
48. *Apologie,* Paris, pp. 257–258; Amsterdam, p. 179.
49. *Apologie,* ch. 11, Paris, p. 267.
50. *La Vie et L'Esprit de Spinosa* (1719), pp. 93–95. Berti, "The First Edition," p. 186, writes "In addition they wrote a wholly new chapter on a legislator they thought deserved more attention, Numa Pompilius." Ancient sources on Numa include Livy, l. 18–21; 40. 29 (on stone chest of Numa's laws discovered and burned); Plutarch, *Numa, Dion. Halic.* 2. 58–76, Cic. *De Rep.* 2.13–15. Montaigne is Naudé's precedent in favoring Plutarch on Numa: see his *Essays,* trans. Frame (Stanford: Stanford University Press, 1958), 2.12:381; 2.16:477; 2.31:539; 2.32:549. Plato's *Laws* 8.837–38 recommended the use of divinely founded law to socialize a population.
51. *Apologie,* Paris, p. 255. See also *La Vie et L'esprit de Spinosa,* p. 94.
52. *La Vie* cites Naudé on *coups d'états* on pp. 123 and 126.
53. *Apologie,* ch. 11, Amsterdam, p. 184, note.
54. Mr. De la Monnoye, "Sentiments Concerning the Treatise of the Three Impostors," in Abraham Anderson, *The Treatise of The Three Impostors and the*

Problem of the Enlightenment (Lanham: Rowman & Littlefield, 1997), pp. 46, 49.

55. King James, *Daemonologie* (1597) (New York: Barnes & Noble, 1966), pp. 24–26. *Apologie,* Paris, pp. 38–40.
56. *Apologie,* ch. 12, Paris, pp. 300–303; Amsterdam pp. 210–211. Also, ch. 3, 27–29.
57. *Apologie,* Ch. 3, Paris, p. 52, London, p. 26 quoted above; Amsterdam, p. 34. Also see ch. 16, Amsterdam, pp. 328–329 on Mohammed and Savonarola, with long editorial note at p. 328 finding Naudé biased against the heretic Savonorala and seeing consistency in Naudé's support for Catholic burning of heretics and the St. Bartholomew Day's Massacre. That reading follows from Naudé, *Considérations politiques sur les coups d'Etat* (Rome, 1639) (reprinted Hildesheim: Verlag, 1993), pp. 109–111.
58. *Apologie,* Paris, pp. 53–54.
59. Maryanne Horowitz, "The 'Science" of Embryology before the Discovery of the Ovum," *Connecting Spheres,* ed. Marilyn J. Boxer and Jean H. Quataert, second ed. (Oxford: Oxford University Press, 1999), p. 107.
60. *Apologie,* ch. 6, Amsterdam, p. 67.
61. *Apologie,* ch. 6, Amsterdam, p. 61.
62. *Apologie,* ch. 7, Amsterdam, pp. 68–75. The London edition is more striking on the contrast between books read and books ignored.
63. *Apologie,* London, pp. 305–306, Amsterdam, p. 470. Paris, p. 615. His statement of belief in the reality of Demonic magic appears also in Amsterdam, pp. 22–23.
64. Gabriel Naudé, *Advice on Establishing a Library,* trans. John Evelyn (1661; reprinted Berkeley: University of California Press, 1950), p. 26.
65. *Apologie,* ch. 14, pp. 246–285, especially p. 284, London, p. 187.
66. *Apologie,* ch. 16, Amsterdam, pp. 330–31.
67. *Apologie,* Ch. 5, London, p. 41, Amsterdam, p. 56
68. *Apologie,* ch. 5, Amsterdam, p. 51
69. *Apologie,* ch. 10, London, p. 109.
70. *Apologie,* ch. 10, London, pp. 103–109.
71. *Apologie,* ch. 21, Paris, p. 448
72. *Apologie,* ch. 15, Paris, pp. 428–429.
73. *Apologie,* ch. 14, Amsterdam, pp. 273, 255
74. *Advice,* pp. 27–28. See Gabriel Naudé, *Bibliografia Politica* (Rome: Bulzoni, 1997), pp. 277–284. A list of Naudé's books in his Paris study appears in Bianchi, *Rinascimento e libertinismo,* Appendix l, pp. 253–270.
75. *Apologie,* ch. 12; Amsterdam, pp. 208–211.
76. Cicero, *De. Div.* 2., London, p. 207.
77. *Apologie,* Paris, London, A 5v; Amsterdam, 7V.
78. *Apologie,* London, p. 9, Amsterdam p, 11.
79. *Apologie,* London, p. 8; Amsterdam, p. 10.
80. My translation of *Apologie,* ch. 7, Amsterdam, p. 76. The magic turns out to be the mechanics of imprinting fake flower borders and trim.

81. Paul Nelles, "The Library As an Instrument of Discovery: Gabriel Naudé and the Uses of History," *History and the Disciplines: Reclassification of Knowledge in Early Modern Europe,* ed. Donald Kelley (Rochester: University of Rochester Press, 1997), pp. 41–57.
82. Horowitz, *Seeds of Virtue and Knowledge,* pp. 225–226, 249. 337n.10, illust. 10.1. See also ch. 8 on Bodin, Epilogue, and illust. 7. 6, E.1, E.2, E.4.
83. *Apologie,* 13, Amsterdam, pp. 235–236.
84. *Apologie,* ch. 13, London, p. 159; Amsterdam, p. 237.

4

Hobbes on Heresy

Martyn P. Thompson

Thomas Hobbes wrote a lot on the subject of heresy, including two histories. The first of these was *An Historical Narration Concerning Heresy* (probably written in 1668; first published posthumously in 1680). The second was in Latin verse, the *Historia ecclesiastica* (begun in the late 1650s, completed around 1671–72, first published in 1688).[1] Elswhere Hobbes wrote an immense amount about religion in general and, in particular, about Christianity, church and state, and a "Christian Commonwealth." Important recent scholarship has uncovered in all of this Hobbesian arguments in favor of religious toleration.[2] Certainly, as John Marshall, Mark Goldie, and others have shown, some of Hobbes's first readers also found intimations of some sort of religious toleration in Hobbes's writings.[3] In particular, they have pointed to the controversies between Anglican latitudinarians and their High Church critics during the Restoration.[4] But I shall argue that these suggestions, then as well as now, misconstrue Hobbes's arguments. The bold advocate of that "mortall God," the "Great Leviathan," was no champion of "tender consciences." All that Hobbes said in numerous places about the *intellectual* or rational indeterminacy of theological and ecclesiological controversies never led him to support any kind of religious toleration.

Quite the contrary, from *The Elements of Law* (1640) to *Behemoth* (posthumously published in 1682), Hobbes insisted that both the nature of public worship and the logic of civil society required that sovereigns establish *uniformity* of religious worship. That Hobbes's Anglican critics should have suspected him of saying one thing while meaning another is not surprising. To them, Hobbes was at best a heterodox Christian, at worst an atheist, and with atheism anything goes. To his Anglican critics during the Restoration, "Hobbism" was a term with which to damn any of the opponents of the recently restored Anglican hierocracy and discipline. But Hobbes explicitly stated, admittedly in retrospect,[5] that he wrote *Leviathan* (1651) to defend the "rights of the king in things temporal as well as spiritual." And he did so because he "considered differences of opinion over theological questions as the cause of the civil war that was then being waged throughout England, Scotland and Ireland." In these circumstances, Hobbes was most unlikely to have supported any arguments for religious toleration. Freedom of religious expression and practice had simply fueled the fire that had almost destroyed civil society altogether.[6]

Given these very well-known positions adopted by Hobbes (one set arising from his exploration of the *logic* of Leviathan, the other from his contemporary *historical* diagnosis of the causes of the English Civil War[7]), readers in search of Hobbes as an advocate of toleration might hope to find more promising material in his writings on heresy. After all, Hobbes was constantly hounded by accusations of atheism and heresy. Despite the patronage of Charles II and Arlington, he was understandably fearful of moves by bishops and others to have him burned as a heretic. In 1662, he published both a rather contrite *Apology for himself and his Writings* and a much more aggressive defense, *Considerations upon the Reputation, Loyalty, Manners, and Religion of Thomas Hobbes.*[8] Despite Charles II's refusal to allow him to publish anything on politics or religion after 1662, Hobbes continued to write extensively on both subjects. He even published his Latin *Leviathan* in Holland in 1668, removing the old concluding chapters, adding new ones plus an appendix discussing the Nicene Creed, the concept of heresy, and a section concerned to rebut 13 accusations of heresy that had been leveled against the English *Leviathan.* He wrote what amounts to a legal brief on the English law of heresy. He added a chapter on the law of heresy to his manuscript that was later published as *A Dialogue between a Philosopher and a Student of the Common Laws of England.* He wrote a lengthy, now lost letter in 1669 about Daniel Scargill's recantation of his admitted "Hobbism." He completed the Latin poem in dialogue form, translated in 1722 as *A True Ecclesiastical History From Moses to the time of Martin Luther.* And in 1668, he completed a lengthy rejoinder to Bishop John Bramhall's book *The Catching of Leviathan* (1658), to which he appended another booklet, *An Historical Narration Concerning Heresy.*

There is considerable variation in style, tone, and form in all of these writings on heresy. There is considerable repetition, too. There can be no doubt that Hobbes was preoccupied during the 1660s and early 1670s with the question of heresy. He was also very much on the defensive. But his defense never led to arguments for religious toleration, even though

Hobbes was as keenly aware as anyone that his own religious ideas were scarcely orthodox. To be sure, at one point in his aggressive defense of himself against the attacks of John Wallis, Hobbes says: "if tenderness of conscience be a good plea [for the toleration of dissenters], you must give Mr. Hobbes also leave to plead tenderness of conscience to his new Divinity, as well as you."[9] But the statement is dripping in sarcasm. There certainly were people claiming "tenderness of conscience," but no matter how sincere such people might be, this was no argument for freedom of religious expression or freedom of public worship.

So what was Hobbes's understanding of heresy? If it did not lead him to construct an argument for toleration, what did it do? And how is it that some sensitive readers, both in the seventeenth century and today, have taken Hobbes to be, in Richard Tuck's term, a "radical tolerationist"?[10] Let me begin with the *Historical Narration Concerning Heresy* and the *True Ecclesiastical History*. The works have been relatively neglected.[11] They obviously serve, at least in part, the very limited purpose of defending Hobbes from current accusations of heresy. So they may seem of merely marginal historical interest, of relevance for a biographer of Hobbes but not for anyone interested in the coherence of his thought. Hence it seemed to Alan Ryan that the *Historical Narration* was "Hobbes's undistinguished little history of heresy," and to George Wright this rather peripheral work contained "Hobbes's first and last intentions in these areas."[12] Both judgments need amending.

On the surface, they are slight works. The main point of the *Historical Narration* is captured in the few lines from Lucretius's *De rerum natura* that Hobbes adopted as the book's motto:

> For just as children tremble and fear all
> In the viewless dark, so even we at times
> Dread in the light so many things that be
> No whit more fearsome than what children feign,
> Shuddering, will be upon them in the dark.[13]

So the argument will be suitably Epicurean and heresy will be shown to have been conjured up (as it emerges by the twin forces of Gentilism and priestcraft) in order to frighten and control the unsuspecting and the gullible. The first two-thirds of the booklet do exactly this. The word heresy, in the original Greek, meant "a taking of anything, and particularly . . . an opinion." So it "originally signified no more than a private opinion, without reference to truth or falsehood."[14] It was a name attached to schools of philosophy that, under the leadership of lesser intellects, turned the pursuit of truth into rhetoric and dogma. With the rise of

Christianity, heresy first acquired its derogatory associations. Because of their rhetorical skills, many of the pagan philosopher-rhetoricians who converted to Christianity were chosen "pastors of the primitive church." They imported their pagan philosophical attachments into their understandings of the Scriptures, and dissension proliferated. The only sanction against doctrinal dissenters was "ignominy," and it was to convey this that the term "heretic" was coined to accompany the already well-established term "heresy." "Catholic" and "heretic" became relative terms and heresy became the name of a doctrine condemned by the whole Church.[15]

The violent unrest of the Arian controversy proved both a milestone and a turning-point. The first Christian Emperor, Constantine, intervened to restore civil peace. His initiative inaugurated the series of great Church councils in Nice, Constantinople, Ephesus, and Chalcedonia, which established the Creed, identified and rejected heresies, and determined appropriate punishments. All of this was initiated and authorized by Constantine "not so much," Hobbes says, because he was concerned with "*the truth*" of Christianity but because he was interested in "*uniformity of the doctrine*" since the "*peace of his people . . . depended on it.*"[16] So Hobbes concluded: "And thus did heresy, which at first was the name of private opinion, and no crime, by virtue of a law of the Emperor, made only for the peace of the church, become a crime in a pastor, and punishable with deprivation first, and next with banishment."[17]

Having pursued the story thus far, Hobbes clearly felt he could wing the next few centuries. As all good Protestants knew, the history of the Roman Church from the later Empire to the eve of the Reformation was one of increasing power and corruption. Through the "negligence or weakness" of civil rulers, the pope came to do "what he pleased in religion." Ecclesiastical authority invaded civil authority. The identification of heresies expanded and were "*punished arbitrarily by the Emperors with banishment or death.*" Christian values were perverted. Serious Christians concerned to understand for themselves the word of God were threatened as heretics; "the careless cold Christian was safe, and the skilful hypocrite [became] a saint."[18]

At this point, Hobbes stops abruptly and turns to consider the English law of heresy. This occupies the remaining third of the booklet. Hobbes's sketch of English statute law from Richard II to the restoration of Charles II is hardly unobjectionable.[19] But the point is abundantly clear. It is designed to show that England no longer had laws in force against heresy and hence *Leviathan* could not be prosecuted by either its Anglican or Presbyterian opponents. The violence of these opponents, Hobbes concluded, was both unlawful and a contravention of St. Paul's injunction that Christians peacefully settle their disagreements with one another by persuasion alone.[20]

This second part of the *Historical Narration* clearly serves Hobbes's pressing need to defend himself from prosecution. But it also displays his conviction that only statute law, the express will of the sovereign, is genuine law. The first part is more interesting for present purposes. Hobbes's argument displays that common seventeenth-century view that origins, rational and/or historical, determine right or value, such that *quod initio non valet, tractu temporis non convalescit.*[21] What Hobbes uncovered, then, in the *historical* origins of the crime of heresy coincided exactly with what *reason* required. Heresy, that is, was a matter of contingent belief, not objective truth. Faith, not reason, determined the minimal Christian creed. Any clear-thinking Christians could find out for themselves in Scripture the simple beliefs God required of them for salvation. So faith and opinion, not truth, established Christian orthodoxy. The Imperial will, and neither God nor the priesthood, authorized both uniformity in religion and the civil punishment of heresy. And uniformity and civil punishments were pursued, not in the interest of true religion, but instead in the interest of civil peace. From then on, the history of Christendom from Constantine to the seventeenth century was one of increasing confusion in the various relations between Church and State, religion and philosophy, faith and reason, opinion and truth.

The *True Ecclesiastical History* fills in many of the gaps and rehearses many of the arguments of the *Historical Narration*. Near the beginning Hobbes notes: "Pure Nature points the God to whom we bow,/But don't the Method of true Worship show."[22] Reason tells us there is a God who we must honor in worship. Revelation tells us that God is Jesus Christ and contains the minimal doctrine for salvation. But men establish churches and devise the manner of public worship. In sketching what men in churches have done, Hobbes reveals much more clearly even than in his other writings that the villains of ecclesiastical history were the clergy: priests and pastors, official and self-styled. To be sure, there were some heroes: Moses, Aaron, St. Paul, St. Peter (Cephas), Apollos, Peter Waldo, Wycliffe (the last two being "Faint Morning Beams of Gospel Glories"[23]), and Martin Luther "(Great Man)," for example.[24] But from their first appearance in ancient Libya and Egypt, through Greece and Rome, the Middle Ages, the Renaissance, the Reformation, and into Hobbes's contemporary world, soothsayers, magicians, priests, pastors, popes, and self-styled saints were all more or less slaves to the sins of the flesh. Pride, greed, ambition, and avarice governed their conduct. In Juvenalian manner, Hobbes lashes them as "Knaves" concerned to advance themselves by deceiving and frightening the "Fools" that constituted the rest of humankind.[25] Among these knaves, on the nonclerical side, were a number of philosophers. Although these included Aristotle, Hobbes was much more hostile to Aristotelians: the clerical Aristotelians, who clouded the simplicity of God's word; and the secular Aristotelians, "the *Hogs* of ARISTOTLE'S *Sty,*" who undermined monarchical authority by stirring up popular rebellions, often with Church backing.[26]

As for the fools, these included all those secular rulers who either willingly subordinated themselves to the Church or who, like Charlemagne, thought they could use the Church's authority for their own ends. Both ended up advancing the power of the papacy over civil sovereigns.[27] Once again, there were some noble exceptions. In Hobbes's histories, Constantine loomed by far the largest. Hobbes's main point was not so much that Constantine converted to Christianity. It was rather that he insisted, for the sake of civil peace, that the Church define heresy and establish uniformity of religious worship and that all this acquired the force of law only by being made public and willed by the Emperor.[28] With few other exceptions (elsewhere Hobbes points to parallels between Elizabeth I, Charles I, and Constantine[29]), foolish civil sovereigns were the playthings of clerical knaves.[30] Even the Reformation had not improved matters much. There had been far too few genuine reformers. Most clerics, Protestant as well as Catholic, were simply concerned with the malicious pursuit of personal gain.[31] But at least since the rise of national churches, Christian civil sovereigns could once more lay claim to supremacy over their churches, if not over all of their church's doctrines.[32]

Now all of this concerning the origins of religious heresy, its relationship to religious orthodoxy and truth, its status as law, the corruption of the clergy, and the rational grounds for uniformity in religious worship, is perfectly consistent with everything else Hobbes wrote about the logic of sovereignty and the character of a "Christian Commonwealth." He never hesitated to insist that the civil ruler even in a Christian Commonwealth must impose uniformity in public worship. Yet diversity in public worship (as well as in private conscience) was a core demand of contemporary religious tolerationists like Roger Williams, Samuel Fisher, John Milton, William Walwyn, John Humfrey, and, eventually, John Locke.[33] So how is it that Hobbes seemed to some contemporaries and to some modern critics to be "a radical tolerationist?"

Hobbes's insistence on liberty of conscience is an obvious source. This certainly disturbed the Anglican opponents of the latitudinarians. To many of these, for a variety of reasons, as Alan Ryan has pointed out, Hobbes seemed to have "subverted *all* intellectual authority and religious authority along with it."[34] But two things need to be remembered. First, the latitudinarians (like Edward Stillingfleet and John Tillotson) were far from being radical tolerationists themselves; they were more often than not opponents of toleration. Theirs was a very limited concern to overcome some of the acrimonious disputes between Anglicans and Nonconformists by allowing scope for some Nonconformists to conform.[35] Second, and more important here, Hobbes's many references to liberty of conscience were not advocacy, but analysis. As a matter of fact, Hobbes always noted, you cannot know what another person is thinking, so it is self-defeating to require that they think other than they do. Thus for Hobbes, the fact of liberty of conscience would only become an argument for toleration if he were to conclude that people had a *right* to act in public on their private consciences. But this, of course, he always denied.[36] In Christian commonwealths, at least outward obedience was always required; in anti-Christian commonwealths, Christians might at most engage "without resistance," Hobbes said, in passive noncompliance and suffer the consequences, even death.[37] No doubt, the coupling of Hobbes's strict Erastianism with these frequent references to liberty of conscience might have led some readers to assume that they were dealing with a covert advocate of some kind of religious toleration.[38] But such a reading could only be sustained by ignoring Hobbes's repeated insistence that civil sovereigns establish uniformity in religion.[39] So to argue from Hobbes's notion of freedom of conscience to even a minimalist program for religious toleration is to mistake Hobbes's point.

More problematic are the conclusions of modern scholars. I shall look briefly at three of these. First, Alan Ryan's discovery of tolerationist arguments in Hobbes arises from a philosophical critique of Hobbes's texts. Ryan accepts all of the points I have made so far. But he detects an unresolved tension in Hobbes's thought between the lawfulness of a sovereign's commands (which are absolute and definitional) and their moral obligation (which depends on the willing compliance of individual subjects, and this in turn is a matter for individual consciences). Ryan acknowledges that the logic of sovereignty in Hobbes requires that the sovereign alone determines the application of legal and moral language. But since on Hobbes's own admission conscience cannot be coerced, Ryan sees a discrepancy between the sort of utilitarian commitment subjects might have to a sovereign's determination of law and justice and the non-utilitarian commitment that is necessary to standards of what is morally good and evil. He suggests that one way of closing the gap would be to see in Hobbes a theory of moral conditioning. Education, exclusively under the sovereign's control, is designed to condition the subjects into internalizing the moral rules established by the sovereign. And this is certainly evidenced in the "Review and Conclusion" to the English *Leviathan*.[40]

So far, Hobbes looks like a very illiberal and anti-tolerationist thinker. But this is a mistake, Ryan argues. We must recall that Hobbes dismissed the possibility of determining people's consciences. The sovereign might condition them, but he could never control them. Further, we must recall that Hobbes consistently warned that a sovereign's determinations, by the law of nature, were only legitimate when they furthered civil peace. Any sovereign that overstepped these bounds would fuel civil disturbance, not secure civil peace. So Hobbes required minimal government. And finally, if we bear in mind that Hobbes valued peace, not just for security's sake, but more importantly for the "commodious living" it permitted, then we can detect some sort of commitment to "intellectual liberty and toleration" in Hobbes. For commodious living involved enjoying "civilized pleasures," pursuing the "virtues of a gentleman," and these included "the pleasures of having a quiet conscience and an easy mind."[41] These would only be possible where sovereigns tolerated considerable diversity in intellectual pursuits and commitments. Ryan admits that this is still only a utilitarian argument. The non-intervention of the sovereign in the "inner lives" of subjects is a matter of "technique not principle." The only principle involved is that sovereigns refrain from intervening in their subjects' lives until the public peace is endangered.[42]

This is a coherent reading of Hobbes. But it seems to me to be historically inaccurate. As Ryan admits, Hobbes has next to nothing to say about anyone's "inner life" that is not exclusively physiological.[43] Furthermore, there is no evidence to suggest that Hobbes anywhere supported religious toleration or that any significant number of his first readers read him as advocating toleration. Some libertines, latitudinarians, and other Erastians, no doubt, extrapolated from his writings a certain permissive potential or "laissez-penser" (as Ryan calls it), which they then reworked.[44] Yet hardly anyone seriously argued that Hobbes himself actually recommended toleration.[45]

But most important for the historical inaccuracy of Ryan's interpretation, Hobbes does not leave an unresolved tension between legal and moral commitments. Both exist for Hobbes only in civil society and both arise from the same act of authorization. The sovereign is authorized by its future subjects to determine the proper meaning and use of both legal and moral language. The result is that the sovereign's acts in all these respects are the subjects' own.[46] This argument works in Hobbes because Hobbes can still invoke a concept of authority that we have all but lost. If sovereigns' decisions are genuinely authorized and hence authoritative (as they are in Hobbes through his notions of authorship, covenant, and representation), then they are unquestioningly recognized as such by their subjects. This is the meaning of *auctoritas,* as Hannah Arendt and others have reminded us.[47] If authoritative pronouncements are questioned (as they always might be), if tender or rebellious spirits refuse this recognition, then the fault, the self-contradiction, lies in the questioners.[48] Hence the short shrift Hobbes gives to claims of tender or inspired consciences.[49]

Hobbes's account of the nature of authority in *Leviathan* is far from unproblematic.[50] By attempting to theorize the concept of *auctoritas* in a manner appropriate for the conditions of a modern state, Hobbes can certainly be seen as contributing to the complex set of intellectual and practical political changes that eventually led to the decline of classical authority in the modern world.[51] With hindsight, Arendt interprets both Luther and Hobbes as doing precisely this.[52] So, too, in a rather different way, does Horst Rabe.[53] But neither Arendt nor Rabe doubt that Hobbes was theorizing a classical conception of *auctoritas.*[54] Since we have all but lost this concept of *auctoritas,* however, we find Hobbes's argument either difficult to follow or simply unacceptable.[55] And it is this, perhaps, that sent even so sensitive a reader of Hobbes as Alan Ryan off looking for the tolerant "conditioning" of recalcitrant subjects' moral sensibilities. Hobbes himself had no such need.[56]

Richard Tuck's discovery of arguments for toleration in Hobbes is very different. His account is exclusively historical, exclusively contextual. The Hobbes that emerges is much more radically tolerationist than Ryan's. Tuck's argument is that in all his writings, Hobbes makes a fundamental distinction between "reason" and "faith." Reason, the province of philosophy and truth, discovers that there is a God, a first cause, and that God must be worshipped. It also discovers the laws of nature and identifies them as God's laws. This natural religion was the basis of all particular religions but their forms varied and they were purely matters of "faith." The relationship between natural reason and Christian faith occupied much of Hobbes's attention in the *Elements of Law* (1640), *De cive* (1642), and *Leviathan* (1651).[57]

So far this is uncontroversial. But then Tuck notes in chapter 47 of *Leviathan* a dramatic change from Hobbes's earlier views. In the earlier works, Hobbes was "a reasonably orthodox Anglican."[58] In chapter 17 of *De cive,* for example, Hobbes insisted that a Christian sovereign is obliged to interpret the Scriptures "through properly-ordained *Clergymen.*"[59] But in *Leviathan* much of this disappears. Now the sovereign alone decides matters of faith and, according to Tuck, Hobbes added to this a "passionate defense of toleration" when "he praised the decentralized ecclesiastical order of the new republic in England."[60]

The crucial text is the famous one from the English *Leviathan* where Hobbes argues "and so we are reduced to the independency of the primitive Christians, to follow Paul, or Cephas, or Apollos, every man as he liketh best which, if it be without contention, and without measuring the doctrine of Christ by our affection to the person of his minister . . . is perhaps the best."[61] This certainly sounds like a defense of religious pluralism and toleration. Tuck calls it a "defense of Independency" that proved so embarrassing to Hobbes after the Restoration that he "eliminated" it from the 1668 Latin *Leviathan*. And Tuck's conclusion is that "*Leviathan* is a defense of toleration, and *De cive* a defense of religious repression."[62]

But both internal and external evidence suggest that Tuck's conclusions cannot be sustained. First, *De cive* is hardly a defense of orthodox Anglican intolerance and repression. Hobbes's point in the passage from *De cive* that Tuck quotes is the familiar Hobbist one of insisting that civil sovereigns authorize Christian belief. The full quotation makes this clear.

> As a Christian, therefore, the holder of sovereign power in the commonwealth is obliged to interpret holy scripture, when it is a question about the *mysteries of faith,* by means of duly ordained *Ecclesiastics.* And so in *Christian* commonwealths judgement of *spiritual* and *temporal* matters belongs to the civil authority. And the man or assembly which holds sovereign power is the head of both the *commonwealth* and the Church; for a *Christian Church* and a *Christian Commonwealth* are one and the same thing.[63]

Certainly, what Hobbes has to say about church government in *De cive* is compatible with much of the theory and practice of Anglicanism, and this did disappear from *Leviathan.*[64] But by then, of course, the Church of England had disappeared from England.[65] From its establishment in the Reformation and into its exile in France, the Anglican Church had always acknowledged the monarch as its head and as the civil sovereign. So it might seem that Hobbes was suggesting in the first part of the passage just quoted that the Church was authoritative in spiritual matters (as Anglican bishops tended to claim), whereas the king was the authority in civil matters (as Anglican bishops always conceded).[66] But the second part makes clear that this was not his meaning. For having said that a Christian sovereign was obliged to interpret the "*mysteries of faith,* by means of duly ordained *Ecclesiastics,*" Hobbes immediately concluded that the judgment of *both* spiritual and temporal matters belonged exclusively to the civil sovereign.[67] And Hobbes consistently maintained this position from the *Elements of Law,* through *De cive* and *Leviathan,* to the *True Ecclesiastical History,* the *Answer to Bishop Bramhall's Book,* and *Behemoth.*[68]

Hobbes always insisted that the Church's authority to interpret scripture was derived from the civil authority and this for the sake of civil peace.[69] He always opposed Anglican claims for a *jure divino* episcopacy. Yet this was a prominent theme in orthodox Anglican thinking especially in the Laudian Church.[70] So what, exactly, did Hobbes mean, in *De cive* and elsewhere, when he noted an obligation of the civil sovereign to ecclesiastics?

It was an obligation with respect, specifically, to the "*mysteries of faith.*" In all of Hobbes's writings about Christian doctrine, he distinguished between matters essential for salvation and those that were inessential. The exact content of the essential and the inessential varied somewhat from work to work. But Hobbes always maintained that the fundamentals of Christian faith were not mysterious and that they alone were essential for salvation. Disputes about the mysteries were matters of religious "SUPERSTRUCTION."[71] The fundamentals always included the belief "*that Jesus is the Christ* . . . and all the explications thereof . . . as also all such as be evidently inferred from thence."[72] The "mysteries," by contrast, always concerned such questions as "how it may be made out with wit, *That there be three that bear rule in heaven, and those three but one? How the deity could be made flesh? How that flesh could be really present in many places at once?* . . . And other metaphysical doctrines."[73] Disagreements about these superstructural mysteries were often the cause of violent conflict between Christians, as they had been under Constantine. And as Constantine had recognized, such conflicts could only be resolved "through properly ordained *Ecclesiastics.*" They were inevitably matters of considerable complexity, indeed obscurity. No sovereign who was not ordained would have the requisite expertise to resolve them alone.[74] The practice of ordination was designed to guarantee a clergy with that requisite expertise. The choice of candidates for ordination was, however, a matter exclusively for the Church, that is, for the Christian commonwealth represented by the civil sovereign. The ordination itself, through prayer and "the laying-on of hands," was a matter for the "Teachers" in the Church, that is, the bishops.[75] But the authority of the ordained clergy was only an authority to teach. It was never an authority to command.[76] Yet maintaining civil peace in times of religious contention might well require turning the expert advice of the clergy into authoritative commands.[77] So the obligation of the sovereign with respect to ecclesiastics was an obligation to accept their expert opinions about scriptural interpretation.[78] It was also an obligation to command obedience to them, if and only if, in the civil sovereign's judgment, doing so would serve the interest of civil peace.[79]

This relationship between a Christian king and a Christian church was the same in *De cive* as in Hobbes's defense of *Leviathan* in answer to Bishop Bramhall. In reply to Bramhall, Hobbes wrote:

> Where the king is a Christian, believes the Scripture, and hath the legislative power both in church and state, and maketh no laws concerning Christian faith, or Divine worship, but by the counsel of his bishops whom he trusteth in that behalf; if the bishops counsel him aright, what clashing can there be between the Divine and human laws? For if the civil law be against God's law, and the bishops make it clearly appear to the king that it clasheth with Divine law, no doubt he will mend it by himself, or by the advice of

> his parliament; for else he is no professor of Christ's doctrine, and so the clashing is at an end. But if they think that every opinion they hold, though obscure and unnecessary to salvation, ought presently to be law, then there will be clashings innumerable, not only of laws, but also of swords, as we have found is too true by late experience. But his Lordship [Bramhall] is still at this, that there ought to be, for the Divine laws, that is to say for the interpretation of Scripture, a legislative power in the church, distinct from that of the King, which under him they enjoy already. This I deny.[80]

To the extent that Hobbes understood Bramhall here as representing orthodox, mid-seventeenth-century Anglicanism, he was opposed to it. He was opposed, it should be added, for substantially the same reasons in both *De cive* and *Leviathan*.[81]

No matter whether Hobbes was correct in detecting this disruptive tendency in Laudian and Restoration Anglicanism, there is strong external evidence that he rejected much of the practice of the Anglican Church in the reign of Charles I. At the time he was writing *De cive,* he supported many of the "root and branch" criticisms of Anglicanism. He suggested that the Anglican bishops were partly responsible for pushing the country toward civil war.[82] Writing to the third Earl of Devonshire in summer 1641, he argued that the "abondance of abuses committed by Ecclesiasticall persons and their Officers . . . can neyther be denyed nor excused." Their "Couetousness and supercilious behauior" had brought episcopacy into disrepute and hence it was perfectly reasonable that ecclesiastical jurisdiction should have been taken from them and placed in the hands of lay commissioners. "I am of the opinion," he continued:

> that Ministers ought to minister rather then gouerne; at least that all Church gouernment depend on the state, and authority of the Kingdome, [without which] there can be no vnity in ye church. your [Lordship] may perhaps thinke this opinion, but a *fancy of Philosophy.* but I am sure that Experience teaches, thus much, that the dispute for [precedence] betwene the *spirituall* and *civill power,* has of late more then any other thing in the world, bene the cause of *ciuill warres,* in all *places of Christendome.*[83]

So the model of church-state relations at work in *De cive* was not that of the intolerant Anglicanism of the Laudian Church. But it may have been that of the severely chastened Anglican Church immediately following the "root and branch" agitation, the reforms of 1641, and the impeachment of Archbishop Laud. Momentarily, perhaps, but significantly for Hobbes, ecclesiastical power was indisputably subordinated to civil power. It was this church that earned Hobbes's highest praise, at least in retrospect in 1668. After the dismantling of the "great power of the bishops" in 1641, he claimed, the Church of England had become "equal to that of the church at the time of Constantine with respect to purity of doctrine but superior to it with respect to the fairness of ecclesiastical laws."[84]

The evidence, then, seems not to support Tuck's first claim that *De cive* was a reasonably orthodox defense of Anglican intolerance. What of his second claim that chapter 47 of *Leviathan* contains a "passionate defense of toleration" with its praise for the "decentralized ecclesiastical order of the new republic in England?" This, too, is untenable. Chapter 47 is not a defense of Independency, though Clarendon apparently attacked it as such and Tuck and others have followed suit.[85] Rather, as Hobbes suggested in various places, it is an attack on the religious chaos that followed the temporary fall of Anglicanism when there "emerged every type of sect of men, writing and publishing whatsoever theology" they wished.[86] Read this way, Hobbes's argument in chapter 47 is not a surprising, uncharacteristic, and short-lived defense of religious pluralism by an otherwise consistent defender of religious uniformity and absolute sovereignty. It is rather a perfectly characteristic attack upon religious chaos by a consistent defender of religious uniformity and absolute sovereignty. It is certainly a passionate attack, but it is heavily laced with sarcasm.[87]

What Hobbes was doing in leading up to the famous passage is clear. In a couple of paragraphs, he sketched the ecclesiastical history of Christianity from the Apostolic Church to the present. His point was to show, in the terms of the *True Ecclesiastical History,* how clerical knaves and lay fools had conspired to tie three "knots" around "Christian liberty." Originally, Christians obeyed the Apostles out of reverence for their "wisdom, humility, sincerity, and other virtues." These primitive Christians' "consciences were free," but their outward doings, "their words and actions," were not. Their outward doings, however, were "subject to none but the civil power."[88] All this changed as the Church expanded and "presbyters" (priests) acquired power over their congregations. An ecclesiastical power was added to the civil. This was the first "knot" tied around Christian liberty. Subsequently, the chief presbyters or bishops acquired authority over lesser priests. This was the second "knot." And finally, the Bishop of Rome "took upon him" authority over all other bishops. This was "the third and last knot . . . of the pontifical power."[89] Then came the Reformation. In England, each of the knots was untied, but in reverse order. The ecclesiastical power of the pope was removed by Elizabeth I. *Jure divino* episcopacy was dismantled in the 1640s. The third knot was almost immediately untied when power was taken "from the presbyterians." And so, Hobbes continues:

> we are reduced to the independency of the primitive Christians, to follow Paul, or Cephas, or Apollos, every man as he liketh best: which, if it be

> without contention, and without measuring the doctrine of Christ, by our affection to the person of his minister (the fault which the apostle reprehended in the Corinthians), is perhaps the best. First, because there ought to be no power over the consciences of men, but of the Word itself, working faith in every one, not always according to the purpose of them that plant and water, but of God himself, that giveth the increase. And secondly, because it is unreasonable in them, who teach there is such danger in every little error, to require of a man endued with reason of his own, to follow the reason of any other man, or of the most voices of any other men, which is little better than to venture his salvation at cross and pile. Nor ought those teachers to be displeased with this loss of their ancient authority. For there is none should know better than they, that power is preserved by the same virtues by which it is acquired; that is to say, by wisdom, humility, clearness of doctrine, and sincerity of conversation; and not by suppression of the natural sciences, and of the morality of natural reason; nor by obscure language; nor by arrogating to themselves more knowledge than they make appear; nor by pious frauds; nor by such other faults, as in the pastors of God's Church are not only faults, but also scandals, apt to make men stumble one time or other upon the suppression of their authority.[90]

At first glance, this may look like a defense of religious freedom and praise for the decentralized ecclesiastical order of England's new republic. To the extent that it was directed against Roman Catholicism, *jure divino* episcopacy, and Presbyterianism, Hobbes was indeed saying that the situation in England was better without them. But this was a far cry from advocating religious freedom and from celebrating the religious pluralism of the late 1640s and early 1650s.[91] In Hobbes's view, men were in fact stumbling in all directions. Although "the sting" had been taken out of Presbyterianism, "Fifth-monarchy-men," other "sectaries" and other "fanatics," had risen in their stead.[92] There is no reason to believe that when Hobbes wrote *Leviathan* he had changed his mind at all about the prevalence of clerical "knavery," lay "foolishness," and the dangers of ecclesiastical disorder.[93] In fact, whenever he cited the *First Epistle of Paul to the Corinthians* (especially I Corinthians 1:12 and 3:18–23), his point was to remind readers of just such knavery, foolishness, and the dangers they posed to civil and ecclesiastical peace.[94] For this, of course, was precisely Paul's point in his blistering attack on the "sectaries" and "fanatics" in Corinth with their spurious claims to be followers of Paul, Cephas, or Apollos. They were destroying the unity of the primitive Corinthian Church.

So Hobbes's parallel between the "independency of the primitive Christians" to choose to follow "Paul, or Cephas, or Apollos" and the current religious scene in England cannot have been meant as praise for the latter. There are at least three reasons why not. First, that would have required Hobbes to reject Paul's ideal of Church unity based on the simple fundamentals of Christian faith,[95] an ideal that Hobbes otherwise shared. Second, Hobbes believed that the primitive Apostolic Church was inevitably vulnerable to disorders caused by the blending of pagan philosophy, religion, and Christian faith, precisely the kind of blending that Paul identified as the cause of sectarianism and irreligion in Corinth. This kind of disorder, Hobbes wrote, "was at the greatest in the times of the Apostles, and in the Primitive Church, till the time of the Nicene Council, and somewhat after."[96] So neither Corinth nor the historical Apostolic Church could function as models of good ecclesiastical order. Third, for the parallel between decentralized Corinth and the decentralized ecclesiastical order of 1650 to work as a favorable gloss on the latter would have required a radically different understanding of the current state of religious affairs in England from the one Hobbes actually held.

But precisely here is the clue to Hobbes's meaning. The parallel was highly appropriate, but for the opposite reason from the one that Tuck and others have assumed. It served to condemn the current disorder. Its heavy sarcasm has been completely overlooked. It is not a defense of religious pluralism and toleration but an attack on religious chaos. The religious scene, in the absence of hierarchy and without a sovereign, was brimful of "contention," private ambition, and personal "affection." It was also completely devoid of those virtues of "wisdom, humility, clearness of doctrine, and sincerity of conversation" that Hobbes claimed were necessary if any of the competing religious doctrines were to carry authority. So Hobbes was not saying that the contemporary religious scene in England was like that of St. Paul's ideal for the primitive church before Constantine. For that would have been a church without contentions, without cults of personality, without "divisions," where "all speak the same thing," and where all are "perfectly joined together in the same mind and in the same judgment."[97] Rather, Hobbes was saying it might have been like that if it were not so fraught with contention, so divided along sectarian lines and so dominated by so many scoundrels.

The passage, furthermore, was not so embarrassing to Hobbes after the Restoration that he expunged it from the 1668 Latin edition. He certainly removed the passage, but by then, of course, the Anglican Church had been restored and was well on the way to cleaning up the ecclesiastical disorder that Hobbes had found so reprehensible before. Equally to the point, Hobbes rewrote the whole of the last part of *Leviathan* for the 1668 edition because he directed that Latin edition to a different, non-English audience. In these circumstances, it is hardly surprising that he omitted many remarks of specific interest only to England.[98]

My third example is Richard Flathman's discovery of a tolerant Hobbes that emerges neither from close philosophical analysis of Hobbes's writings nor from their historical contextualization. His is rather an exercise in normative political theorizing in which Hobbes is enlisted to serve Flathman's own purposes in exploring the creation and maintenance of "conditions favorable to individuals' self-making and to their never-ending struggle to achieve felicity."[99] Flathman is perfectly clear about the character of his intellectual encounter with Hobbes's writings. He accepts that it involves the "risk of distorting Hobbes's intentions and/or the meaning of Hobbes's texts." It does so because his point is "to wrench [Hobbes] out of his context, to thrust him into ours, to make his texts speak to questions of present philosophical and especially moral and political concern."[100]

So Flathman's Hobbes is at once much easier and much more difficult to criticize than Ryan's or Tuck's. Philosophical and historical distortions of Hobbes's meanings are pre-programmed. But to dwell on them is to miss Flathman's point. He reads out of Hobbes's writings arguments that champion robust individuality, scepticism, and a defense of both limited government and the circumscribed character of political activity. Hobbes's concerns with "commodious living" and the limited aims of governmental activity are said to generate a significant sphere of individual liberty, especially "religious liberty."[101] To be sure, Flathman sees many unresolved tensions in Hobbes's thought between his defense of absolute sovereignty and his defense of robust individuality. He sees Hobbes's preoccupations with the sovereign's role as educator as inadequate to resolve these tensions. Nonetheless the thrust of Hobbes's argument and "the best of Hobbes's thoughts" supposedly point in the direction of religious toleration and more. In *Leviathan,* Flathman claims, Hobbes offers a "celebration of freedom, of pluralism, above all of individuality and individual self-making."[102]

In short, then, up to this point Flathman's selective but sympathetic reading of Hobbes (which, despite his disclaimers, does have some foundation in the historical Hobbes) has much in common with Ryan's. Flathman presents it as a reading inspired by Michael Oakeshott but his annotated bibliography includes appropriate references to Ryan as well.[103] The bibliography also includes appropriate references to Tuck's work. As it happens, *Leviathan*'s extended play on the *First Epistle of Paul to the Corinthians* is as crucial for Flathman's identification of Hobbes as a radical religious tolerationist as it is for Tuck's. But Flathman, too, misreads the passage, though in a somewhat different way.

According to Flathman, Hobbes's quotations from and paraphrasing of St. Paul on the freedom "to follow Paul, or Cephas, or Apollos" are an "endorsement" of the "religious freedom and pluralism that obtained in the earliest days of Christianity."[104] In making this suggestion, Flathman ignores the fact that Hobbes well knew that St. Paul was condemning religious pluralism at Corinth, not endorsing it. He also ignores the fact that by invoking St. Paul and the ideal of Christian unity to which Paul subscribed, Hobbes was condemning the religious factionalism of the late 1640s and early 1650s, not endorsing it. And furthermore, Flathman's suggestion ignores the fact that Hobbes, on occasion, explicitly rejected the idea that the primitive Christian Church before Constantine was a model of good ecclesiastical order.[105]

Flathman's final claim about the extent of Hobbes's commitment to religious toleration is obscure. He suggests that Hobbes's reference to "Paul, or Cephas, or Apollos" indicates that "the freedom Hobbes admires was for more than just the Christians." But why this is so is unclear. Paul, Cephas, and Apollos were all indisputably Christians. The immediate reference was to Christian factions in Corinth and the indirect reference was to Christian factions in mid-seventeenth-century England. Certainly, part of the problem in Corinth sprang from the enduring attachments of some of the Christians to pagan ideas, beliefs, and practices. But Paul in no way endorsed this religious diversity. Certainly, too, there are many unresolved questions in the literary, historical, and theological interpretation of the *Epistles to the Corinthians.* In particular, it is uncertain whether the competing Christian factions in Corinth were actually claiming to be followers of Paul, or St. Peter (Cephas), or Apollos (the Christian convert who once taught at Corinth and who was with Paul in Ephesus when the *First Epistle* was written).[106] Since Paul notes at one point that he was applying the names Paul and Apollos "figuratively," it seems probable that they were not (or at least not only) making this claim.[107] Paul's point, then, in employing figuratively the names of three such prominent Christians may well have been to draw attention to the foundation of Christian unity that Corinthian factionalism was destroying. But however one decides the issue, there is no evidence that Paul was addressing his remarks to any but the Christians at Corinth. And there seems, therefore, no reason to believe that Hobbes was doing anything different with respect to his contemporaries.

It seems, then, that none of the three interpretations of Hobbes as a champion of religious toleration is really sustainable. This does not mean that Hobbes's writings cannot be made to generate tolerationist ideas. Clearly they can. Ryan and Flathman make only this claim. But it should be abundantly clear that the more Hobbes's *writings* are torn from their historical moorings, the less they are *Hobbes's* writings. Let me then conclude by returning to the historical Hobbes. I shall simply summarize my very different interpretation of Hobbes's views on religious uniformity and religious toleration from Richard Tuck's.

My suggestion is that Hobbes's deep and abiding concern with religion, religious controversy, and the notion of heresy and his deep and abiding awareness of the corruption of the clergy of all denominations, never led him to advocate any form of religious toleration. Of course, he never wavered in the belief that private conscience could not be coerced. But equally, he never wavered in the belief that all outward actions could be, if need be. The central question of religious toleration was the right to express outwardly, publicly, the diverse religious beliefs that various groups and individuals held inwardly. Hobbes always denied that this was a right. He denied it for perhaps four main reasons, each reinforcing the others. First, he was convinced as a matter of human experience, that religious dissension was the main cause of civil wars. Second, also as a matter of human experience, to accept diversity in religion would be to unharness the forces of clerical "knavery." Third, any such multiplication of competing authorities would unnecessarily complicate and ultimately contradict the logic of sovereignty in a modern state. And fourth, he always insisted that natural religion required uniformity of public worship.[108] It did so because the concept of God contained the concept of *public* worship. Public worship was human action designed to pay appropriate honor to God as witnessed by other human beings. But one person's conception of honor was another's idea of insult. So, as in everything else that concerned public life, only the will of the Christian sovereign, paying due attention to the advice of Christian clerics, could determine unequivocally what constituted genuine public worship.[109] For a Christian civil sovereign to permit public expressions of religious diversity would be to sanction, in the eyes of fellow citizens, insulting God. Instead of securing civil peace, it would simply nourish the seeds of yet further civil strife.

Notes

1. On the dates of Hobbes's works on heresy, see Philip Milton, "Hobbes, Heresy and Lord Arlington," *History of Political Thought* 14, 1993, pp. 501–46. Hobbes's Latin poem was translated into English as *A True Ecclesiastical History from Moses to the Time of Martin Luther in Verse* (London: E. Curll, 1722). All quotations are from this edition.
2. I shall be particularly concerned with three recent arguments by Alan Ryan, Richard Tuck, and Richard Flathman. See Alan Ryan, "Hobbes, Toleration, and the Inner Life," in David Miller and Larry Siedentop, eds., *The Nature of Political Theory* (Oxford: Oxford University Press, 1983), pp. 197–218, and "A more tolerant Hobbes?" in Susan Mendus, ed., *Justifying Toleration: Conceptual and Historical Perspectives* (Cambridge: Cambridge University Press, 1988), pp. 37–59; Richard Tuck, "Hobbes and Locke on Toleration," in Mary G. Dietz, ed., *Thomas Hobbes and Political Theory* (Lawrence: University of Kansas Press,

1990), pp. 153–71, and in several subsequent publications; and Richard E. Flathman, *Thomas Hobbes: Skepticism, Individuality and Chastened Politics* (Newbury Park: Sage, 1993). See also, amongst many others, the well-informed discussion in Glenn Burgess, "Thomas Hobbes: Religious Toleration or Religious Indifference?" in Cary J. Nederman and J. C. Laursen, eds., *Difference and Dissent: Theories of Toleration in Medieval and Early Modern Europe* (Lanham: Rowman and Littlefield, 1996), pp. 139–61.

3. See Mark Goldie, "The Reception of Hobbes," in J. H. Burns, ed., *The Cambridge History of Political Thought 1450–1700* (Cambridge: Cambridge University Press, 1994), pp. 610–15.
4. See John Marshall, "The Ecclesiology of the Latitude-men 1660–1689: Stillingfleet, Tillotson and 'Hobbism,'" *Journal of Ecclesiastical History* 36, 1985, pp. 407–27; Mark Goldie, "Sir Peter Pett, Sceptical Toryism and the Science of Toleration in the 1680s," in W. J. Shiels, ed., *Persecution and Toleration* (Oxford: Oxford University Press, 1984), pp. 247–73; Jon Parkin, "Hobbism in the Later 1660s: Daniel Scargill and Samuel Parker," *The Historical Journal,* 42, 1999, pp. 85–108.
5. This 1668 comment does not falsify Hobbes's intentions in 1651. Hobbes defended the king's rights in all his writings before *Leviathan* and presented the exiled king with a copy of *Leviathan* as soon as it was ready. He was looking for praise and did not anticipate the hostility of exiled Anglicans at Charles II's court in Paris.
6. George Wright, "1668 Appendix to Leviathan," *Interpretation* 18, 1991, pp. 380–1.
7. By far the best account of the relationship between philosophy and history and reason and faith in Hobbes is J. G. A. Pocock, "Time, History and Eschatology in the Thought of Thomas Hobbes" in his *Politics, Language and Time* (New York: Atheneum, 1972), pp. 148–201.
8. I accept the dates proposed by Philip Milton, "Hobbes, Heresy and Lord Arlington," esp. 542–6.
9. *The English Works,* IV, 431.
10. Tuck, "Hobbes and Locke on Toleration," p. 167.
11. See P. Springborg, "Hobbes on Religion," in T. Sorrell, ed., *The Cambridge Companion to Hobbes* (Cambridge: Cambridge University Press, 1996), pp. 346–80, esp. 349.
12. Ryan, "Hobbes, Toleration, and the Inner Life," p. 204; Wright, "1668 Appendix to Leviathan," p. 340, fn. 24.
13. *The English Works,* IV, p. 386. I have adopted W. E. Leonard's translation.
14. Ibid., p. 387.
15. Ibid., pp. 388–90.
16. Ibid., p. 393.
17. Ibid., p. 399.
18. Ibid., pp. 402–3.
19. Compare Robert Willman, "Hobbes on the Law of Heresy," *Journal of the History of Ideas* 31, 1970, pp. 607–13 and Samuel I. Mintz, "Hobbes's

Knowledge of the Law: A Reply," *Journal of the History of Ideas* 31, 1970, pp. 614–5.

20. Ibid., pp. 407–8.
21. The idea that "what is not right in its beginnings cannot become right through the passage of time," is made much of, for example, by William Atwood, *Jus Anglorum ab Antiquo* (London, 1681), Preface, n.p.
22. *A True Ecclesiastical History,* p. 3.
23. Ibid., pp. 174–8, quotation on p. 178.
24. Ibid., p. 184.
25. Ibid., pp. 5–36.
26. Ibid., pp. 55, 27.
27. Ibid., pp. 140–44.
28. Ibid., pp. 40–53. More fully: *An Historical Narration,* 391–99. See also: "1668 Appendix to Leviathan," 370, 372, 374–5, 378.
29. For example: "1668 Appendix to Leviathan," 372.
30. For example: *A True Ecclesiastical History,* 92.
31. "For deeply rooted in the Clergy's Mind,/The Dregs of Rage and Malice lay confin'd;/Ambition there with Avarice agree,/Pride, and Revenge, and innate Cruelty"; ibid., 127.
32. Ibid., 110, 147–8. The relationship between Christian sovereigns and Christian doctrine is discussed at length in Pocock, "Time, History and Eschatology," esp. 190ff.
33. See John Coffey, "Puritanism and Liberty Revisited: The Case for Toleration in the English Revolution," *The Historical Journal* 41, 1998, pp. 961–85 and Burgess, "Thomas Hobbes," pp. 152–55.
34. Ryan, "A more tolerant Hobbes?," 38.
35. Marshall, "The Ecclesiology of the Latitude-men," passim and R. Ashcraft, "Latitudinarianism and Toleration: Historical Myth versus Political History," in R. Kroll, R. Ashcraft, and P. Zagorin, eds., *Philosophy, Science and Religion in England 1640–1700* (Cambridge: Cambridge University Press, 1992), ch. 7.
36. On the denial of a right to private conscience in determining public worship, see T. Hobbes, *Man and Citizen,* ed. B. Gert (Indianapolis: Hackett, 1991), pp. 300–307.
37. *De Corpore Politico,* ch. 25, para. 14. See T. Hobbes, *The Elements of Law,* ed. J. C. A. Gaskin (Oxford: Clarendon Press, 1994), p. 154.
38. Goldie, "The Reception of Hobbes," pp. 610–15. The problematic relationship between Erastianism and toleration, however, is emphasized in Burgess, "Thomas Hobbes," pp. 144–5.
39. *On the Citizen,* pp. 181, 220–21 and *Leviathan,* pp. 238–41.
40. Ryan, "Hobbes, Toleration and the Inner Life," esp. pp. 210–14.
41. Ibid., pp. 216–7.
42. Ibid., p. 217.
43. Ibid., p. 198.
44. Ryan, "A more tolerant Hobbes?," pp. 37–9.

45. The first mention of *Leviathan* in print that Goldie found involved an attempt to enlist Hobbes's arguments in the cause of toleration for Roman Catholicism. This was an extraordinarily idiosyncratic reading. Goldie, "The Reception of Hobbes," p. 613.
46. *Leviathan,* chs. 13–18.
47. H. Arendt, "What is Authority?" in *Between Past and Future* (London, 1961), pp. 91–141.
48. Exactly this point was also made by Hobbes's friend and otherwise arch-opponent of *Leviathan,* the Earl of Clarendon. See Edward Hyde, "Of Conscience" (1670), in *A Collection of Several Tracts of the Right Honourable Edward, Earl of Clarendon* (London, 1727), pp. 162–7.
49. In the *Elements of Law,* Hobbes makes the point by reference to private conscience. "When a man in a Christian commonwealth hath once transferred his right of judging to another [the sovereign], that which shall be commanded, is no less his judgment, than the judgment of that other; so that in obedience to laws, a man doth still according to his conscience, but not his private conscience. And whatsoever is done contrary to private conscience, is then a sin, when the laws have left him to his own liberty, and never else." *The Elements of Law,* p. 153.
50. See A. P. Martinich, *A Hobbes Dictionary* (Oxford: Oxford University Press, 1995), pp. 37–42.
51. Hobbes's rather clumsy distinction between a person's "public conscience" and "private conscience" in the *Elements of Law* (see note 49 above) reveals something of his own dissatisfaction with the translation of classical *auctoritas* into the circumstances of a modern state. But the tension between the two is resolved in *Leviathan.* Here Hobbes unequivocally shows that "private conscience" is mere ill-considered opinion, not "conscience" at all. He offers a brilliantly succinct account of the modern corruption of the word "conscience." *Leviathan,* ed. M. Oakeshott, pp. 41, 211–2, 224–5, 327–8, 381–2.
52. For Arendt, the coherence and stability of *auctoritas* arose from the mutual reinforcement of the elements of what she called "the Roman trinity" of religion, authority, and tradition.
53. H. Rabe, "Autorität," in O. Brunner, W. Conze, R. Koselleck, eds., *Geschichtliche Grundbegriffe: Historisches Lexikon zur politisch-sozialer Sprache in Deutschland,* vol. I (Stuttgart: Klett-Cotta, 1972), pp. 382–406.
54. *Leviathan,* pp. 105–6.
55. Compare Michael Oakeshott, *On Human Conduct* (London: Oxford University Press, 1975), pp. 149–58 and 189–93 and *Rationalism in Politics* (Indianapolis: Liberty Press, 1991), pp. 265, 445–9.
56. This is not to neglect Hobbes's interest in public education. But his concern here was never to massage private conscience. That would have been pointless. It was rather to teach true, scientific principles instead of false, seditious ones.
57. Tuck, "Hobbes and Locke on Toleration," pp. 160–1.

58. Ibid., p. 162.
59. *De cive,* chap. 17, sect. 28.
60. Tuck, "Hobbes and Locke on Toleration," p. 163.
61. *Leviathan,* p. 456.
62. Tuck, "Hobbes and Locke on Toleration," pp. 164, 165.
63. *On the Citizen,* ed. R. Tuck, trans. M. Silverthorne (Cambridge: Cambridge University Press, 1998), p. 233.
64. Tuck, *Leviathan* (revised edition, Cambridge: Cambridge University Press, 1996), pp. xxxviii-xliv.
65. In reply to Bishop Bramhall's criticism that Hobbes had defended Anglican episcopacy in *De cive* but not in *Leviathan,* Hobbes wrote: "Being at Paris when there was no bishop nor church in England, and every man writ what he pleased, I resolved (when it should please God to restore the authority ecclesiastical) to submit to that authority, in whatsoever it should determine. This his Lordship construes for a temporizing and too much indifferency in religion," which in the circumstances is hardly fair. John Bramhall, *Castigations of Mr. Hobbes, 1658* (New York: Garland, 1977), 483–5, and Hobbes, *An Answer to a Book Published by Dr. Bramhall* (1668) in *The English Works,* IV, 366–7.
66. Tuck, *On the Citizen,* xxx-xxxiii.
67. Hobbes repeated his point toward the end of the next and final chapter, *On the Citizen,* p. 245.
68. *Elements of Law,* 162, *Leviathan,* 306, 355–7, *True Ecclesiastical History,* 147–8, *Answer to a Book Published by Dr. Bramhall,* 330, 364–5, *Behemoth,* ed. S. Holmes, 51–3.
69. *Elements of Law;* 142, *On the Citizen,* 232, 219; *Leviathan,* 355–61; *Behemoth,* 51–3.
70. William M. Lamont, *Godly Rule: Politics and Religion 1603–1660* (London: Macmillan, 1969), 56–77.
71. *Elements of Law,* 144.
72. Ibid., 144–5.
73. *Behemoth,* 55.
74. *True Ecclesiastical History,* 147–8.
75. *On the Citizen,* 224
76. Ibid., 208–10, 224; *Leviathan,* 325–6.
77. *On the Citizen,* 219, 230–1, 232.
78. Everyone had this kind of prudential obligation, not just sovereigns. *Elements of Law,* 69.
79. This is also the doctrine of *Leviathan,* 355–6.
80. *Answer to a Book Published by Dr. Bramhall,* 364–5.
81. *On the Citizen,* 218–9; *Leviathan,* 356–7, 360.
82. Tuck, *Leviathan,* p. xl, fn. 53.
83. *Thomas Hobbes: The Correspondence,* Vol. I, 120.
84. "1668 Appendix to *Leviathan,*" 372.
85. E. Hyde, Earl of Clarendon, *A Brief View and Survey of the Dangerous and Pernicious Errors to Church and State, In Mr Hobbes's Book Entituled Leviathan*

(Oxford, 1676), 308–9. Flathman, *Thomas Hobbes,* 154. Jeffrey R. Collins, "Christian Ecclesiology and the Composition of *Leviathan:* A Newly Discovered Letter to Thomas Hobbes," *Historical Journal* 43, 2000, pp. 227–8.

86. "1668 Appendix to *Leviathan,*" p. 381. *An Answer to a Book Published by Dr. Bramhall,* p. 366. *Historical Narration,* p. 407.
87. The passage must be read as an example of the kind of Hobbesian irony and sarcasm which Quentin Skinner discusses in *Reason and Rhetoric in the Philosophy of Hobbes* (Cambridge: Cambridge University Press, 1996), esp. ch. 10.
88. *Leviathan,* p. 455.
89. Ibid., pp. 455–6.
90. Ibid., p. 456.
91. See also Hobbes's *Behemoth,* passim.
92. *Behemoth,* pp. 169, 181–3.
93. No evidence points in this direction. The story Hobbes later told in *Behemoth* is replete with examples of greed, ambition, hypocrisy, foolishness, and deception, both clerical and lay.
94. *Elements of Law,* pp. 144, 147, 150; *An Answer to a Book Published by Dr. Bramhall,* p. 303, *True Ecclesiastical History,* p. 77.
95. I Corinthians 1:10 and I Corinthians 3.
96. *A Dialogue between a Philosopher and a Student of the Common Laws of England,* ed. Joseph Cropsey (Chicago: University of Chicago Press, 1971), p. 124.
97. I Corinthians 1:10.
98. See the introduction by George Wright to the "1668 Appendix to Leviathan," pp. 323–47.
99. The words are Morton Schoolman's in the "Series Editor's Introduction" to Flathman, *Thomas Hobbes,* p. xiii.
100. Flathman, *Thomas Hobbes,* p. xxi.
101. Ibid., p. 152.
102. Ibid., p. 154.
103. Ibid., p. 174.
104. Ibid., p. 154.
105. *A Dialogue between a Philosopher and a Student,* p. 124. Everything Hobbes wrote in praise of Constantine was a critique of the primitive Church as a model of good ecclesiastical order.
106. Acts 18: 24–28, Acts 19: 1 and I Corinthians 16: 12.
107. I Corinthians 4: 6.
108. *On the Citizen,* pp. 181, 220–21; *Leviathan,* pp. 238–41.
109. *On the Citizen,* pp. 175–83; *Leviathan,* pp. 236–7, 240).

5

Between History and Politics

Philipp van Limborch's *History of the Inquisition* [1692]

Luisa Simonutti

With a decree of the Holy Office of May 1690, Charles Dellon's *Relation de l'Inquisition de Goa (Relation of the Inquisition in Goa)*[1] (1687) was brought under the hammer of the Inquisition by Pope Alexander VIII. An equally severe decree of the Inquisition of May 1694 laid down "that no one in any role or condition, or in any language or version whatsoever, should in any manner dare to print, or arrange to have printed, in any place" Jacques Marsollier's *Historie de l'Inquisition et son origine (History of the Inquisition and its Origin)*[2] (1693) and Philipp van Limborch's *Historia Inquisitionis (History of the Inquisition)*[3] (1692). These books, "sic prohibitos, et damnatos," were not merely subjected to the censorship of the Roman clergy, but were forbidden and condemned so that "under no pretext could they be legally printed, kept, or read."[4]

In 1709, going to press with a new, revised, and more complete edition of the *Relation de l'Inquisition de Goa,* physician and traveler Dellon broke out of anonymity, placing the work under the patronage of the Conti princes and the influential Bishop of Meaux, Bossuet, in his preface. Intending to bring his work back within the confines of Catholic legality, Dellon specifically distanced himself from the work of Limborch.

> A Latin folio volume titled *Historia Inquisitionis Tolosanae,* printed at Amsterdam and dedicated to the archbishop of Canterbury, fell into my hands. I have tried to rebut the charges that the author has wrongly brought against the Roman Church concerning the irregularity of the procedures of the Holy Office. . . . This author wants to make the Church responsible for the abuses that take place in the tribunals of the Inquisition, as if she authorized those abuses. But besides that they were not established by the Holy Office, and they were not approved by any General Council, it is also very certain that the Popes, who are the visible heads of the church, have tried at different times all of the imaginable ways of stopping those abuses and of obliging the Inquisitors of Spain and Portugal to rectify their procedures and to regulate them at least according to those that the Holy Office observes at Rome.[5]

Limborch had in fact not only cited the Frenchman's work among his sources, but had also made use of his account of the Inquisitorial procedures and had utilized many of the engravings present in his work.[6]

From the moment it appeared, Dellon's work aroused great interest by reason both of its content and of the discursive and exotic nature of the narrative. The account of his travels in India, the description of the influential tribunal set up for the protection of Catholicism in such remote lands, and the documentation of its inquisitorial procedures and of the tragic experiences of two gentlemen, Joseph Pereira De Meneses and Luís Pessoa and his family, garnered wide interest. Jansenists and Gallicans found here an illustration of the brutality and hypocrisy of the judicial processes of the Holy Office. Pro-Protestant and anti-Spanish Frenchmen who abominated the Inquisition—whether Spanish or Roman—for a variety of political and religious reasons were also avid readers of Dellon's work. Accused of pro-Calvinism, but above all identified with the Gallican policy championed by Louis XIV, to whose court Dellon had been admitted, the text was subject to a variety of interpretations and underwent significant alterations in the course of the author's revisions.

Dellon's work was marked by the critical independence that characterized the attitude of the French Court and Church toward the Papacy. It is not surprising that the influential Cardinal Ciampini was entrusted with drawing up the report for inclusion of the work on the Index. The Cardinal did not hesitate to express his incontestable judgement: "In the first place I execrate this fallacious, perjured, and impious writer, undeserving of any justification. . . . Secondly I consider that this book should be banned without hesitation on the grounds that it is highly dangerous to religion."[7]

On 31 March 1694, the Cistercian monk and Chief Inquisitor, Giovanni Maria Gabrielli, sent to the cardinals of the Congregation of the

Holy Office a long report on the work of another French writer who, under the cloak of anonymity, had presented a critical interpretation of the origins and development of the Inquisition that appeared to draw its inspiration from the Jansenist position.

Recalling the teaching of the apostles and the political and religious figures of the Early Church in his *Historie de l'Inquisition et son origine,* Jacques Marsollier upheld the independence of the judgment of ministers of the Church and magistrates in trials against heretics. He emphasized that the Church had had numerous opportunities for action against heretical ideas, such as in the case of the persistent Pelagian heresy, but that it had always expressed itself through ecclesiastical penalties. These were excommunication in the case of laymen, and banishment from the ministry and excommunication in the case of the clergy. The ecclesiastical power did not intervene directly or indirectly in the deposition of monarchs who had fallen into heresy or apostasy, nor could such errors imply the dissolution of the bonds of obedience of subjects.

To support his arguments, Marsollier drew on the historical report of the Inquisition in the Venetian Republic by Paolo Sarpi, as well as numerous examples drawn from the history of the Roman Empire. He stressed that in all historical situations in which magistrates had to intervene against heretics, they did so not in fulfillment of any ecclesiastical mandate, but only in obedience to Imperial law. As a result, penalties inflicted on the heretics by the civil court were limited to fines or exile of the heretic, never the death penalty. Cases of social disruption or of violation of the laws of the state by heretics were different. They resulted in rigorous intervention by the state, whose fundamental duty was the safeguarding of social peace.

This argument represented for Marsollier the justification for toleration. He reiterated the classic argument according to which toleration undermined religious belief and resulted in indifference, not only toward the dogmatic apparatus, but toward religion itself. But he also affirmed that freedom of worship and mutual toleration, far from being a source of confusion, actually was an effective instrument for the maintenance of social peace. It is for the prince, insists Marsollier, to promulgate laws that are appropriate for his own country. In cases where the ecclesiastical power has made its weight felt in the sphere of temporal power, such as concerning heresy, this has disrupted the ancient order. That was increasingly evident over the years and, according to the author, was clearly visible now in territories where the Inquisition held power.

Having mapped out the principal arguments of the four sections that composed Marsollier's work, the Inquisitor concluded by calling upon the "most wise and most prudent" judgement of the Congregation of a book the insane contents of which could not be effectively annihilated either by censorship or by just refutation. To Gabrielli, the *Historie de l'Inquisition et son origine* appeared full of historical distortions and was permeated by a malevolent interpretation of the Holy Institution, convincing him to condemn it.[8]

At the meeting of the Congregation held on 19 May 1694 (as had happened in May 1690), Giulio M. Bianchi, Secretary of the Holy Congregation of the Index, drew up the *Protocolli* and the *Diari,*[9] thus passing down to the modern reader a record of the procedures to which books were subjected, among them the *Historia Inquisitionis* of Philipp van Limborch. The Index specified the extreme danger of this text because of its attack on the procedures of the Holy Office, which, according to the Protestant theologian, placed people under a terrible yoke.[10] Nor was the *Historia Inquisitionis* the only of Limborch's works to suffer the censure of the Holy Office. During the first half of the eighteenth century his *Theologia Christiana*[11] and his *De veritate Religionis Christianae, amica collatio cum erudito judaeo*[12] were condemned.

"The Indubitable Evidence of the Manuscript Itself"

In his works, and in his unpublished manuscripts, Limborch frequently tackled the question of the relationship between political and ecclesiastical power. One of the crucial features of the Remonstrant creed, of which he was at this time the most eminent representative, was a concern with the problem of heresy, the defense of liberty of conscience, and political toleration of the different Christian doctrines. In his massive *Historia Inquisitionis,* Limborch made his voice heard in the lively debate on the question of liberty of conscience and the persecution of heretics not by using the well-known weapons of anti-Catholic controversy, but philological reasoning and historical criticism.

Stories of the brutality of the ecclesiastical tribunals had been recounted during these years by Dellon and Marsollier in the strong language of polemic, or, as in Gregorio Leti's *L'Inquisizione processata, opera storica e curiosa,*[13] using the sharp edge of sarcasm. The latter, a prolific and adventurous writer of Italian origin who had settled in Amsterdam, in the early 1680s set out in the form of a dialogue the protests of the Christian religion before the throne of Christ, and its exhortations that the deity should intervene to free it from the yoke of the Inquisition and the tyranny of Rome. Leti then recounts how God's endeavor was to no avail, since when he sent Justice to Rome to settle the controversy, she was denounced by the Inquisition and sent to languish in pontifical prisons.

Adopting a different approach, and nevertheless succeeding in focusing the attention of a wide readership on this still extremely delicate matter, Limborch chose to present in its entirety, and in a philologically correct edition, the precious medieval codex containing the sentences delivered by the Toulouse Inquisitors between 1307 and 1323. The Inquisitorial reports, the harshness of the original language, and the crudeness of the descriptions, were, Limborch believed, more effective than any ironic account or polemic in illustrating to the contemporary reader the illegitimacy of an Office that took upon itself the title of Holy, while at the same time forcing men to the faith by means of violence.

> My History of the Inquisition gives Light to the Book of Sentences. My Design in it was to give a Representation of that Tribunal, not in a false Disguise, nor deform'd by unnatural and hideous Colours, but in living and genuine ones; I mean, to draw the Picture of that horrible Court, which makes its principal Boast of the Title of Sanctity, to the Life, not from the Writings of those who separate from the Church of *Rome,* but that there may be no Room for Calumny, from those of the Popish Doctors, and even Inquisitors themselves; that hereby the vast Power granted to the Inquisitors, the most cruel Laws of it, and the injust Method of Procedure, quite different from the Usage of all other Courts, might appear to the whole World, and that hereby the Papacy it self might be known to all Mankind to be what it really is.[14]

In Limborch's opinion the word "heretic" was also illegitimate; it was widely abused and expressed only the point of view of a faction or religious doctrine. He concluded his introduction with the following explanation:

> The Reader may, perhaps, wonder at one Thing, that I have always called those Hereticks that have been proceeded against by the Inquisition: But, as I was relating the Popes Bulls, and the Decrees of Popish Councils, I could not help using the same Words I found in them. By an Heretick, therefore, I understand one condemned for Heresy by the Church of *Rome.* I could not rehearse their Decrees but in their own Words, and was therefore forced always to use them, unless I would have interrupted the Course of the History, by repeated and innumerable Alterations, and thereby rendered it less pleasing and acceptable. Let it therefore suffice, once for all, to say, that, by the Word *Heretick,* when I ever speak of the Inquisition against Hereticks, I do not mean one who is truly an Heretick, but accounted an Heretick by the Church of *Rome,* taking the Word in the Popish Sense of it. In the mean while, those who are Hereticks in their Account, are not so in mine; and I sincerely believe, that those which the Church of *Rome* hath condemned for Heresy, have died, and gloriously endured the Punishment of Fire for the Testimony of Jesus Christ, and the maintaining a good Conscience.[15]

The theme of heresy and the non-punishability of heretics was present in Limborch's thought from his earliest historical and polemical works and in the course of his dispute with Orobio. From his very first polemic, an attack on the rigid Calvinism of Jacobus Sceperus in the early 1670s, and subsequently in his pamphlets against Bredenburg and his edition of the works and letters of the Fathers of Arminianism, Arminius, Courcelles, and Episcopius, Limborch proposed an anti-dogmatic and anti-theocratic reading of Christian doctrine, bringing to the forefront rational ethics and a balanced distinction between the roles and the aims of ecclesiastical and political systems. From *De veritate* to the historical works on the Synod of Dordrecht, and even in the unpublished works, the themes of the impunity of heretics, the irreconcilability of consciences, peace, and religious liberty remain pivotal aspects of Limborch's theological-political thought. In the unpublished manuscript entitled *Reformatorum de poena haereticorum,*[16] he proposed an analysis of the heresy-toleration dialogue from different points of view through the works of the major Christian and Reformed authors. These themes of the final chapters of his major theological work, the *Theologia Christiana,* along with the *Historia Inquisitionis* and the manuscript repertory of the opinions of ecclesiastical commentators, theologians, and writers who had written upon the matter, were the themes that accompanied the last years of Limborch's work of historical criticism.

Presenting the reader with the precious medieval codex, Limborch considered it necessary to provide a context for the documentary material, the reports, interrogations, and accounts of the execution of political figures and ordinary people of Toulouse who had fallen into the net of the Holy Office. They were accused of harboring the Albigensian heresy (the name given to the Cathar heresy in the area of Toulouse). For this purpose, he prefaced the edition of the *Liber Sententiarum Inquisitionis Tholosanae* with a history of the medieval Inquisition in the various European countries. In the introduction he also analyzed the canon law upon which the Holy Office based its judgement of the nature and gravity of the crimes attributed to the heretics, and supplied a full description of the role of the vicars, notaries, and Inquisitorial judges, and of the methods and procedures they used.

What was originally intended as a mere introductory discourse became a broad and thorough history of the Inquisition. Limborch divided his historical study into four main sections, the first two dedicated to the origins and evolution of the Inquisition and an examination of the tasks and powers of the Inquisitorial officials and of the prison sentences. The third part consists of an analysis of the various crimes and punishments, from the principal ecclesiastical censure of excommunication to the principal civil penalty represented by the confiscation of property, and to the most serious punishments, which handed the offender over to the secular authorities. Finally, in the last section of his introduction, Limborch described the Inquisitorial procedures for denunciation and accusation, the role of witnesses and the superfluous role of the defense lawyers, dwelling on the risks that witnesses and lawyers ran of being implicated in the accusations of falsity and heresy.

"The primitive Christians opposed with the greatest Vigour, all Cruelty and Persecution for the Sake of Religion"[17] claimed the Amsterdam theologian. He took up ideas expressed by the Doctors of the Church, and in particular by Lactantius, affirming that Christian doctrine must be defended with the sword of reason and confirmed only by evidence and the truth. The conversion of Emperor Constantine marked the delivery of political power into the hands of the ecclesiastical apparatus and the beginning of an intermingling of religion and politics that was to have a negative influence on posterity. The Remonstrant theologian identified another crucial precedent in the figure of St. Augustine. Faced by the Manichean heresy, the Bishop of Hippo had been against forcing consciences. Later, however, in the struggle against the Donatist heresy, he became a relentless supporter of the persecution and execution of heretics, in the conviction that this represented "the chief duty towards God." Limborch concluded: "But the Minds of Christians have been perverted from his branch of Equity through the Prevalence of Self Love; so that when they could prevail with the Civil Power to assist them, they have pronounced all that differed from them Hereticks, and then exercised all Kinds of Cruelty against them."[18]

In the following chapters, analyzing the Albigensian and Waldensian doctrines and the history of the Tribunal of the Holy Office in various European countries up to the period of the Reformation, Limborch placed special emphasis on the relationship, sometimes conflictual but more often of mutual support, which existed between the civil and ecclesiastical authorities. This relationship not only accompanied the establishment of the Inquisition, but favored its consolidation, as emerged most significantly in the relations between Emperor Frederick II and Pope Gregory IX, and in the introduction of the Inquisition in Spain. In these pages, Limborch continues to focus attention on the question—close to the hearts of the Remonstrant branch of Dutch Calvinists—of the interrelation and relative boundaries of the spheres of politics and religion.

Finally, Limborch entrusted to the Toulouse document itself—compiled by four notary Inquisitors and based on the activities of the notorious Bernardo Gui—the task of illustrating, directly and with all its incisive power, the sentences issued and the execution of the condemned.[19] The *Liber* is made up of 16 sermons containing rules of inquisitorial procedure and penalties for the heretics, most of which were pronounced by Bernard

Gui, in collaboration with other prelates, in the Cathedral of Toulouse. The manuscript also contains an alphabetical index of names, places, and dates relating to individuals tried by the Inquisition in the period between 1307 and 1323. The codex, a fundamental document in the history of the medieval Inquisition, for which even today there are very few documentary sources, was for a long time thought to be lost, existing only in the printed edition that came to us from the Arminian theologian. However, it was found at the British Library in the early 1970s.[20]

True to the principles of philological and historical rigor, Limborch used as sources for his historical account almost exclusively Catholic writers and historians, ecclesiastical annals, histories of the church written by clerics, and the manuals of famous Inquisitors, "who are so fare from having written any Thing untrue, out of Hatred to the Inquisition, that they every where cry up the Sanctity of it."[21] He provided the reader with a bibliography of works used before beginning his historical introduction. These authors included Bernardo of Como, Simancas, Zanchini, Juan de Rojas, Emerycus, Pegna, Ludovicus da Paramo, Sousa, Fulcodi, Carena, and Sarpi, as well as the contemporary work of Dellon.[22] The only exception was the work of Reformed inspiration by the pseudonymous author Montanus.[23]

Providing his readers with a long account of Limborch's *Historia Inquisitionis,* the editor of the erudite journal *Bibliothèque Universelle et Historique,* the Arminian polyhistor and philologist Jean Le Clerc, underlined the criteria and ideals that inspired the edition:

> M. Limborch has had the scruple to employ, as much as he could, the very words of the authors he uses; even if the style is sometimes barbarous, and almost unintelligible. He felt bound to do so, so no one could impute anything to him. It is not necessary to exaggerate in order to cause a just horror of the procedures of the Inquisition. The injustice and cruelty are so visible that it is enough to explain things simply and naturally as they were, to make any reasonable man conclude that this Tribunal could never have been established, except by the inspiration of the declared enemy of humanity from the beginning of the world.[24]

Limborch chose to fight in defense of liberty of conscience and toleration and against the persecution of heretics not by means of invectives but through the presentation, in a philologically accurate edition, of a precious historical document that would reveal, beyond the reach of partisan commentary, not only the cruelty and irrationality of the Catholic Church, but of any form of persecution whatsoever on the grounds of faith.

This editorial choice was fully supported by John Locke, who in correspondence with the theologian, and in particular in letters referring to this editorial enterprise, wrote: "Theological zeal, as I see it, is always and everywhere the same and proceeds in the same way."[25] Precisely by virtue of their common interests and their profound friendship established during his Dutch exile, Locke wished to entrust to Limborch the realization of this project.[26]

The English philosopher had consulted the medieval codex at Montpellier in 1677, during his stay in France. Ten years later he revealed his admiration and envy of his friend and host in Rotterdam, Benjamin Furly, when the latter came into possession of the precious codex—initially temporarily, and then Furly purchased it for his library.[27] Together with Furly, who financed the copying of the text and sponsored its publication, Locke oversaw the different phases of the publication of the work, as the correspondence with Limborch illustrates. He encouraged Limborch to obtain the patronage of the Archbishop of Canterbury, John Tillotson, and to present the work to the Latitudinarian clergy and the British public.

Limborch commented that "many of the Contra-Remonstrants are buying and reading it; so far however I have not heard any unfavorable opinions."[28] Nevertheless, even in his homeland some Calvinist theologians perceived, amidst the explicit accusations against the Church of Rome, an implicit reference to the persecution suffered by the Remonstrants during the Synod of Dordrecht and the condemnation of their creed. Nor had Limborch, moreover, ever relinquished his role as a theologian who was not only committed to the defense of the original Christian message, but also engaged on the wider front of the political-religious debate that involved Remonstrants as well as Huguenot exiles and English dissidents in favor of religious liberty and toleration.[29]

Preparing the Second Edition

A cultural enterprise of such importance could not pass unobserved in Rome. Limborch, moreover, was conscious of the reactions that his work provoked in the Catholic world. He transcribed in full on his own copy of the *Historia Inquisitionis,* which he annotated with great care with a view to a second edition, the decree of prohibition of the work issued by the Holy Office in May of 1694. In the course of his painstaking revision Limborch made lengthy handwritten additions to the historical treatise preceding the edition of the Toulouse codex.[30]

If the philological rigor with which the codex was published was unequalled (and remained so for a long time[31]), the history of the Inquisition that preceded it required, in Limborch's eyes, revisions based on improved use of historical sources, explication of doctrine of the Fathers, Catholic manuals of Inquisitorial history and practice, and papal bulls and documents.

The additions, enhanced by textual references, would highlight the rules and procedures of the judicial system and the ideological aims of the Holy Inquisition. This is probably the motive behind the transcription of the bull of Pope Paul V that he inserted in the chapter dealing with an analysis of the different crimes on the part of heretics indictable by the Inquisition.[32] Issued in 1616, this bull redefined the jurisdictional sphere of papal excommunication. The incontestability of this document would, therefore, have provided an illustration of the effective extension of the temporal authority of the Church and its significance in relation to the power of the state.

Similarly, in the chapters dedicated to the Toulouse Inquisition, Limborch, through the addition of passages and sometimes entire handwritten pages interleaved in the working copy, drew attention to the enormous power that the pope placed in the hands of the Dominicans, and the role of a veritable "holy militia" which the latter adopted in the war on heresy.[33] He then provided a meticulous description of the functioning of the Holy Palace, the tasks of the various congregations, and the roles performed by the cardinals.[34]

Recalling Ludovico da Paramo's description of the activities of the Inquisition in the territories of Central America governed by the Spanish crown, the Amsterdam theologian emphasized the historical weakness of this pro-government account, which not only failed to account for the oppression, but rather attempted to justify the cruelty of the Inquisitorial fathers through supernatural events.[35] But the iron pact drawn up by Ferdinand and Isabella with the tribunals of the Inquisition, which frustrated all efforts against the Inquisition in defense of the freedom of Aragon,[36] set itself firmly against any prerogative of freedom, both in the colonies and in the homeland.[37]

Retracing the history of the Spanish and Portuguese Inquisition and its influence on Renaissance Europe, in a lengthy note Limborch described the attempt made in Italy to supplement the procedures of the Roman Inquisition with the abhorred "modum Hispanicum." He emphasized the hostility that the Papal-Hispanic understanding had immediately aroused in numerous regions of the peninsula, and in particular the opposition of the people and the Senate of the Duchy of Milan, which, while it accepted the Tridentine directives of the Holy See, was certainly not disposed to bow before this new Spanish interference or to adopt the tyrannical methods used against the Moors and Jews.[38]

Nor could there fail to be a long indictment of that censure in which Limborch himself had been directly involved, "that this may be more distinctly explained, and that it may be known what and how great the Power is which is committed to the Inquisitors, concerning the Prohibition of Books."[39] In ancient times, in fact, the condemnation of heretical books came under the power of the emperor. But in the course of the Middle Ages the pope gradually usurped many branches of civil government, among them the right to ban books suspected of heresy.[40] To bring a little order to the tangled confusion of sentences that had accumulated against forbidden books, authors, and even readers, it was decided to publish the catalogue of the works that had been placed on the Index. At this stage the censure was extended not only to the works of authors condemned for heresy, but also books by authors who belonged to the legalized Christian world, sometimes printed in Rome and already approved by the Inquisition or the pope. Limborch indicates a significant example: "Such are the Annotations of *Erasmus* on the *New Testament,* approved by *Leo X* after he had read them over, by a Bull dated at *Rome, Sept. 10, 1518.*"[41] And he comments:

> But above all this is remarkable, that under Colour of Faith and Religion, the Books of such Authors are prohibited and condemned with the same Severity, in which the Authority of Princes, and the Civil Magistrate is defended against the unlawful Usurpations of the Ecclesiasticks; and in which the Power of Councils and Bisshops is maintained against the unjust Encroachments of the Court of *Rome:* And finally, all those which plainly detect their Hypocrisy and Tyranny, by which, under the Disguise of Religion, they impose on the People, or persecute them. In a Word, Religion was never before made subservient, by such an admirable Mystery, to bereave Men of their Senses.[42]

Concluding the chapter dedicated to an analytic description of the Inquisitorial processes, which were marked by torture, Limborch supplied the account of the trial and torments suffered by Orobio De Castro, "as I had it from his own Mouth."[43] This was the Jewish doctor with whom he had an intense but unprejudiced debate, published in the second half of the 1680s.[44] Now, reexamining the chapter of the *Historia Inquisitionis* dealing with the trials of persons suspected of heresy, the Remonstrant theologian decided that the most exemplary manner of closing this chapter was to make available to the reader the sentence issued against Galileo Galilei and his abjuration, an outstanding example of illegitimate and unjust persecution.

> What the Manner is of proceeding at this Day in the Inquisition general at *Rome,* against Persons suspected of Heresy, we have a most clear Instance, in the Sentence pronounced against *Galileus Galilei,* a famous Astronomer, because he taught, that the Sun stood immovable in the Midst of the Universe, and that the Earth moved round it as about its proper Center. I have thought it worth while to transcribe it here intire, that it may appear, that Opinions purely Astronomical, and that have nothing to do with Religion, and which

can neither be of any Advantage or Disservice to Piety, which soever side of the Question is defended, are a sufficient Argument to the *Roman* Inquisitors, to render any one vehemently suspected of Heresy, after the Prelates of that Church have once determined on one side.[45]

In drafting the notes for the revised edition, Limborch remained faithful to his original intention. He provided a historical description and contextualization of the origins and development of the Inquisition, illustrating its procedures through incontestable Catholic sources. He skillfully combined philological rigor with close attention to ancient and modern sacred historiography, and to significant aspects of the contemporary scene.

The second Latin edition of the *Historia Inquisitionis* never saw the light. But in the early 1730s the nonconformist divine Samuel Chandler offered English readers a new edition and translation of Limborch's work. Contrary to what the Dutch theologian would have wished, Chandler placed primary emphasis on the critical force of the arguments. In fact, he did not publish the *Liber Sententiarum* at all but merely a precise translation of the long historical introduction, enriched with all the additions that Limborch had prepared over many years of revision. Chandler recounted how he came into possession of the precious notes:

> Mr. *Lock* often mentions, in his Letters, several Additions which Mr. *Limborch* had prepared, and promised to transmit to him, that he might insert them in their proper Places in the Margin. I know not whether he ever had the Pleasure of seeing them; 'tis certain the Publick hath never hitherto been favour'd with them. When I first began my Translation of the History, the late ingenious *Anthony Collins,* Esq. informed me, that he had some M.S. Papers of Mr. *Limborch* relating to it, and generously sent them to me for my Perusal. After this, I was informed by a worthy Friend, that there was a Gentleman in *Holland* who had a large Number of Corrections and Additions; and, upon my Application to him, he very kindly ordered them to be transcribed out of the Copy Mr. *Limborch* kept by him, which he had corrected and enlarged with his own Hand, and transmitted them to me from the *Hague.*[46]

In Chandler's opinion, these notes made Limborch's historical account even more incisive and relevant. Chandler prefaced the work with a lengthy introduction that traced the history of religious thought from its classical Greek and Latin origins up to the early decades of the seventeenth century, with the intention of defending the legitimacy of liberty of conscience and demonstrating the need for toleration on the part both of the political apparatus and the Church. This author, eager for controversy, launched the *Historia Inquisitionis* into the midst of the eighteenth-century religious and political battle in favor of toleration and of free thought.

Notes

* I thank the Director of the Vatican Archive of the Congregation for the Doctrine of the Faith, Alejandro Cifres, for valuable suggestions, and I am grateful to the library staff for assistance. I also thank Jos A. A. M. Biemans, Curator of manuscripts at the University of Amsterdam, for his help; Gianni Paganini for his support; and Ugo Baldini for advice. This study was supported by Inter-University Ministerial funding and the CNR-Milan.

1. Charles Dellon, *Relation de l'Inquisition de Goa* (Leiden, 1687; Paris, 1688; Cologne: Marteau, 1709).
2. Jacques Marsollier, *Historie de l'Inquisition et son origine* (Cologne: Marteau, 1693).
3. Philippus van Limborch, *Historia Inquisitionis. Cui subjungitur Liber Sententiarum Inquisitionis Tholosanae ab anno Christi 1307 ad annum 1323* (Amsterdam: Wetstenium, 1692).
4. *Decretum Feria IV, die 19 Maij 1694,* drafted during the papacy of Innocent XII, and issued by the Congregation of the Holy Inquisition in Rome in 1694. Città del Vaticano, Archive of the Congregation for the Doctrine of the Faith (ACDF), Index, *Protocolli* (EEE 52, Bianchi 1694), vol. X, c.362. On the Inquisition, see Gustav Henningsen, John Tedeschi, and Charles Amiel, eds., *The Inquisition in Early Modern Europe. Studies on Sources and Methods* (Dekalb: Northern Illinois University Press, 1986); *The Prosecution of Heresy: Collected Studies on the Inquisition in Early Modern Italy,* ed. John Tedeschi (Binghamton: Medieval and Renaissance Texts, 1991); Francisco Bethencourt, *L'Inquisition à l'époque moderne. Espagne, Italie, Portugal, XVe-XIXe siècle* (Paris: Fayard, 1995); *L'apertura degli Archivi del Sant'Uffizio Romano. Giornata di studio* (Rome: Accademia Nazionale dei Lincei, 1998); and the introduction by Andrea Del Col to *L'Inquisizione nel Patriarcato e Diocesi di Aquilea, 1557–1559* (Trieste: Edizioni Universita di Trieste, 1998).
5. Charles Dellon, *Relation de l'Inquisition de Goa,* Preface. See also the recent edition by Charles Amiel and Anne Lima, *L'Inquisition de Goa. La relation de Charles Dellon (1687)* (Paris: Editions Chandeigne, 1997).
6. Francisco Bethencourt, "The 'auto da fé': ritual and imagery," *Journal of the Warburg and Courtauld Institute,* vol. 55, 1992, pp. 155–168, shows Limborch's debt to Dellon's engravings, but underlines the enrichment brought to the engravings by his description of the images, scenes, and ceremonies of the Spanish Inquisition.
7. ACDF, Index, 1690 (IIa, 47, cc. 305*r*–308*v*: cc. 306*v*–307*r*).
8. ACDF, S.O., *C.L.,* 1693–1695, n. 20.
9. See ACDF, Index *Protocolli* (EEE 52, Bianchi 1694), vol. X, c. 296, c. 353 e c. 355 See also ACDF, Index, *Diari,* (Bianchi, 1692–96), vol. X, c. 751. The report censuring Limborch's work does not appear to have been preserved.

Moreover, documentation relating to several of the Ferie of 1694 is incomplete. It is possible, in the case of an important work, that the reading was entrusted to a cardinal of the holy Congregation so that he could express his opinion verbally.

10. See ACDF, *INDEX LIB. PROHIB.* vol. IV, c. 2975. A large number of works subjected to censure, including the texts discussed here, are preserved in the Biblioteca Casanatense in Roma. See the catalogue *Inquisizione e Indice nei secoli XVI-XVIII* (Rome: Biblioteca Casanatense, 1998).
11. Amsterdam, H. Wetstein, 1686. See ACDF, Index, 1724–1725 (F4, IIa, 73, c. 945); 1726 (G 4, c. 747), 1727–28 (IV, 76, c. 340).
12. (Gouda: J. van der Hoeve, 1687). See ACDF, Index, 1740–48 (IIa 82, c. 365, c. 369).
13. (Cologne, Paolo della Tenaglia, 1681).
14. Philippus van Limborch, *The History of the Inquisition,* translated into English by Samuel Chandler (London: J. Gray, 1731), *The Dedication to John Tillotson.*
15. Ibid., *Preface to the Reader.*
16. Unpublished manuscript entitled *Reformatorum de poena haereticorum sententia ex variis autoribus et ecclesiarum responsis collecta et exemplis aliquot demonstrata,* with preface dated October 1710. (UBA, MS, III G 28, 199 ff.) The internal organization of the text and the presence of the preface lead one to think there was a plan for publication. See also Abr. des Amorie van der Hoeven, *De Joanne Clerico et Philippo a Limborch dissertationes duae* (Amsterdam: F. Muller, 1843), in particular Appendix *E.*
17. van Limborch, *The History of the Inquisition,* B.I, chap. II, p. 4.
18. Ibid., B.I, chap. VI, p. 40.
19. For a detailed examination of the work and a reconstruction of the phases of its editing, see L. Simonutti, "Limborch's *Historia Inquisitionis* and the pursuit of toleration," in A. P. Coudert, S. Hutton, R. H. Popkin, and G. M. Weiner, eds., *Judaeo-Christian Intellectual Culture in the Seventeenth Century* (Dordrecht: Kluwer, 1999), pp. 237–255.
20. M. A. E. Nickson provides a history of the manuscript's passage from France to Holland and then to England, where it was eventually acquired by the British Museum in 1756, in "Locke and the Inquisition of Toulouse," *The British Museum Quarterly* 36, 1972, pp. 83–92. See also Simonutti, "Limborch's *Historia Inquisitionis* and the pursuit of toleration."
21. van Limborch, *The History of the Inquisition, Preface to the Reader,* p. IX.
22. See the bibliographical references in E. van der Vekené, *Bibliographie der Inquisition* (Hildesheim: Olms, 1963).
23. See B. A. Vermaseren, "Who was Reginaldus Gonsalvius Montanus?," *Bibliothèque d'Humanisme et Renaissance,* XLVII, 1985, pp. 47–77.
24. Jean Le Clerc, *Bibliothèque Universelle et Historique* (Amsterdam: A. Schelte, 1692) t. 23, pp. 360–409; pp. 361–362.
25. John Locke, *The Correspondence,* edited by E. S. De Beer, 8 vols. (Oxford: Clarendon Press, 1976–1989), letter 1473.

26. Locke echoes Limborch: "On my suggestion and incitement our friend Furly has sent you through van Helmont a book in manuscript, with a view to its remaining in your keeping whilst arrangements are made for printing it. When you see the contents I think you will agree with us that it deserves to see the light; for it contains authentic records of events in that rude age, of which historians have given us either negligent or faulty accounts. I would prefer, and, unless I am mistaken, all lovers of truth would be with me in preferring, such artless narratives, drawn from the very fountain-head, to the flowers of rhetoric offered by historians, which are ornaments in the writer, indeed, but deceive the reader and lead him into errors." Locke, *Correspondence,* letter 1023, p. 393.
27. The codex appears not to have been sold at the auction held after Furly's death, probably on account of its high value. See *Bibliotheca Furliana, sive Catalogus Librorum Honoratiss. et Doctiss. Viri Benjamin Furly* (Rotterdam: Fritsch and Bohm, 1714). When the German bibliophile and academic Zacharias von Uffenbach visited Furly and his library, he insisted on seeing the precious codex. Furly expressed to the visitor his disillusion at the generic acknowledgements of Limborch. See Zacharias Conrad von Uffenbach, *Merkwürdige Reisen durch Niedersachsen Holland und Engelland* (Frankfurt: Wagner, 1753–1754), vol. I, pp. 279–80. In the introduction to the "Liber Sententiarum," Limborch wrote: "I have had [the manuscript] in my possession for four whole years thanks to the generous consent of its owner[; it] was written by notaries of the Inquisition and is an authentic exemplar taken from the archives of the Inquisition of Toulouse." See van Limborch, *Historia Inquisitionis, Praefatio* to the *Liber Sententiarum Inquisitionis Tholosanae.* While preparing the second edition, he expressed his intention of giving further prominence to the role played by Furly and to the originality and integrity of the codex, quoting a declaration to Furly in 1695 by the Catholic prelate C. Bontius.
28. Limborch recounted to Locke the critical reactions of Triglandius toward the *Historia Inquisitionis* and de Volder's defense: "Not long after the publication of my *Historia* Professor Trigland among others was in company with Mr. de Volder. He saw my *Historia* and after he had looked at it a little de Volder asks him whether he has seen and read it. He says no, and adds, 'These men believe that they were wronged in the year '18 of this century, and therefore are wont to insist on it and repeat it in season and out of season.' De Volder replies that nothing of that, not so much as a single word, is to be found in this whole *Historia,* unless perhaps he supposes that readers who see here the proceedings of the Inquisition painted to the life will believe that the acts of the year '18 are similar to them in all respects, and for that reason these acts are condemned in this book. Many of the Contraremonstrants are buying and reading it; so far however I have not heard any unfavourable opinions." Locke, *Correspondence,* letter 1640, p. 695.

29. Notes on reactions to Limborch's work can be found in L. Simonutti, "Limborch's *Historia Inquisitionis* and the pursuit of toleration," pp. 248–250.
30. This volume is preserved in the Universiteit Bibliotheek of Amsterdam (UBA).
31. Historians believed that the precious codex was lost. Consequently, for almost three centuries Limborch's edition of the *Liber Sententiarum* was considered a documentary source of exceptional historical value. They also believed that Limborch compiled his edition from a copy, most likely an imperfect one, of the *Sentences;* they were able to explain neither how Limborch had come to possess the codex nor whom the generous donor had been. See Celestin Douais, *Les sources de l'histoire de l'Inquisition dans le midi de la France aux XII et XIV siècles* (Paris: Palme, 1881); Id., *Documents pour servir à l'histoire de l'Inquisition dans le Languedoc* (Paris: Renouard, 1900); H. C. Lea, *A History of the Inquisition of the Middle Ages* (Cambridge: Heffer, 1956); H. Kamen, *The Spanish Inquisition* (London: Weidenfeld & Nicolson, 1965). For a historical and social analysis of the Inquisition and its social impact, based on a Vatican manuscript from Provence from the first decades of the fourteenth century, see Emmanuel Le Roy Ladurie, *Montaillou, village occitan de 1294 à 1324* (Paris: Gallimard, 1975).
32. Mss additions, *HI (Historia Inquisitionis),* B. III, chap. I, p. 178. CE (Chandler's edition), vol. II, pp. 6–13.
33. Mss additions, *HI,* B. I, chap. IX, p. 42. CE, vol. I, p. 63, "For this Fraternity *Dominick* framed certain Constitutions, by which they were to preserve and govern themselves. The first was, that such who entered into this Warfare, should take a solemn Oath, that they would endeavour with all their Might to recover, defend, and protect the Rights of the Church against all who should pretend to usurp them; and that in Defence of the Ecclesiastical Effects they would expose themselves and their own Estates, and take up Arms, as often as they should be called on to do it by the Prelate of the War, who was then *Dominick,* and afterwards the Masters General of the *Dominican* Order."
34. Mss additions, *HI,* B. I, cap. XXIX, p. 100 and p. 101. CE, vol. I, pp. 154–155.
35. Mss additions, *HI,* B. I, chap. XXIV, p. 81. CE, vol. I, pp. 125–26: "What *Paramus* adds farther is ridiculous: That when the aforesaid Inquisitor of the holy Office eagerly desired some Sign from the blessed Virgin *Mary,* of *Guadalup,* in those Days of the Inquisition, to exalt the Faith, and for the Confusion of Heresies and Errors, there were so many evident Miracles wrought by piously invocating the holy Virgin, that Doctor *Francis Sanctius de la Fuente,* one of the said Fathers, who had undertaken the Care of writing those Things, was quite tired out with writing them, through the Multitude of the Miracles. For there were sixty Miracles counted up for the Confirmation of the Faith, the Approbation of the Office of the holy Inquisition, the Edification of many, and the Display of the Divine Power."

36. Mss additions, *HI,* B. I, chap. XXIV, p. 81. CE, vol. I, p. 126.
37. Ibid.
38. Mss additions, *HI,* B. I, chap. XXVII, p. 96. CE, vol. I, pp. 146–147.
39. Mss additions, *HI,* B. II, chap. XXIV. CE, vol. I, p. 222.
40. Mss additions, *HI,* B. II, chap. XVI, p. 149. CE, vol. I, p. 223.
41. Mss additions, *HI,* B. II, chap. XVI, p. 149. CE, vol. I, pp. 224–225.
42. Mss additions, *HI,* B. II, chap. XVI, p. 149. CE, vol. I, p. 225.
43. van Limborch, *The History of the Inquisition,* vol. II, B.IV, p. 221.
44. van Limborch, *De veritate Religionis Christianae amica collatio cum erudito judaeo* (Gouda: J. ab Hoeve, 1687).
45. Mss additions, *HI,* B. IV, chap. XXX, p. 329. CE, vol. II, p. 229.
46. van Limborch, *The History of the Inquisition,* translated into English by Samuel Chandler (London: J. Gray, 1731), vol. I, *Preface,* p. XII.

6

The Public Context of Heresy

Bayle, Maimbourg, and Le Clerc

Sally L. Jenkinson

The stake and the gibbet, the terrifying tribunal of the Inquisition, Crusades, Papal bulls to incite subjects to rebel, seditious preachers, conspiracies and assassinations of princes were the ordinary methods employed [by Christians] against those who did not submit to their commands.

—*Bayle,* Dictionary, *"Japan," Remark E.*[1]

There was no issue more embarrassing for writers of the Protestant communion in the sixteenth century than the necessity, to which they thought themselves reduced, of urging the magistrate to punish heresy; whilst thinking it strange at the same time that Catholic princes should persecute the Protestants.

—*Bayle,* Dictionary, *"Sainte-Aldegonde," Remark L.*[2]

Heresy in the Baroque State

This chapter examines the ideas of three authors—a Catholic, a Huguenot, and a Dutch Remonstrant—who engaged one another in debate about heresy at the turn of the eighteenth century. By heresy in the "baroque" state, I mean the status of a sect designated "heretical" in the criminal law of Christian society in post-Reformation Europe. Following the breakup of Catholic Christendom into a network of sovereign states, each state continued to support an official Church responsible for instruction in a body of "orthodox" ideas, and for excluding instruction in "heretical" ideas.

"The baroque state" denotes the society evolving in Europe during the period after the age of Reformation and before the age of Enlightenment. It denotes equally the territorial unit of the age recognized by Europe's diplomatic community as "sovereign," that might of course be enlightened or despotic, republican or monarchical, Catholic or Protestant, large or small, efficient or ramshackle. Because this essay is concerned with the public notion of "heresy," the main conceptual point is that at this period the state acquired the authority to define Christian orthodoxy and heretical departures from it in its own terms, and under its own laws.

Ideas about heresy in the baroque society reflected a transitional stage in European thought. They were the product of a culture poised between the theory and institutions of the Middle Ages and those of the Liberal Enlightenment. Though the historic loyalties and immediate concerns of our authors are very different, they responded to changes that were occurring throughout Europe. In different degrees all three authors were innovative in their concerns and methods of argument.

The first section of this chapter describes the background of each of our authors. The second section distinguishes between two paradigms of theologico-political government. These two paradigms were rivals for the allegiance of minds throughout the age of the baroque state. Propagandists for each paradigm sought to caricature the other's failings, creating pitfalls for the historian of ideas. The third section interprets texts from two works by philosopher and critic Pierre Bayle. The first is his *Critique Générale de l'Histoire du Calvinisme de M. Maimbourg* (1682),[3] in which he examines the way in which a Gallican Catholic perceives the history of his community. The second is his pamphlet *Entretiens de Maxime et Thémiste,* published posthumously in 1706,[4] in which he replies to an attack from co-religionist and fellow man of letters Jean Le Clerc. Bayle says he is obliged to defend his reputation as a detached scholar in response to Le Clerc's accusation of "heretical" writing. Finally, we draw from the reciprocal involvement of our three authors some conclusions about the evolution of the ideas of heresy and intellectual freedom.

Maimbourg, Bayle, and Le Clerc: Background

Louis Maimbourg (1610–1686), of a Catholic family from Nancy in northern France, was sent in 1626 to pursue a career with the Jesuits.[5] Following studies in Rome he returned to France to teach humanities at a Jesuit academy. In consequence of Louis XIV's policy, intensified after 1660, Maimbourg was drawn into the project that sought to bring the Huguenots back into a single official Church. Writers were encouraged to try to convert the "heretics" to the official religion.[6] The writing of history was deemed helpful to this project, and between 1673 and 1686, Maimbourg published eleven works of history, of which seven were histories of heresy.[7] In due course Maimbourg broke with the Jesuits, receiving an income directly from the Gallican regime that enabled him to continue his historical writing. *L'Histoire du Calvinisme* of 1682, to which Bayle and others responded, was followed by a reissue in 1686 of all of his histories. Through his family he had close contacts with Anglicanism.[8]

Pierre Bayle (1647–1706), born at Le Carla in the southwest of France, was the son of a Huguenot pastor. He attended a local Huguenot college, moving in 1668 to the Catholic University of Toulouse. He converted briefly to Catholicism in 1669 but reconverted in 1670 to the Reformed Church.[9] He was then swiftly dispatched by his family to Protestant Geneva, for he could have been prosecuted under a French law that punished converts to Catholicism who rescinded their conversion.[10] In 1675 Bayle became professor of philosophy at the Huguenot academy of Sedan, in northern France.[11] On the academy's suppression in 1681 in anticipation of the Revocation of the Edict of Nantes, Bayle moved to the Netherlands to become professor of Philosophy and History at Rotterdam's Ecole Illustre.[12] He took with him several completed manuscripts that reflected his scholarly interest in the movement of the *libertins érudits,* or erudite sceptics,[13] including his *Pensées diverses sur la comète* (1681). In 1682 and at great speed, Bayle wrote and published the *Critique générale de l'Histoire du Calvinisme de M. Maimbourg,* which we shall discuss below.[14]

To criticize Maimbourg, Bayle drew on the perspective of both critical philosopher and humanist historian. He used his erudition in the Ancients and the Moderns to question the Gallican *politiques,* whose position Maimbourg had adopted. This school was wary of sectarian diversity in the state, but less from the perspective of theology than from the perspective, adopted by Hobbes some decades before, of pragmatic opposition to the proliferation of sects. Following this work, Bayle produced many further publications, including his celebrated *Commentaire Philosophique* of 1686.[15] All of Bayle's writings favored intellectual diversity on the general grounds that one man's orthodoxy was another's heresy, so that persecution for holding one metaphysical belief rather than another was as paradoxical as it was unjust.[16]

Bayle's work reflects his awareness that the advent of the post-Reformation state had modified but not overturned received beliefs about the importance for orderly society of a public religion. Though enlightened minds—whether Catholic, such as Maimbourg, or Protestant, such as Le Clerc—had no love of persecution, they still believed that it was the duty of the state in the baroque age to maintain an official religion. Jean Le Clerc (1657–1735) differed from both Bayle and Maimbourg in that he came from Protestant Geneva and not religiously divided France. His learned Calvinist family encouraged him to study theology, philosophy, and biblical exegesis, and to travel in France during the last years of the Edict of Nantes. He took Protestant orders in Geneva, traveled to England, and in 1682 became minister to the Savoyard and Walloon congregations in London.[17] Two years later Le Clerc settled in Calvinist Amsterdam, where he joined the Remonstrant sect, receiving patronage from Remonstrant theologian Philipp van Limborch and making friends with the exiled John Locke, whose writings he translated.[18] Successively Le Clerc was professor of Hebrew, Belles lettres, and Church History in the Remonstrant Seminary. In 1686 he became editor of the journal *Bibliothèque Universelle et Historique* (later the *Bibliothèque Choisie*). During a lifetime devoted to religious journalism and letters, he published 79 books.[19] Though of course an opponent of Louis XIV's persecution of the Huguenots, Le Clerc, surprisingly, became Bayle's adversary. One historian suggests that their dispute arose from professional rivalry.[20] Le Clerc's antagonism toward his Huguenot colleague surfaced after 1700, when he made objections from the perspective of Protestant rationalism, as he understood it, to two propositions that Bayle had consistently defended. Le Clerc expressed his criticisms in a pseudonymous work entitled: *Parrhasiana, ou Pensées diverses sur des matières de critique, d'histoire, de morale et de politique.*[21] Bayle responded in various essays,[22] to which Le Clerc replied in the pages of *La Bibliothèque Choisie.* The discussion continued after Bayle's death, when the work which is discussed below, *Entretiens de Maxime et de Thémiste,* was published posthumously in 1706 with a foreword by Bayle's colleague, Basnage de Beauval.[23]

Two Paradigms of Theologico-Political Theory

Some who write about European institutions at the end of the seventeenth century suppose that a given political theorist—especially in the age of Louis XIV, William III, Hobbes, Locke, and Bossuet—should be pressed into service of either "absolutism" or "toleration." However, this dichotomy has few advantages in the context of ideas about heresy. For the political writings of the age that followed the Reformation—for example, those of Bellarmine, Hobbes, Grotius, Locke, Spinoza, Leibniz, Bossuet, Maimbourg, Le Clerc, or Bayle—indicate that notions of "heresy," "toleration," and "absolutism" are used with very different intentions that must be interpreted in the context in which they are deployed.[24]

Throughout Europe the large fault lines were drawn in terms of two opposing paradigms of authority in armed competition for the allegiance of peoples. Both paradigms required society to define orthodoxy, and to deal appropriately with those who emerged, by intention or default, as "heretics." These two paradigms are named here "The Paradigm of Ultramontane Catholicism"[25] and "The Paradigm of the Sovereign State."[26] That theorists of these paradigms disagreed profoundly—to the point of resorting to arms—about who defined heresy, but without wishing to abandon the concept of heresy as such, indicates the importance attributed to the label "heretic." From the time of Augustine such labeling had been used by ecclesiastics both as an instrument for maintaining social control and as a means of integrating peoples into mainstream society.

The first and more ancient of these theories of authority—opposed by all three of our authors—was that of Ultramontane Catholicism. This doctrine—known sometimes as the theory of the Two Swords—was the received wisdom of the Western Church more or less from the age of Saint Augustine until the age of the Reformation.[27] It asserted that a Christian ruler should treat as heretics and rebels those parties and sects that attempted to challenge the authority of the Church. To secure the support of the secular arm, the Medieval Church taught that allegiance was always due to the Christian sovereign. However, the Church could absolve peoples from allegiance to a ruler it declared to be "heretical," such as Elizabeth I of England or Henry III of France, who had failed to extirpate heresy with sufficient vigor. It could even allow their assassination.

In his *Dictionary* article on Michel de l'Hôpital, the first layman to be Chancellor of France, Bayle illustrates how Ultramontane Catholicism abused the notion of "heresy" in the sixteenth century to impede political and religious innovation. To illustrate his point he cites a letter that was sent to the Pope on 14 June 1562 by Cardinal Hippolyte d'Est, Cardinal Legate in France. "Among other difficulties," so the Cardinal had written,

> "not one of the least is to dislodge from court, as your Holiness desires, the chancellor and several other relevant persons. For your Holiness puts in this number both heretics and those who are suspected of heresy. But if all the latter were to be banished from court it would undoubtedly be deserted, these new opinions having already made such an impression on the minds of courtiers that there are few to be found that have not some small tincture of them."[28]

By "The Paradigm of the Sovereign State," I mean the theory of authority that sought to put limits on attempts by Ultramontane Catholicism to control a nation's affairs. Some of its supporters, such as Machiavelli, were republican thinkers who joined with religious reformers of the age of Renaissance and Reformation to contest the claim of the Roman Curia that it could define heresy throughout Christendom. Constitutional thinkers, Bodin above all, denied that the civil authority should share its power with either pope or ecclesiastics.[29] But their movement was not necessarily on the side of egalitarian reform. After the Wars of Religion, advisors to princes insisted that former Catholic Christendom was now a network of independent states whose religious politics were their own concern. Encouraged by the new doctrine—compressed into the ambiguous slogan *cuius regio eius religio*[30]—the rulers and elites of Europe's post-Reformation communities agreed reciprocally that within each state, prince and people were to adopt their preferred version of Christianity and conform to it.

Fearing that religious sects might constitute a fifth column, civil elites reciprocally agreed that "heretics"—defined as communities of sectarians not of the orthodox confession—should be encouraged to leave a territory, even though they should be allowed to take their possessions with them and not be molested. Civil authorities likewise accorded exemptions from expulsion for certain sects—Catholic, Lutheran, and Calvinist minorities in the German states; Socinian minorities in Transylvania and Poland; and Jewish minorities in the Netherlands and elsewhere—provided that these communities were law-abiding and conducted their devotions in private. In short, post-Reformation treaties and conventions did not abolish notions either of "heresy" or "schism." Enlightened minds of the baroque age prided themselves nevertheless on having dissociated themselves and their laws from the stake, the gibbet, Crusades, Papal bulls to incite subjects to rebel, the Inquisition, and so on.[31]

Bayle, who opposed penalties for heresy, whatever their severity and whatever regime sought to impose them, drew attention to the "embarrassment" of the best Protestant societies on the question of sectarian freedom. Once they had secured their independence from the Catholic Curia, newly independent rulers often oppressed sectarians, using the arguments formerly used against them. The Dutch patriot Phillipe de Marnix, Bayle noted in his *Dictionary,* was "strangely perplexed" when it was complained that he had "exhorted the Estates to persecute the sects."[32] His inconsistency exemplified a paradox that lay at the heart of Europe's independent republics:

> . . . there was no issue more embarrassing for writers of the Protestant communion in the sixteenth century than the necessity, to which they thought themselves reduced, of urging the magistrate to punish heresy; whilst thinking it strange at the same time that Catholic princes should persecute the Protestants.[33]

Hobbes, Bayle noted, had provided a quintessentially robust account of the new doctrine.[34] Bayle perceived that Hobbes—like Bodin and Sainte-Aldegonde—sought to secure the new republic from troubles by making no concessions to incitement to political violence, whether from Ultramontane Catholics or from "heretical" sects. A theory of government advocating "two swords" in which civil power was supposedly shared with a church, Hobbes declared, was a usurped and absurd doctrine. No sovereign authority could share power with a church, national or international. For this reason the independent civil ruler carried the sword of state in one hand and the pastoral crook in the other. A truly Christian commonwealth had no reason to rely on dogma constructed by Ultramontane ecclesiastics. The sovereign state, the theory insisted, must have its own official creed, ensure that clerics teach its tenets, and protect it from "heretical" sectarians.[35] But who in fact defined "heresy" under the new doctrine of *cuius regio eius religio,* and where was orthodoxy's ultimate court? According to Ultramontane theory, orthodoxy meant the body of Christian doctrine, sanctioned by the hierarchy of Catholic ecclesiastics, at whose apex was the pope. The theory of *cuius regio eius religio,* on the other hand, called for defining orthodoxy in one of two ways: by the civil authority alone, as Hobbes taught; or by the civil authority acting in partnership with society's rational theologians, as Le Clerc probably supposed.

Assessed in the context of these two paradigms, our three thinkers, as opponents of the first paradigm, may be regarded as "advanced" minds. Assessed in the context of the second paradigm, Maimbourg and Le Clerc are harder to place. Maimbourg's Gallicanism was "radical" to the extent that he opposed Ultramontane and Papal authority and dismissed the claims of the Catholic League to take up arms against the Calvinists in defiance of the royal edicts of toleration. Le Clerc's Protestantism, by the same token, was "radical" to the extent that it condemned the stake, the gibbet, and the Catholic Inquisition. Yet, it was intolerant to the extent that Le Clerc would like to suppress certain doctrines taught by Catholics, Socinians, Jews, *esprits forts,* and others. Le Clerc—like his colleagues Locke and Limborch—takes for granted that the best commonwealth is a Christian commonwealth whose doctrines can be legitimately determined by reasonable exegesis of the Scriptures. According to Bayle's critique, Le Clerc was naive enough to suppose that the Protestant exegete possessed a special entitlement—not available to other scholars—to declare which opinions were "orthodox" and which should be silenced.

Bayle's Reply to Maimbourg

In refuting Maimbourg's *Histoire du Calvinisme,* Bayle invoked great subtlety and bite, notwithstanding his acknowledgement of Maimbourg's erudition, historical knowledge, and rhetorical and logical skills. First, Bayle called attention to the ex-Jesuit's financial dependence upon the income he received from Louis XIV and to his reputation for accommodating himself to political circumstances. The *Histoire du Calvinism* was published later than expected, he observes wryly, perhaps because, having promised a work on the "birth, progress, and annihilation of Calvinism" by Louis XIV, its author "could not honorably have published his book before the entire extirpation of the Huguenots" (p.7).[36] Maimbourg wrote, Bayle continues, because he perceived that the court of Louis XIV was "determined to ruin Calvinism in as little time as he could write his history" (p.9). He therefore prepared an apology for all the violence that would be used to accomplish this grand design.

Could any reader doubt, Bayle asks, that they were about to read a biased history? The author would have to represent the Calvinists as supporters of "heinous ideas" (p.9), always poised to "revolt against their legitimate sovereigns and always ready to plunge their country into lamentable devastation" (p.9). Was it not evident that Maimbourg intended to derive from history the maxim that "a prince who loved the glory of God," and who wished "to ensure the safety of his kingdom and his throne," should always be ready "to exterminate these monsters, to cut off every head of the hydra and to crush those infernal enemies of God and state" (p.9)?

Bayle's thesis was that Maimbourg was misusing the art of history to support government policy. The first purpose of the book was to justify the policy of oppressing the Huguenots and give new impetus to the king and his ministers, in case their "campaign [against heresy] was not going as fast as the ecclesiastics might wish" (p.9). The book, Bayle warned, would be a justification for the religiously unified state.

> You will not find a page in the *Histoire du Calvinism* where the following notion is not thrust before the reader: namely, that one is an enemy of the state and one's King if one puts up with heretics, and a king who suffers them will be guilty of a negligence that loses both king and kingdom" (p.9).

Bayle will not refute Maimbourg point by point or check his footnotes (p.10). He will demonstrate that the skilled historian can make of the past what he wishes. Given his record, experience, and position, it would be surprising had Maimbourg *not* used his *Histoire du Calvinism* to propagate political doctrines that conformed to government policy.

Maimbourg's thesis against tolerating "heresy," as Bayle grasps, relied on a version of early-modern political philosophy, not on medieval political theology. The politicians and the ecclesiastics of the age of Louis XIV had united to assert in a new way that the civil ruler was obliged to extirpate heresy. Maimbourg insists in the manner of the *politique* of the age of *cuius regio,* that the ruler, to prevent civil strife, is obliged to prevent the proliferation of sects by bringing together prince and citizen within the public religion: *une foi, une loi, un roi.* His status as a cleric notwithstanding, Maimbourg advanced a civilian's version of the theory, supported by *politiques* and theorists such as Hobbes, and by the doctrine of "the Paradigm of the Sovereign State," that the very presence of sectarians within a state was necessarily a threat to public order.

Nor does Bayle misread Maimbourg by supposing that the ex-Jesuit supported the theocracy of the ecclesiastics, the divine right of the monarch, or absolutism of the civil authority. He acknowledges the apparent plausibility of the premise from which Maimbourg begins and sets out to demolish it. Was it not evident, Bayle asks, that Maimbourg would select for his *Histoire du Calvinisme* the appropriate instances of disorderly events in history that would be suitable for supporting his position? However, Bayle points out, any competent historian schooled in logical argument could disprove that proposition by showing from history that there was often disorder in states united in religion. There was often peace in societies where there was a diversity of religion. There was no necessary connection between religious diversity in a state and an orderly society. Additionally, by making his own critical technique transparent, Bayle draws attention to the fact that if a party can select instances from history to justify a public policy that happens to be mistaken, so a critic of that policy can select instances from history to justify a better policy.

Bayle's *Critique Générale* draws attention to the arsenal of such devices that could be put to use by the historian. For example, he brings out the irrelevance to present public policy of historical accounts of massacres and similarly disputed events. He assesses with evidence the certainty or uncertainty of facts and propositions. He stresses the need to read a source with care. He searches out alternative documents for evidence that could refute a mistaken proposition. Though a full account of Bayle's rigor cannot be given here, we can give some illustrations.

On Disputed Events in History

Though important to the parties to a dispute, Bayle proposes, disputed events in history are not useful in making judgements about the truth of one's religion. For, he says,

> When I read a history of the civil wars composed by our authors I find that the Protestants were never in the wrong . . . but when I read about the same

> wars in the history of the opposition, above all if they were written by the monks or churchmen, I find myself in unrecognizable territory. . . . The former claim that the Protestants were never aggressors. . . . The monks reverse this argument; [saying that] the Huguenots were the first to rebel. They conspired against the King. They burnt and ransacked the country . . . and the Massacre of St Bartholomew Eve was a necessary and legitimate act of prudence to forestall Admiral de Chatillon who had resolved to butcher the Catholics (p.10).

As a scholar, he could not resolve these contradictions, but as a Huguenot he knew that conflicting accounts would not justify "any legitimate prejudice against the divinity of my religion" (p.11).

The Certainty or Uncertainty of Facts and Propositions

Bayle argued that when scholars read history, they should see it as a field of experience for testing their propositions. They must, nevertheless, remain on their guard against "the clever writer" who could pervert "the most innocent facts" (p.11). As in advocacy, Bayle argues, there are no facts which, in the hands of opposing parties, "could not take on a completely contrary appearance" (p.10). Bayle supposes that it is neither possible nor desirable for a historian to be completely without bias. But having a point of view, he insists, does not prevent the historian from investigating the evidence well or badly. A good historian can distinguish between degrees of evidence, types of evidence, and corroboration of evidence, and so assess whether an event was nearly certain, or so uncertain that judgement about it should be completely suspended (pp.11–12).

A sound historian can also judge aims and motives if all the parties to a dispute agree, though he would need to be more cautious when parties disagree.

> If, on the one hand, good sense requires that I suspect that a Huguenot historian has failed to penetrate the pernicious designs of his party because of a want of discernment or because he is blinded by prejudice, or that he dissimulates in order to save the honour of his party; the same good sense requires that I similarly suspect a Catholic historian (p.11).

Despite Bayle's familiarity with the school of thought called "skepticism," it would be wrong to suppose that as a historian he agreed with those who said there was no possibility of establishing a historical fact, or extending objective knowledge about the past. The "serious historian," he insists, by examining the sequence of events, the character of the actors, weighing the whole with what has been said "by one party and the other," can illuminate events well enough both "to uncover many impostures" and "to refute many calumnies" (p.12). A more difficult problem for a historian lay in ascertaining contested events when those who knew the facts concealed them, as with broken treaties, secret conspiracies, and brutality against the innocent. In such matters, the sound historian—able to confirm neither one position nor the other—was obliged to suspend judgement (p.11).

On Reading a Source with Integrity

Throughout his critique, Bayle applies to the historical text the methods increasingly used by the natural scientist, the legal investigator, or the logician. Maimbourg's *Histoire du Calvinisme,* accordingly, is always treated with respect. When he refutes his adversary, Bayle first restates the historian's thesis. Having recovered the proposition in a form with which the author would agree, he applies the evidence. The initial step—giving an account of an adversary's position—is always the most important. It is time wasted, as well as bad logic, to attempt to refute a proposition that has been incoherently asserted in the first place. As an example of a contestable assertion, he cites Maimbourg's report of an uprising of villagers followed by the massacre of the whole community that had occurred during the French Wars of Religion (p.27). The full facts had long been a bone of official contention between Catholics and Protestants. Bayle objects that events could not have been as Maimbourg describes them since his account contained internal inconsistencies. Where there are unresolved contradictions in a report, Bayle teaches, the source is suspect. It is then appropriate to propose an alternative theory of what might have happened. He suspects, he says, that these supposed heretics never used violence in the first place, and that their supposed uprising was invented by the persecutors to excuse the brutality of the troops toward the villagers and their attempt to remove eyewitnesses (p.51). Bayle then shows how Maimbourg's report of this disputable event in the history of France's Wars of Religion affected recent policy. It was invoked to defend state violence toward the Huguenot community. The regime of Louis XIV—citing the supposedly factual proposition that Huguenots were heretics, and that heretics were violent rebels—could then send in dragoons to intimidate Huguenot families. By criminalizing the victim, the regime exonerated itself from past and future accusations of using violence against the innocent (pp. 51–53). The "infamous practice," Bayle concludes, "of attributing crimes to a party, in order to excuse the injustice one has done to them, is the principle cause of uncertainty in history" (p. 51).

Discovering Alternative Documents

The technique of citing documentary evidence that refutes a historian's errors, Bayle remarks, can test the truth of any proposition about past or contemporary affairs. It could be applied in particular to the *Histoire du Calvinisme.* Had Maimbourg not written in his Preface that the treatment of the Huguenots by the regime of Louis XIV had been "gentle, charitable . . . zealous for the salvation of souls . . . and opposed to brutality and force?" (p.32).

No less than any other historian, Bayle could show that there were four different sorts of testimony that could refute that statement. First, there was the verbal testimony of whole families that had left France, obliged to live in countries they found disagreeable (p.32). Second, there was the Jansenist Catholic author Antoine Arnauld who, in his *Apologie pour les Catholiques,* unwittingly conceded that violence was indeed being used against Huguenots when he justified its use on the grounds that it would be in the interests of the heretics' eternal salvation (p.32). Third, Bayle points out, there were the formal complaints made to the government by different communities of Huguenots, coupled with the formal refusals of the administration to investigate their complaints (p. 32). Fourth, international documents provided evidence that foreign princes had offered asylum to Protestant refugees from France. Bayle cites at length from published letters from Charles II, King of England, dated 22 July 1681, to the Bishop of London and the Lord Mayor of London asking them to accord asylum to French Protestants "because of the persecution to which they had been exposed." Was not this document, Bayle concluded rhetorically, "a thousand times more authentic as evidence that in France they do not treat us kindly, than a thousand volumes by Mr. Maimbourg declaring that they do?" (pp.51–2).

From the standpoint of the evolution of ideas about the definition and treatment of heresy at this period, the dialogue between Maimbourg's *Histoire du Calvinism* and Bayle's *Critique Générale* is uniquely significant. It reveals the shift in reasoning used by public intellectuals to justify the presumed need to impose an "orthodox" religion. It reveals that the age of the baroque state recognized new standards based on international treaty to regulate the treatment of those expelled and exiled for "heresy." It shows us that Bayle, concerned to uphold these new standards—though seeking to push them further—was ready to accord as much credit as possible to the fact that Maimbourg condemned cruelty toward heretics, and that he was motivated by political realism and the doctrine of *raison d'état.* Turning Maimbourg's pragmatism to his own account, Bayle seeks to show that Maimbourg's error—like that of the regime's other *politique* realists—is to begin from the pragmatic case for religious unity and then select evidence that supports the case.

It remains common today for liberal critics of Louis XIV's suppression of the Huguenots to presume that the cause of Louis XIV's policy was the "absolutism" and "theocracy" of the regime, including belief in the supposed obligation of the Catholic ruler to convert heretics and "save souls." In fact, as Maimbourg is aware and as Bayle recognizes, the civilian thinkers of the age relied on more pragmatic doctrines to support their policy. By the turn of the seventeenth century, Europe was already a patchwork of states committed to different Christian confessions, whose publishing houses not only sent banned books across frontiers, but knew how to exploit clandestinity to promote sales. Protestant states were to be found on almost every frontier of France: Switzerland, Geneva, the United Provinces, and England. Zealots in these communities preached that their own souls were "saved" but those of other confessions "damned." Nevertheless, politicians and the diplomatic community, while making use, when necessary, of rival religious loyalties, advanced more pragmatic reasons for opposing sectarians within a state. The theory that Maimbourg advanced, therefore, and that Bayle contested, asserts that peace in society requires a common religion and a firm policy for ensuring religious conformity.

A better theory, Bayle believed, would have recognized Europe as it was capable of becoming; namely, as a network of religiously plural states.[37] Its rulers would be committed not only to restoring toleration to its formerly persecuted communities, they would "persecute no one for his opinions in religion," nor "use the right of the sword against conscience."[38] Bayle thereby opposes public policy concerning "heresy" as reflected in both the paradigm of Ultramontane Catholicism and the paradigm of the Sovereign State. For this latter institution, devised to enable the early reformers to protect themselves from the Inquisition, had ceased to be on the side of reform. In Bayle's day, privileged elites in the baroque state, Protestant and Catholic, found new reasons for fearing innovators, and new generations of dissenters to control.[39]

Bayle's Reply to Le Clerc

There could hardly have been more difference between Bayle's reply to Maimbourg of 1682, and his response to Le Clerc more than 20 years later. As Basnage de Beauval remarked in the foreword to Bayle's *Entretiens de Maxime et Thémiste,* Bayle's *Critique de l'Histoire du Calvinisme* and *Nouvelles de la République des Lettres* had been written with "a courtesy and charm that are rarely found among authors" (p.2).[40] Yet in the dispute with Le Clerc, Basnage de Beauval continued, Bayle poured scorn upon his adversary, and had perhaps gone too far (p.2).

Le Clerc claimed to have found in Bayle's writings two notions that seemed to him, as a Christian theologian, incompatible with orthodox Christian teaching. The first notion to which he objected was Bayle's proposition, asserted consistently throughout his work, that a society of atheists, whose members believed in neither the existence nor the providence of God, would, if it had good civil laws, be capable of sustaining itself.[41] The second was Bayle's treatment of the Manicheans. Bayle had pointed out that this sect had taught that the goodness and power of God were difficult to reconcile with reasoned argument. Bayle's observation as a philosopher was to say that though the position of which he gave an account was "heretical" from a Christian perspective, and though Manicheanism undoubtedly entailed morally repugnant consequences, no system of theology had so far been able to resolve the question satisfactorily. In consequence of these speculative writings Bayle's work was judged by Le Clerc and other Protestant theologians to be incompatible with Christianity. Bayle's religious criticism, Le Clerc concluded, was therefore a matter for grave public concern in a Christian commonwealth.

Given the erudition of the age, no scholar, Christian or otherwise, would have supposed that Bayle himself subscribed to the beliefs concerning God of either the "atheist" or the "Manichean."[42] The metaphysics of the matter were undoubtedly difficult, but a careful reader could see that Bayle had distanced himself from the position he described. Le Clerc, despite being a scholar himself, chose to argue that Bayle treated these heretical notions as if they deserved to be discussed philosophically, failing to associate them with sufficient horror.[43] It was "against the rules of every religion," Le Clerc said, "to find problems in Providence," and especially to cast them "in language that outrages Divinity" (p.31). That sort of critical practice undermines "civil society," of which one of the most important fundamental principles is that there is a God who is holy and well intentioned (p.31).

In short, Le Clerc was an opponent of Catholic intolerance, but intolerant in turn of those who criticized Christian teaching as his own sect defined it. Bayle's perspective was different. Though a Christian layman, he analyzed human psychology as a natural philosopher and defended the hypothesis that belief in a deity seemed to be unconnected with moral conduct. According to Basnage de Beauval, Bayle regarded Le Clerc's question-begging attack on his demonstration as a slur on his honor as an honest scholar, and it was for this reason that he used such strong language in his reply.

It was equally the case, however, that Bayle used Le Clerc's attack to make some points in public about another matter that profoundly interested him. This was the differing psychology of the polemicist and the scholar, regardless of the era in which they lived. The "prejudices of a party produce, among other faults," he explains, "a violent inclination to favor everything that can render the other sects odious." A man who has this fault, he continues, "cannot bear it to be said that other sects have anything about them deserving of praise," or "that anyone should refute the falsehoods that blacken their name" (p.35). Bayle's literary and historical research on societies that governed effectively without an official religion apparently brought out violent emotions in Le Clerc. Accordingly, Bayle makes his character Maxime speculate that Le Clerc cannot bear having Bayle assert that atheism is not the "worst of the states into which the soul of man can fall" (p.35–6). Nor can he bear to read that a society that had no religion could, nevertheless, sustain itself.

By replying to Maimbourg on his own ground, through history and political theory, Bayle, by implication, is criticizing not just Maimbourg and Gallicanism but also the many other humanist writers of the age, such as Lipsius, Grotius, or Hobbes, for example, who assumed that sects within the state were always a threat to public order. By replying to Le Clerc, on the other hand, Bayle showed that a colleague had made foolish charges. He judged that in Le Clerc he was confronting a minor instance of Protestant intolerance in metaphysical debate, which he found embarrassing in a co-religionist. The question that he finds absurd in a post-Cartesian age was whether a scholar might raise and discuss in a journal an idea he acknowledges as "heretical," if a co-religionist judges that in so doing, he scandalizes Christian contemporaries and undermines their faith. Through his response to Maimbourg's history, on the one hand, and to Le Clerc's accusation, on the other, Bayle demonstrably moved the debate about "heresy" in the direction of the liberal enlightenment.

Conclusion

From the debates among Bayle, Maimbourg, and Le Clerc we learned that to evaluate the ideas of these writers we had to point out that heresy could not exist without orthodoxy, and that rival ideas about orthodoxy were embodied in the rival paradigms of authority. All three authors opposed the persecution of heresy by Ultramontane Catholicism. Maimbourg and Le Clerc, according to the received wisdom of the time, apparently saw no better alternative than imposing new orthodoxies through the relatively preferable institution of the Sovereign State.

Bayle's counter-assertion, that defining orthodoxy and suppressing heresy should be condemned in all contexts, was many decades ahead of its time, in an age that feared (or pretended to fear for propaganda purposes) a return of Ultramontane rulers, the Inquisition, and a train of cruel punishments. Bayle's supporting premise, that a public religion was less essential for an orderly society than the age supposed, was not what the baroque age wished to hear. It was rather a prerequisite for the future liberal state that would make "no law concerning the establishment of a religion." Few contemporaries understood Bayle's position, and those who did preferred a civil religion to a plurality of sects. Despite the differences between our three thinkers, all defended freedom from papal persecution, made possible by the paradigm of the "Sovereign State." However, Maimbourg and Le Clerc were engaged, from different perspectives, with the way in which Christian orthodoxy should be managed by the sovereign. In contrast Bayle, between the lines of his historical criticism, asked whether orthodoxy needed to be "managed" at all.

Despite his background as a French ecclesiastic and his education as a Jesuit, Maimbourg represented the Gallican tradition within the second paradigm. Because of the *politique* and prudential dimension to his thought, Bayle finds Maimbourg's position more worthy of criticism than that of Le Clerc. For Maimbourg was inclined to perceive "heresy," as did Hobbes, less as an error against truth, than as an unacceptable challenge to public authority. Le Clerc represented the perspective of the theologian of the Protestant state who found that the political doctrine of *cuius regio eius religio* tended to curtail the political influence of the theologian. Le Clerc, like his Remonstrant co-religionist Limborch and their English colleague John Locke, believed that a Protestantism supposed reasonable is the sole means of safeguarding Europe from a return to Ultramontane Catholicism. In reality, the Protestant theologian in the Europe of sovereign states had inherited little of the political authority that had accrued to the theologian of Ultramontane Catholicism, or had been acquired by the lay politician of the Sovereign State. This hidden conflict of authority between ecclesiastics and laymen within the Protestant state was, without doubt, reflected in the antagonism of Le Clerc toward Bayle. Le Clerc hoped that the Christian scholar, through his skill in exegesis of texts, would, in due course, recover a rational Christianity capable of reuniting all well-intentioned Christians. Bayle, as a critical teacher of science and epistemology, had no more interest in restoring authority to Protestant theologians than to Catholic theologians. His position is that of the philosophic layman and the scholar of the republic of letters, who writes within the paradigm of the baroque state while aiming to transcend it.

Given his interest in the skeptical philosophers of the ancient world and in the institutions that preferred critical teaching to dogmatic teaching, we might say—using today's vocabulary—that Bayle would have "abolished orthodoxy" and "decriminalized heresy." Such phrases, wholly familiar to the ear of a modern liberal, would have bewildered Bayle's contemporaries. Bayle can praise Bodin, Grotius, Hobbes, and other theorists of earlier decades to the extent that they had insisted that the civil sovereign's right of the sword was the prerequisite of intellectual freedom.[44] He transcends these theorists by adding that the prince abuses the right of the sword if he wields it to extirpate heresy *even when he himself defines it*. The sovereign's duty is not to fight a holy war against heresy; it is to keep the public peace and to respond to emergencies.

Bayle did not dismiss out of hand Maimbourg's premise that the presence of "unorthodox" sects in a given society constitutes a threat to public order. He examined the proposition from a historian's perspective, asking if the doctrine—despite being the received wisdom of many honorable thinkers—can be substantiated in the present age. His alternative proposition is that the Huguenots of France under the Edict of Nantes, like the Socinians in Poland[45] under their initial edicts of toleration, had never constituted a danger to public peace. Undoubtedly the doctrine seemed valid to those, such as Bodin, who lived through France's bloody Wars of Religion, or Hobbes, who survived the Civil War in England, but it was a refutable theory with reference to the broader historical canvas of the ancient and the modern world.

Neither Maimbourg, who promoted Gallican unity, nor Le Clerc, who promoted a Christian society based on a Protestantism supposed reasonable, would have expressed sympathy with Bayle's opinions even had they understood them. Maimbourg, as a Gallican Catholic, and Le Clerc, as a Dutch Remonstrant, supported an official Christian orthodoxy, even though the particular orthodoxies each supported, and the particular heresies they condemned, were not substantively the same. Bayle is of the avant-garde in the context of the baroque state because he proposed, as would a liberal constitutionalist in our own day, that no harm would befall civil society if there were no established orthodoxy. The notion, with its logical corollary that heresy in religion should be no crime, secured support only in the late eighteenth century with the coming of the liberal constitution.

Notes

1. Pierre Bayle, *Dictionnaire Historique et Critique par Monsieur Bayle* [1697] (Rotterdam: Leers, 1702). English translation: *Bayle: Political Writings,* ed. Sally L. Jenkinson (Cambridge: Cambridge University Press, 2000), p. 132.
2. Bayle: Political Writings, p. 242.
3. Pierre Bayle, *Critique générale de l'Histoire de Calvinisme de M. Maimbourg,* 1682, third edition, in Pierre Bayle, *Oeuvres Diverses* [1727–31], ed. Elisabeth Labrousse (Hildesheim: Georg Olms, 1965–1984), vol. II, pp. 1–160.

Hereinafter abbreviated as *OD* I-V. The work he criticized was Louis Maimbourg, *Histoire du Calvinisme,* 2 vols. (Paris: Sebastien Marbre Chamoisy, 1682).

4. *Entretiens de Maxime et de Thémiste où réponse à ce que Monsieur Le Clerc a écrit dans son X. tôme de la Bibliothèque Choisie, contre M. Bayle,* 1706, *OD* IV, pp. 1–36; hereinafter referred to as *Entretiens.*
5. See Solé, Jacques, *Le Débat entre Protestants et Catholiques Français de 1598 à 1685* (Paris: Amateur de Livres) 1985, vol. I, pp. 100–4. For biographic and bibliographic information about Maimbourg and Le Clerc, see Michaud, *Biographie Universelle* (Paris: 1859) and Bayle's article "Maimbourg" in the *Dictionnaire Historique et Critique.*
6. Maimbourg published three such works: *La Méthode pacifique pour ramener sans dispute les protestans à la vraie foi sur le point de l' Euchariste* (1670); *Traité de la Vrai Eglise de Jesus Christ pour ramener les enfans égarez à leur mère* (1671); *Traité de la vrai parôle de Dieu pour réunir toutes les societez chretiènnes* (1671).
7. Maimbourg's histories included the following: *Histoire de l'Arianisme . . . , avec l'origine et le progrès de l'hérésie des Sociniens* (1673); *Histoire de l'hérésie des iconoclastes* (1674); *Histoire des croisades* (1675); *Histoire du Luthéranisme* (1680); *Histoire du Calvinisme* (1681/2).
8. See article "Maimbourg" in Micheaud ed., *Biographie Universelle* (Paris: 1859) and article "Maimbourg" in Bayle, *Dictionnaire Historique et Critique,* Remarque E.
9. Solé, *Le Débat,* pp. 100–1.
10. See E. Labrousse, *Pierre Bayle* (The Hague: Nijhoff, 1963–64), vol. I, pp. 63–71.
11. Ibid., pp. 94–130.
12. Ibid., p. 166.
13. See Richard H. Popkin, *The History of Skepticism from Erasmnus to Spinoza* (Berkeley: University of California Press, 1979), pp. 87–109.
14. See *OD.*II, pp. 1–160.
15. *Commentaire philosophique sur ces paroles de Jesus Christ 'Contrain-les d'entrer'* (1686). In *OD.* II, pp. 357–496. For an article on the treatment of heresy in Bayle's *Dictionnaire Historique et Critique,* see Hubert Bost, "Regards critiques ou complices surs les hérétiques: de l'Alphabet de Pratelous au Dictionnaire de Bayle," in Pierre Bayle, *Citoyen du Monde: De l'enfant à Carla à l'auteur du Dictionnaire. . . . réunis par Hubert Bost & Philippe de Robert* (Paris: Champion, 1999, pp. 199–214).
16. They included *Nouvelles de la République des Lettres* (1684–7); *Nouvelles lettres critiques sur l'histoire du Calvinisme* (1685); *Ce que c'est que la France toute catholique sous le règne de Louis le Grand* (1686); *Dictionnaire Historique et Critique* (1697); and *Entretiens de Maxime et de Thémiste* (1706).
17. Le Clerc, through his patron, Philipp van Limborch (1633–1712), became associated with John Locke during the latter's stay in the Netherlands from 1684 to 1689. On van Limborch, see the essay in this collection by Luisa Simonutti.

18. See Micheaud, op. cit.
19. See Micheaud, op. cit; Solé, op. cit. vol 1, pp.130–1.
20. See E. Labrousse, "Introduction Historique," in Bayle, *OD,* IV, pp. vii-xvi, viii-ix.
21. Amsterdam (1699; 2nd ed. 1701), 2 vols. Translated as *Parhasiana, or thoughts on several subjects* (London: 1700).
22. In *Réponses aux questions d'un provinçial,* 4 vols., 1703–7, in *OD.* I-V, vol III, pp. 601–1083.
23. Bayle, Pierre *OD.* I-V, vol. IV, pp. 1–36; hereinafter referred to as *Entretiens.*
24. See Melvin Richter, "Absolutism," in D. Miller et al. eds., *The Blackwell Encyclopaedia of Political Thought.* (Oxford: Blackwell, 1988), pp. 1–3. John Dunn cites the "absolutism" of Louis XIV as a reason for Locke's hostility toward the French regime in *John Locke* (Oxford: Oxford University Press, 1984), p.12.
25. Defended, for example, by Cardinal Robert Bellarmine, in *Tractatus de potestate summi pontificis* (1610).
26. Machiavelli, Bodin, Grotius, and Hobbes. These authors, respectively Florentine, French, Dutch, and English, were laymen concerned to show that the civil law should override the canon law, above all in defining sovereignty, orthodoxy, and heresy.
27. The notion of an orthodox Christian religion can be dated from the establishment of Christianity as the Roman Empire's de facto official religion consolidated by the Church's condemnation of the heresy of Pelagius in the fourth and fifth centuries. The Fathers of the Church, such as Augustine, taught that the Christian ruler should maintain Christian orthodoxy.
28. Cited in *Bayle: Political Writings,* "De l'Hôpital," Remark H, p. 102.
29. On an early ideal of mutually tolerant societies in Europe in the work of Nicolas of Cusa, see Cary Nederman, "Natio and the Variety of Rites: Foundations of Religious Toleration in Nicolas of Cusa" in J. C. Laursen, ed., *Religious Toleration: "The Variety of Rites" from Cyrus to Defoe* (New York: St. Martin's Press, 1999), pp. 59–74.
30. The earliest legal formulation of the doctrine of *cuius regio* was in the Religious Peace of Augsburg of 1555. Its theory was expanded in the Treaties of Westphalia of 1648, which brought to an end the Thirty Years War.
31. Bayle's words: see *Bayle: Political Writings,* "Japan," Remark E, p. 132.
32. Cited in *Bayle: Political Writings,* "Sainte-Aldegonde," Remark L, p. 242.
33. Ibid., p. 242.
34. See *Bayle: Political Writings,* "Hobbes," Remark E, pp. 84–87.
35. Thomas Hobbes, *Leviathan,* ed. R. Tuck [1651] (Cambridge: Cambridge University Press, 1991), Parts II and III, respectively, "Of a Christian Commonwealth" and "Of the Kingdom of Darkness."
36. References in the text refer to the relevant page numbers in *La Critique générale de l'histoire du Calvinisme de M. Maimbourg,* in *OD.* II, pp. 1–159. Translations are my own.

37. On this theme see J. C. Laursen, "Pierre Bayle on Citizenship and Cosmopolitanism in the Republic of Letters" in E. Banús and A. Llano, eds., *Razón pública/Public Reason* (Pamplona: Universidad de Navarra, 2001).
38. *Bayle: Political Writings,* "Macon," Remark C, p. 176. See editorial introduction in ibid., pp. xxxiv-vii.
39. See *Bayle: Political Writings,* "Socinus (Faustus)," Remark K: "Those who say that the Socinians have a full liberty of conscience in the United Provinces are poorly acquainted with history," pp. 267–8.
40. Henri Basnage de Beauval, Preface to *Entretiens de Maxime et de Thémiste,* 1706, *OD* IV, p. 2.
41. Bayle devoted many pages in his *Pensées diverses sur la comète* (1681), *OD.* III, pp. 1–160 and *Continuation des pensées diverses,* (1704), *OD.* III, pp. 189–415 to the viability of a society without religion.
42. In his *Theodicy* (1710), Leibniz, who, like Le Clerc, found Bayle too open in discussing the Manichean heresy, did not assume that Bayle supported the view. See my "Bayle and Leibniz: Two Paradigms of Tolerance and some Reflections on Goodness without God" in J. C. Laursen, ed., *The Variety of Rites,* pp. 173–89.
43. *Entretiens,* p. 31.
44. On this apparent paradox, see J. C. Laursen, "Baylean Liberalism: Tolerance Requires Non-Tolerance," in Laursen and C. Nederman, eds., *Beyond the Persecuting Society: Religious Toleration before the Enlightenment* (Philadelphia: University of Pennsylvania Press, 1998). Bayle goes further than his contemporaries in resolving the dilemma. See *Bayle: Political Writings,* "Bodin," Remarks P and Q, pp. 17–28; also the introduction, pp. xxxv-vii.
45. On Bayle's comparison between the revocation of toleration of the Socinian sect in Poland and the Huguenots in France, see *Bayle: Political Writings,* "Socinus (Faustus)," Remark A, pp. 254–8, p.258.

Part Three

Radical Heretics on the Offensive

Introduction

John Christian Laursen

The Socinians did not have to sit back and let others defend themselves by attacking Socinians. As Martin Mulsow shows in his chapter below, some of them wrote counter-histories as counter-attacks. And some were more subtle than simply to insist that their heresy was orthodoxy and the other side's orthodoxy was a heresy. An example Mulsow explores attempts to show instead that the heretic of the orthodox was actually one of them, a pillar of orthodoxy; but that orthodoxy had indeed gone astray.

Some heretics were too cautious to publish anything, but we know what they thought because they circulated manuscripts that have come down to us. They consisted of at least two kinds. Some, like the writers reviewed in Antony McKenna's chapter, were atheists, deists, or nonbelievers who deliberately tried to make trouble for believers by bringing out the inconsistencies in orthodox accounts of the history of the Church and of heresy. Many of them pushed Bayle's insights into the incoherences of theology much further than Bayle had; but then, they did not dare publish.

Others who did not publish, like Isaac Newton, sincerely held a religious position that was, however, considered heterodox. Newton wrote literally thousands of pages of historiography of heresy. He evidently felt he had to suppress publication of his own manuscripts on heresy partly because of the outraged response to writings of others such as Richard Simon.[1] His massive manuscript histories of heresy portrayed the Catholic Church as the inheritor of many heresies, including idolatry and sorcery. But this did not mean that he did so as a partisan of the Protestants; rather, even the Reformation was a corruption of the pristine early church.[2] On the one hand, he rewrote the history of heresy such that almost all contemporary manifestations of organized religion represented a heresy. On the other hand, he belongs in the history of late-seventeenth-century and early-eighteenth-century heresy as a representative of the actually quite widespread heresy of Arianism. Although he did not publish, his histories of heresy were influential: he shared them with John Locke and others.

It is important to understand that theology and natural philosophy (what we now call science) were not independent spheres. To Newton, the Cartesians and Leibnizians were modern Cabbalists, Gnostics, and Platonists. Their heretical emanationist cosmogony contradicted Newton's voluntarism, contributing to the corruption of philosophy at the same time as they corrupted religion.[3]

Since most of Newton's manuscripts were never published, Leibniz may not have known that he was a heretic in Newton's eyes. Leibniz's own attitude toward heretics such as mystics, quietists, and millenarians—some of whom had been burned by the Inquisition or other authorities—was that something could be learned from all of them.[4] He also pointed to the personal and professional reasons for some of the heresy-hunting: theologians needed heretics in order to draw attention to a threat that would justify their work.[5]

A writer's ideological intentions are important, but one factor in the production of all literary outputs that should never be overlooked is the economic factor. Writers have rarely written and printers have rarely published without an eye for some sort of pecuniary gain. Maimbourg wrote his orthodox histories of heresy in return for subsidies, as Bayle pointed out. Others wrote for the market. And paradoxically, one reason for not publishing a history of heresy was to cater to the clandestine market. Part of that market was the people like Queen Christina of Sweden who offered $1,000,000 for a copy of "The Three Impostors." Part was people like Baron Hohendorf, who collected manuscripts for Prince Eugene of Savoy and spent good money doing so.[6] So it occurred to impecunious writers and printers to produce for that market.

Marketing clandestina included such strategies as fake medievalization, that is, efforts to make the manuscript appear very old, even if it cited

Descartes! A plausible reason for the suppression of most copies of the printed version of "The Three Impostors" was price maintenance both for the manuscripts and for the few surviving printed volumes.[7] And these marketing techniques apparently worked.

The "Three Impostors" theme had become such a commonplace by the 1730s that someone could entitle a medical dissertation in critique of caffeine *The Three Impostors: Coffee, Tea, and Chocolate!*[8]

Articles in the francophone scholarly journal *La lettre clandestine [Clandestine Letters],* which has been appearing annually since 1992, explore a lot of this literature and its markets. It is regrettably too little known in anglophone scholarship. Not all of it remained unpublished. Some of Robert Darnton's forbidden bestsellers mined the theme of the history of heresy for scandal and sales.[9]

Newton's strategy of turning nearly all contemporary positions into heresies was followed by others. One of the most influential histories of heresy in eighteenth-century Germany was Gottfried Arnold's *Nonpartisan History of Churches and Heresies* of 1699–1700.[10] It was anything but nonpartisan. Arnold was a mystical Pietist who even quit his position as professor of history at Giessen because of the sense that it interfered with his pursuit of religious salvation. Paradoxically, his attacks on every other religious position were influential in at least two directions. Generations of atheist materialists such as Johann Conrad Dippel, Theodor Ludwig Lau, Johann Christian Edelmann, and Johann Lorenz Schmidt (who wrote of "baptism by the fire of Arnold") took to heart Arnold's defense of Spinoza, but turned it against all religiosity.[11] But Goethe admired Arnold, and the Göttingen historians could also draw on him in pursuit of an orthodox religiosity.[12]

One response to Arnold was Jakob Friedrich Reimmann's effort to capture the history of heresy for literary history, removing it from the exclusive property of the church historians.[13] Arnold's influence is also indicated by a serious and self-proclaimed orthodox namesake volume, Johann Gottfried Hering's *Compendious Church and Heretic Lexicon* (1731, 1734).[14] Konrad Füessli has been called "the Swiss Gottfried Arnold" for his *New Impartial Church and Heretic History* of 1770.[15]

Jewish Heresies

If anyone is inclined to think that suffering persecution is likely to limit the persecutorial impulses of a community, Richard Popkin's chapter in this volume will dispel that notion. We have already mentioned Spinozism, which was a heresy not only to Christians but also to Jews. Sabbatianism was another heresy internal to the Jewish community, and Popkin's chapter traces its rise, fall, and continued influence today.

The impostorship literature that has already been discussed several times in this volume was reproduced in a myriad of connections, both Christian and Jewish. U.C.L.A. Special Collections has a copy of a pamphlet entitled *Pantheon of Baptists and Enthusiasts and Spiritual Rust-house against the old Quakers and the new Free thinkers* (1702) in which James Naylor of the Quakers appears together with Shabbatai Zevi as a pair of great impostors.[16] By the time Ludvig Holberg wrote the first complete history of the Jews in 1742, he could cite more than a dozen Jewish impostors who had duped the people.[17]

Notes

1. Justin Champion, "'Acceptable to Inquisitive Men': Some Simonian Contexts for Newton's Biblical Criticism" in James E. Force and Richard Popkin, eds., *Newton and Religion: Context, Nature, and Influence* (Dordrecht: Kluwer, 1999), pp. 77–96.
2. See Rob Iliffe, "Those 'Whose Business it is to Cavill': Newton's Anti-Catholicism," in Force and Popkin, eds., *Newton and Religion,* pp. 97–119 and *Priest of Nature: Persecution, Heresy, and the Private World of Isaac Newton* (forthcoming).
3. See Matt Goldish, "Newton's 'Of the Church': Its Contents and Implications," in Force and Popkin, eds., *Newton and Religion,* pp. 162–4.
4. Compare Donald Rutherford, "Leibniz and Mysticism" and Allison Coudert, "Leibniz and the Kabbalah" with Daniel J. Cook, "Leibniz on Enthusiasm," in Coudert, Richard Popkin, and Gordon Weiner, eds., *Leibniz, Mysticism and Religion* (Dordrecht: Kluwer, 1998), pp. 22–46, 47–83, and 107–135.
5. Cook, "Leibniz on Enthusiasm," p. 133.
6. For these examples, see S. Berti, F. Charles-Daubert, and R. Popkin, eds., *Heterodoxy, Spinozism, and Free Thought in Early Eighteenth-Century Europe* (Dordrecht: Kluwer, 1996), pp. ix, 293, 407ff, etc.
7. See Berti, et al., eds., *Heterodoxy, Spinozism, and Free Thought,* pp. 293ff.
8. Cited in Ludvig Holberg, *Selected Essays,* tr. Mitchell (Westport: Greenwood, 1976), p. 68.
9. Robert Darnton, *The Forbidden Bestsellers of Pre-Revolutionary France* (New York: Norton, 1995); *The Corpus of Clandestine Literature in France, 1769–1789* (New York: Norton, 1995).
10. Dietrich Blaufuss and Friedrich Niewohner, eds., *Gottfried Arnold (1666–1714): mit einer Bibliographie der Arnold-Literatur ab 1714* (Wiesbaden: Harrassowitz, 1995).
11. Stefan Winkle, *Die heimlichen Spinozisten in Altona und der Spinozastreit* (Hamburg: n.p., 1988), pp. 27, 29, 35.
12. Peter Reill, *The German Enlightenment and the Rise of Historicism* (Berkeley: University of California Press, 1975); Winkle, *Die heimlichen Spinozisten,* p. 14.

13. See Hanspeter Marti, "Litterärhistorie und Ketzergeschichte," in M. Mulsow and H. Zedelmaier, eds., *Skepsis, Providenz, Polyhistorie: Jakob Friedrich Reimmann (1668–1743)* (Tübingen: Niemeyer, 1998), pp. 60–75.
14. *Compendieuses Kirchen- und Ketzer-Lexicon.*
15. *Neue unpartheyische Kirchen- und ketzerhistorie* (Frankfurt, 1770). See Günter Muhlpfordt, "Pietistische Ökumene: Die Bogomilen als östliche Pietisten in der Sicht Gottfried Arnolds," in Blaufuss and Niewohner, eds., *Gottfried Arnold (1666–1714),* p. 244.
16. *Anabaptisticum et enthusiasticum pantheon und geistliches Rüst-Hauss wider die alten Quacker, und neuen Frey-Geister.*
17. See J. C. Laursen, "Ludvig Holberg's History of the Jews: Arguments for Toleration" in H. Méchoulan, R. Popkin, G. Ricuperati, and L. Simonutti, eds., *La formazione storica dell'alterità: Studi di storia della toleranza* (Florence: Olschki, 2001).

7

Histories of Heresy in the Clandestine Philosophical Manuscripts

Antony McKenna

Discord and controversy among Christian churches and sects has been identified as a source of free thought by a number of critics.[1] Various clandestine manuscripts seem to bear out this analysis. The anonymous philosophers of the early eighteenth century take much pleasure in enumerating different interpretations of historical testimony and sacred texts and recording the multiplication of consequent heresies. The complexities of sacred history and exegesis are such that the boundaries between orthodoxy and heresy become blurred and our authors underline that the orthodoxy or heterodoxy of a given belief depends simply on the balance of ecclesiastical and political power at a given moment.

Orthodoxy appears in these manuscripts as the affirmation of a powerful Church, heresy as the (often intellectually or morally legitimate) questioning of that authoritative affirmation. The reader can only conclude that the Church is governed like a human society according to the requirements, not of inherent moral or doctrinal authority, but of political and economic power. The panorama of churches and sects thus reinforces the philosophical analysis of the political imposture of the Church, which imposes arbitrary superstitions on its submissive servants in an exercise of power rather than in the search for salvation.

Intellectually, the triumphant doctrine appears to be a result of infinite caviling sophistry and scholastic hair-splitting, and the persistent controversies between Christian churches and sects are interpreted as proof that authentic doctrine is beyond the reach of the ordinary person. He can no more distinguish between rival interpretations of obscure dogma (the nature of divine beings and the relations between the three persons of the Trinity, for instance, or the exact manner in which the body of Christ is present in the bread and wine of sacred mass), than he can sort out the different claims to doctrinal infallibility founded on historical authority. The history of the Christian Church thus appears as a long process of fragmentation, splitting hairs, and splitting churches, and Roman Catholic orthodoxy thus appears as one sectarian doctrine among many others.

How should the ordinary person choose a belief? Whatever may be the original ambition to examine the foundations of doctrine and authority, he or she inevitably yields to the weight of education and prejudice. Religious identity is determined in the same way as national or cultural identity, by the chance circumstances of birth and education. "We are Christians by the same title we are Germans or Périgordians," declared Montaigne (II, 12), and Pascal had echoed: "custom is our nature, and whoever is accustomed to a faith believes it" (Lafuma 419; Sellier 680). The clandestine philosophers then underline gleefully, with many a quotation of Christian apologetics, the crucial importance of salvation, the urgency with which the question of the authentic religion must be addressed, and the care with which the evidence and the arguments must be weighed up and examined. They conclude, of course, that the crucial question is commonly and inevitably resolved on the basis of trivial and circumstantial evidence: "it is the country and not reason, that decides the religion of almost all men" (*De l'examen de la religion,* ed. S. Landucci, p. 51). Not that this conclusion is specifically philosophical: it is borrowed from the theologians themselves: "Mr. Nicole is convinced that it is chance that decides the religion of almost all men" (ibid); the Protestant Ostervald agrees: "Most people are only Christians because they have been required from the infancy to profess Christianity; but they know little of truth and divinity. They would be Jews or pagans if they were born pagans or Jews. Thus, speaking properly, one cannot say they believe and that they have faith . . ." (p. 53). The would-be believer is thus confronted with an impossible task: "we are like voyagers who, before beginning the voyage, close their eyes and take for a guide a blind instinct that conducts them by chance" (p. 54).

This blind faith is then put in the balance with the claims of philosophical reason. Malebranche is called on to define true faith as founded on the evidence of rational philosophy, as well as on the submission of blind faith.[2] These two paths prove divergent, but the authority of Malebranche is invoked to legitimize the denunciation of all those elements of doctrine that appear incompatible with philosophical reason. The great Christian Philosopher himself had lent priority to reason and authorized this philosophical appropriation of his definition: "the certainty of faith depends on the knowledge that reason has given us of the existence of a God" (ibid, IV.vi. §2). In the framework of anti-Christian philosophical rationalism, born of the special Malebranchist brand of Christian rationalism, reason thus appears as a legitimate criterion by which to judge the truth of doctrine.

The influence of Malebranche on the clandestine philosophers in France is well known. In other circles, particularly in the Netherlands, the fragmentation of Protestant churches and sects played an important role in the emergence of Enlightenment philosophy: the "inner light," the light of conscience, is heralded as that of reason. Recent studies have underlined the contribution of the "collegiants."[3] Socinianism can also be seen as a crucial standpoint that was denounced by theologians and greeted by philosophers as an important contribution to the rational evolution of Christian philosophy—or, as some would have it, the subservience of Christian faith to philosophical reason. Here, I would like to draw attention to the position of Pierre Bayle, who exerted enormous influence on clandestine philosophy. Bayle explicitly adopted a Socinian stance in his *Philosophical Commentary.* He first launches a disclaimer:

> God forbid that I accept the jurisdiction of natural light and metaphysical principles, like the Socinians, who claim that any sense given to the Scripture that does not conform to that light and those principles must be rejected, and by virtue of that maxim refuse to believe the Trinity and the Incarnation. No, no, that is not what I claim to do, without limits. . . ."[4]

But, as so often happens, Bayle immediately transgresses the limits of the rejection of Socinianism that he has just offered:

> I well know that there are axioms against which the most evident and express words of the Scripture gain nothing, such as *the whole is greater than its parts* [. . .]. If one showed in the Scripture the contrary of these propositions; if one were to accomplish thousands and thousands of miracles, more than Moses and the Apostles, to establish a doctrine opposed to these universal maxims of common sense, no one would believe them at all; and he is more likely to persuade himself either that the Scripture speaks in metaphors and irony, or that the miracles come from the devil, than to believe that natural light is false in its maxims. (ibid, p. 87)

Both Catholics and Protestants admit that "neither Scripture, nor the Church, nor miraclers can prove anything against the evident light of reason" (ibid).

The disclaimer appears, therefore, as a smoke screen, and the definition given here of the status he attributes to self-evident rational maxims will be a stepping stone to his analysis of Christian doctrine in the article "Pyrrhon," Remark B. After another digression, Bayle returns to the question of the status of self-evident ethical maxims, underlining the Socinian nature of the position he is adopting:

> I repeat it again. God forbid that I understand this principle the way the Socinians do; but if there may be certain limitations with regard to speculative truths, I do not think that there should be with regard to practical and general principles concerning morals. I want to say that without exception one must submit all moral laws to that natural idea of equity that, just as much as metaphycial light, illumines every man on earth (ibid, p. 89–90).

In so far as the *Philosophical Commentary* defines Bayle's doctrine of the rights of individual conscience against persecution, this text must weigh heavily in the interpretation of his intention in the article "Pyrrhon," Remark B, where he explicitly underlines the contradiction between Christian doctrine and our rational conception of justice.

Let us take this question a step further and consider the text of the article "Socin (Fauste)":

> It is a heavy yoke for reason to captivate its understanding to faith in the three persons of the divine nature and in a God-man: one relieves the Christians infinitely when one delivers them from that yoke; and consequently it is believable that a lot of people adhere to them. . . . That is why these Italian refugees transplanted to Poland rejected the Trinity, the hypostatic union, original sin, absolute predestination, etc. ("Socin [Fauste]," Remark H)

Bayle takes up this point against Jurieu, who had attributed such motives to the Socinians in his violently polemical work *L'Esprit de M. Arnaud* (tom. I, chap. 6, p. 211). This analysis, Bayle retorts, is founded on a miscomprehension of the nature of faith, which depends on the obscurity of its formulæ:

> The speculative mysteries of religion do not bother the people: they certainly bother a professor of theology who meditates on them with attention in order to explain them and to satisfy the objections of heretics . . . but all the rest of the people are in perfect tranquility on such questions: *they believe, or think they believe,* everything they are told. . . . It would be hallucinating to think that one is delivering the bourgeois and the peasant, the warrior and the gentleman, from a heavy burden if one dispenses them from believing in

> the Trinity. . . . One admires more what one does not understand. . . . All the purposes of religion are found better in objects that one does not understand: they inspire more admiration, respect, fear, and confidence (ibid).

The Socinian effort to elucidate religious mysteries thus goes against the very nature of religion. Despite the rather caustic appraisal of popular faith in this passage, one might be tempted to dismiss it as yet another expression of Bayle's "fideism": *o altitudo!* Faith is by nature incompatible with reason, as Christ and his apostles, and the long tradition of the Church after them, had declared. But the expression "*they believe or think they believe*" seems to me to exclude this interpretation. Bayle is asking: What do popular believers believe? They can only repeat mysterious formulæ and cannot define them. What does it mean to believe in formulæ that, all admit, have no rational sense and that can only be repeated like a parrot?[5] This declaration of war on obscurantism in the very domain of faith seems to me to be an invitation to reconsider Bayle's attitude to Socinianism.

The relevance of these points for the philosophical clandestine manuscripts will soon be clear. Let us return to our subject via another crucial passage in Bayle, a passage that furnished the clandestine philosophers with some of their most dangerous weapons. Believers, he suggests, are unable to clearly define what they believe and thus unable to define the limits of the heresy that they must seek to avoid. This is in the very nature of religious belief. Indeed, how better determine those limits than by the establishment of a creed of faith, which enumerates the articles of doctrine? But enumeration is not enough: we must then seek to specify the meaning of the formulæ by which we claim to express our faith:

> Thus it is not enough to know that there is a God. The sense of this word must also be determined to attach it to an idea; so, I say, we must understand what is the nature of God, and that is where the difficulties begin. It is a subject that the greatest philosophers have found obscure. (*CPD,* ch. 20)

He reduces these difficulties to a single cause: "It cannot be easy for man to know clearly what is appropriate and what is not appropriate for an infinite being" (ibid). Thus, so long as we are content with the general formula, *God exists,* everyone will agree; but as soon as we try to be more precise, *God created the world,* some philosophers will disagree with popular opinion; if we add *God governs the world and dispenses events,* other philosophical sects will form. Soon, everyone will be a nonconformist (ibid, ch. 21).

In Bayle's analysis, believers are unable to express their faith in unequivocal formulæ, in terms that do not inevitably raise new questions and new doubts, new definitions and new frontiers, new explanations and new heresies. The generation of heresy is imbedded in the very expression of faith—not by some fluke or accident, but because of the very nature of language and of faith.

Let us now examine the clandestine heritage of this very radical stance.

1. *Examen critique des apologistes de la religion chrétienne,* composed around 1733, attributed to Jean Lévesque de Burigny, circulated in manuscript form: seven copies are now known. It was published in the various editions of the complete works of Nicolas Fréret (Londres 1775, 1778; Paris 1792).[6]

   The author insists on the mass of pseudo-gospels and rival apocryphal works circulating shortly after the death of Christ. It is quite impossible, he declares, to discover, among all these texts, an authentic historical testimony. The present Christian canon is simply the result of chance, ignorance, misunderstanding, prejudice. Innumerable sects were born of these rival texts and testimonies (ch. 1, 2). The Christian religion was first chosen by the uneducated; it was later spread by force (ch. 6, 7). Extravagant behavior and superstitious beliefs characterized early Christians and the beginnings of all religious sects, such as the Quakers.[7] All attempts to reform the decadent church are inevitably followed by the decadence of the reformers, as has happened in the Protestant church (ch. 10). The obscurities and contradictions of the biblical text have inevitably given rise to misunderstandings and conflicting interpretations (ch. 11). The text defies literal interpretation and is thus exposed to all kinds of fanciful readings. Exegesis is far beyond the capacity of common folk, as is also the examination of the historical authority of the Church (ch. 12). Sects thus abound because of the very nature of the Christian Bible. Orthodoxy signifies nothing but the momentary triumph of one sect over another for purely circumstantial reasons.

   This brief résumé does not imply that Lévesque de Burigny was extrapolating or simply reciting a polemical piece against Christianity. Each point is abundantly illustrated by quotations from appropriate ancient texts and erudite sources. We have here the result of centuries of bickering between Christian factions and sects. The result is chaos, according to Lévesque de Burigny, policed only by the disciplinary organizations engendered by the Church to control the masses of its ignorant and superstitious community. No authentic texts, no assured and unequivocal interpretation, no authorized ministers, no moral integrity can be found. The proliferation of Christian sects obeys the same laws as those governing other religions: the laws of human fantasy, ignorance, superstition, and political power.

2. *Examen de la religion, ou Doutes sur la religion dont on cherche l'éclaircissement de bonne foi,* text composed by César Chesneau Du Marsais around 1705; circulated in two manuscript versions, of which some 65 copies are now extant; first published in 1745.[8]

   "The scripture is full of modes of expression that are inexact and completely opposite to healthy theology" (Mori, ed., p. 165). Obscure expressions in the Bible are held responsible for the birth of heresy. The ambiguities of the text inevitably lead believers astray. Thus, for example, the Arians: "It is a heresy to say simply and without distinctions that Jesus Christ is less than his Father; nevertheless, that is what the scripture says, and Jesus Christ himself says *My father is greater than I am.* . . . Do not the Arians have reason to maintain, on the basis of this passage, that Jesus Christ is inferior to God the Father?" (ibid).

3. Given the importance lent by Lévesque de Burigny to the proliferation of pseudo-gospels in the birth of various heresies, the exploitation of one of these texts in Catholic and Protestant controversy and thereafter in philosophical anti-Christian polemics is worth mentioning.

   I quote here a little-known manuscript, entitled *De la conception et de la naissance de la Sainte Vierge, et sa généalogie,* discovered in Carpentras and Reims, composed in the 1720s or 30s.[9] The anonymous author cites explicitly his source: a work of Protestant anti-Catholic controversy published by Jean-Baptiste Renoult in 1701, *Les Aventures de la Madone et de François d'Assise.* Renoult claims to be following the example of Pierre Jurieu in his 1685 retort to Nicole entitled *Préjugez légitimes contre le papisme.* In this work, Jurieu explicitly—and very violently—attacks the Jesuit Jean Crasset, *La Véritable dévotion envers la Sainte Vierge établie et défendue* (Paris 1679), a work in which the Jesuit had denounced the French translation of a work by a German Catholic, Adam Widenfeld, *Avis salutaires de la bienheureuse vierge Marie à ses dévots indiscrets* (Lille 1674), where he attempted to discourage and ridicule the excessively superstitious cult of the Virgin Mary in popular Catholic devotional practice.

   The apocryphal text concerning the Virgin Mary, exploited by Crasset, denounced by Jurieu, and copied by Renoult and our anonymous philosopher, is the *Golden Legend* by Jacques de Voragine (dating from 1472–74). De Voragine, whose influence on Renaissance iconography is well known, based his story of the Virgin on the apocryphal *Book of the Nativity of Mary,* composed toward the end of the ninth century. It exploits the *Gospel of the childhood of Jesus,* or *pseudo-Matthew,* a recomposition, dating from the end of the sixth or beginning of the seventh century, of the *Proto-gospel of James,* the oldest of the gospels of the childhood of Christ, mentioned by Clement of Alexandria and Origen in the third century. Our text thus reaches back to the very sources of Christian myth and symbolism.

The cult of the Virgin Mary is quoted by Huguenots and clandestine philosophers alike as an example of superstitious worship practiced by the ignorant masses and cultivated by the Catholic Church. Sectarian cults are thus presented as the result of ignorance and superstition, exploited by the ecclesiastical authority. Once again, religious controversy has given birth to clandestine philosophy.

4. In the *Nouvelles de la République des Lettres,* Bayle draws attention to the great work of religious controversy composed by Pierre Nicole, *Les Prétendus réformés convaincus de schisme* (Paris 1684) and uses it to define his own ambitions as a journalist: "We will give a balanced summary of this work, without partiality and according to the duty of a pure historian."[10] He first gives all the references necessary to understand the ambition of Nicole's work, which is a reply to Claude's book, *Défense de la Réformation,* in which the Protestant theologian retorted in 1673 to the *Préjugez legitimes* of Nicole. Bayle then briefly defines Nicole's aim, which is to convince his readers that the separation of the Protestants from the Catholic Church would be schismatic, even if it were conceded that the community they left was guilty of serious errors. The debate on schism is then brought back to the debate on the status of examination and authority in establishing the truth of the Christian religion. Nicole claims that reason is incapable of finding its way in the labyrinth of biblical exegesis: examination leads only to doubt. Thus, simple folk (and, by implication, the less simple) "have no solid and reasonable way to assure themselves of the truth of most of the articles of faith proposed by M. Claude, nor to know the necessity of them, nor to assure their sufficiency, nor to judge the incompatibility with the articles he excludes." They must therefore, naturally, according to the theologian of Port-Royal, have recourse to authority. Bayle comments:

> Here this author swims in the open sea, or at least he is like a river that rolls majestically in a great and deep channel, but in his chapter 17 he finds himself back between rocky points. . . . One no longer sees that liberty of spirit, that abundance, that confidence . . . of the earlier chapters. (November 1684, article 1)

For the same theologian, who insisted so heavily on the uncertainty of textual interpretation, claims to found on those same texts the authority of the Church and the Church Fathers. Bayle underlines the contradiction and concludes:

> One is forced to admit that God does not require of the ignorant that they know of the infallibility of the Church by an examination of the Scripture accompanied by all the enlightenment and all the research that M. Nicole mentions (ibid).

Incapable of finding true religion by examination of the biblical text or by reliance on Church authority, simple folk (and the less simple) are abandoned to their own devices. This is the starting point of a clandestine manuscript, *De l'examen de la religion,* also attributed to Lévesque de Burigny.[11] He first invokes a number of theological authorities, among them Nicole, to underline how important it is to choose the true religion. Such a religion, he declares, must be universal and must be characterized by obvious signs of its authority. From these premises he concludes that none of the known revealed religions can qualify as true religion. A simple syllogism clinches this argument:

> Any religion the proofs of which are not clear to all reasonable men cannot be the true religion; but, there is no religion, of all those which claim to be revealed, the proofs of which are clear to all men; therefore, none of the revealed religions is the true one.

All revealed religions are founded on books written in obscure languages, reporting miracles and prophecies according to more or less improbable testimony. Not only should we examine all the proofs alleged in favor of the Christian religion, but we should also compare them with those of all the innumerable other religions and sects of the world.

All of Nicole's arguments on the difficulties of interpretation of the biblical text and for the need of an authoritative doctrine are turned back against him. The clandestine author invokes the response of Jurieu, who does not hesitate to declare that Nicole "has argued in favour of the heathens," and he concludes: "If examination is impossible for most men, which cannot be doubted, it must be concluded that they are not obliged to engage in it, and consequently they cannot do better than suspend their judgment." The apologetic method that tries to avoid examination of the texts is driven inevitably to invoke those same texts as a guarantee of its authority: examination is both necessary and impossible. Once again, the birth of heresy is inscribed in the very nature of the biblical text.

5. Many a passage in Bayle's masterpiece of pamphlet polemics, *France Wholly Catholic under Louis the Great,* denounces Catholic theologians and historians who deny the use of violence in the conversion of the Huguenots at the time of the Revocation of the Edict of Nantes in 1685. Everyone knows that violence was employed, he argues, and in particular the military officers who were bound to enforce the violent policy of the political authorities (ed. E.Labrousse, Paris, 1973, p. 79).

   Bayle's passage may well have inspired the reaction of Robert Challe in his *Difficultés,* when he goes through the episodes of his childhood and military career that "opened his eyes" as to the imposture of the Catholic Church:

> The persecution of the Huguenots followed these early times. Ah! M.R.P., what cruelties and what harshness have I not seen! When I remember how, filled with wine, we pulled a miserable old man, suffering from gout, from his bed . . . without letting his pitiable cries and the tears of his two poor daughters pleading at our feet stop us from our barbarity—what a cruel memory!—the pen falls from my hands and my eyes can no longer guide it. (*Difficultés sur la religion proposées au Père Malebranche,* ed. F. Deloffre et al., Oxford, 1983, p. 49)

Challe denounces religious enthusiasm in terms similar to those of Bayle[12] and he inherits Bayle's moral rationalism. Religious bigotry and sectarian prejudice can easily be identified by comparison with a universal moral criterion: "That general charity which we owe to all men, by the indispensable duties of humanity" (ed. cit., p. 72).

One might also suggest that he inherits Bayle's rational method in the denunciation of superstition. Challe's analysis of superstition owes a lot to Bayle's *Pensées diverses,* which systematically put forward the *reasons* for dismissing such beliefs. It is true that Bayle's progression from argument to argument involves many a digression, as he humorously demonstrates the absurd consequences of his adversaries' arguments. It is true also that the critical analysis of astrology gives way, in many chapters, to the question of the status of atheism and to that of the relation between conviction and behavior. Nevertheless, Bayle never loses sight of his *reasons.* The whole structure of his argument depends on the obvious nature of the reasons he advances; he considers it sufficient to demonstrate, according to reason, the absurd nature of a given belief in order to consider that belief ridiculed, refuted, and dismissed. This seems to be the model for the clandestine philosophical rationalism we find in Du Marsais's works and in the *Difficultés* of Challe.

6. What of the heresies that spring up within the Catholic Church? Should they be examined by the pope alone or by a General Council? What is the foundation of Church authority? The *Examen de la religion* imagines a dialogue between "an Indian of good faith" and the Church, and this text had a separate existence in some clandestine manuscript collections:

> An Indian of good faith arrives in Europe, raises his voice, and demands: who can assure me of the truth of divine revelation? Which of you claims to be infallible? The Church of Roman appears: it is I, it says, who is infallible. The Indian prepares to listen, but soon demands proof of the infallibility that is claimed. The Scripture, responds the Church. But what is Scripture, asks the Indian. It's a book inspired by God, the Church responds. How can I tell, asks again the Indian? I assure you it is, answers the Church. If the Indian has good faith and good sense, will he ask anything else? (ed. cit., p. 175)

The same manuscript goes on to quote Fontenelle on the strength of imagination, which has given birth to a myriad of superstitions and sects: "There is no ridiculous belief that has not had its sect [. . .] As ridiculous as a thought may be, all you have to do is find the means to sustain it for a moment; then it becomes an old belief, which is sufficient proof" (ibid, p. 177). Du Marsais further denounces the imposture of Church authority and its tyrannical hold on the biblical text and the tradition of the Church Fathers: the Church wields its power to confirm its authority. And Church Councils are the instruments of Church authority: "Councils are a proof of the falsity of the religion. Indeed, what is a council? It's an assembly of men that after debating, decide among themselves that which they will propose to the rest of men as a truth that God has revealed" (ibid, p. 179). Robert Challe concurs:

> These councils where the decisions are baptized with the name of truth and as oracles of the Holy Spirit, and which have no advantage over the opposite decisions except that they are the most recent and are the result of [. . .] low flattery, sometimes criminal, and of pride and interest. (ed. cit., p. 51; see also, p. 240 *sqq.*)[13]

Heretical sects are therefore simply communities who do not have the temporal power to implement their doctrinal decisions, nor the possibility to escape the instruments of Catholic discipline. The Inquisition is never far from the minds of our clandestine philosophers: Charles Dellon's *L'Inquisition de Goa* (1687) had made its mark and had been exploited at length in the philosophical novel of Simon Tyssot de Patot, *Voyages et avantures de Jaques Massé* (1710).[14]

7. *La Censure du Symbole des Apôtres, par M. xxx, Encyclopédiste* constitutes another excellent example of clandestine philosophy born of religious controversy between rival Christian traditions. It is a systematic criticism, obviously tongue-in-cheek, article by article, of the Apostles' Creed. The title, "par M. xxx, Encyclopédiste" suggests a date after 1752, but fortunately an article by Prosper Marchand reveals a rich bibliography behind this parody and the original circumstances of its composition.

   In 1626, the Jesuit Juan Battista Poza published a defense of the immaculate conception of the Virgin Mary under the title *Elucidarium deiparæ*. From this work were extracted various passages sewn together to form a criticism of the Apostles' Creed, which enabled Francesco Roalès, professor at the University of Salamanca and chaplain to the king of Spain, to have Poza's work censured in Rome and to have the author dismissed from his post as "qualificator" or censor of the Inquisition. Some years later, in 1633, Gaspard Scioppius condemned Poza's work and compared it with a similar satire composed by an English Jesuit and entitled *Censura in Symbolum apostolorum*. In 1665, in an article devoted to Rufinus Tyrannius, priest of Aquilea, known for his *Commentarius in Symbolum apostolorum* composed in the context of his quarrel with Origen, the Jesuit Father Théophile Raynaud defended his English colleague, the author of the satirical *Censura:* the real intention of this theologian "eruditissimus et Catholicissimus" had been to show that it is easy to criticize even the most revered and respectable texts and he had given an audacious example of such ill-founded censorship.

   The work of A. F. Allison has confirmed Raynaud's version and thrown much light on the details of the story. On 12 January 1625, Richard Smith was inaugurated bishop *in partibus* of Chalcedonia with authority over England. He left for England on 28 April/8 May 1625. Two years passed by without any major difficulties, but then Smith's claim to exercise the full authority of a Catholic bishop brought him into conflict with the regular orders in England and in particular with the Jesuits. Smith appealed to Rome on 16 December 1627, but some of his claims were disavowed by the pope. This quarrel drew public attention to Smith, who was threatened with a civil order of arrest. The next year, the pope imposed a pause in hostilities and Smith retreated to France on 24 August 1631. He was then forced to resign and, despite his resistance, the mission of Gregorio Panzani brought an end to the bishopric of Chalcedonia.

Smith's quarrel with the regular orders was then taken up in France. Taking position against a work published by Rudesind Barlow in 1628, Matthew Kellison published his *Treatise on the hierarchy and various orders of the Church* in May 1629 in defense of episcopal authority. Kellison was countered by the Jesuits Edward Knott (*vere* Matthew Wilson, under the pseudonym of Nicholas Smyth) and John Floyd (under the pseudonym of Daniel a Jesu), and their works were submitted to the Sorbonne by Smith's agents in Paris: the works were censured in February 1631, thanks to the interventions of some declared enemies of the Jesuits: Elie Du Fresne de Mincé, Jean Filesac, and Georges Froger, all known to be friends of Jansenism and of Port-Royal. It was in this context, in 1631, that the *Censura in Symbolum apostolorum* was first made public, as a denunciation of the sophistry of the Sorbonne condemnation. It was this same controversy that provoked the composition by Saint-Cyran, the celebrated spiritual director of Port-Royal, of his famous *Petrus Aurelius*. First publication of the *Censura* was therefore a Jesuit protest against Jansenist censorship.

In 1718, François Vautier was imprisoned in the Bastille for his attempt to publish the *Censura* in the context, this time, of the controversy surrounding the *Unigenitus* Bull. Now it was a Jansenist retort to Jesuit censorship.

The manuscript translation to be found at the Mazarine library in Paris[15] is a new "recuperation" of the text, this time in favor of the philosophical authors of the *Encyclopedie*. Prosper Marchand probably derives all his bibliographical information from an unpublished article "Poza" by Bayle, since Bayle refers to such an article but it does not figure in the *Historical and Critical Dictionary*. The *Censure* remains a denunciation of unjust censorship and the author claims to be publishing it "to counter the dangerous artifices of the many unquiet spirits of our century who want, on the pretext of defending the apostolic tradition, to maintain in public that the philosophical works of the moderns are full of false and impious propositions." From Jesuits to Jansenists to philosophers: once again, religious controversy between Church traditions and sectarian interests served the ends of clandestine philosophers.

8. Finally, the *Symbole d'un laïque,* freely translated by d'Holbach from Thomas Gordon's *Creed of an Independent Whig* (London 1720),[16] allows the clandestine philosopher to turn his back on the Church

and its internal upheavals: Christianism is itself nothing but a sectarian religion.

> I believe that one can believe in God without being forced to believe everything that interested priests say, that they do not know any better than honest laymen . . .
>
> I believe that one can disagree with one's priest without being an unbeliever or an atheist. I believe that the religion of priests is not always that of God.
>
> I believe that God is just and will not punish anyone for following his conscience, erring in good faith, and not believing that which is absurd.
>
> I believe that God is wise and reasonable, and not such an enemy of reason as his priests.

Whatever its sources in English, this declaration sounded in France like an echo of Bayle's denunciation of a corrupt clergy and its policy of persecution, and of his appeal to our natural and rational sense of justice.

In conclusion, I would like to underline the clandestine philosophers' debt to Bayle. Without going into the question of Bayle's sincerity or irony, it is easy to see that his works were a permanent source for references to the absurdities of superstitious sects and of analyses of the contradictions of Christian doctrine. He made explicit the lessons of philosophical rationalism, whether or not he adhered to the obvious rationalist conclusions. The clandestine philosophers were among the first to exploit the *Dictionary* in this way and they met with a wealth of material. In its columns, Bayle evokes the golden calf (*Aaron*), the magic wand of Jacques Aymar (*Abaris*), the oracle of the god Bésa (*Abyde*), the figures of Greek mythology (*Achille, Actor, Adonis, Ajax, Alcmène,* etc), various errors concerning the height of Adam (*Adam*), the extravagant behavior of the Adamites (*Adamites*), mystical visions (*Marie d'Agreda, Antoinette Bourignon*), the astrology of Agrippa (*Agrippa, Henri Corneille*) and so on, in alphabetical order. The lesson is plain: each article of the creed and each tenet of popular belief is a source of misunderstanding. All biblical episodes are subject to multiple and contradictory interpretations, which give rise to strange and ridiculous beliefs. Heresy has its origin in the very quest for belief.

But Bayle also provides a rational criterion by which to judge the absurdity of such superstitions and he heavily underlines the contradictions between reason and Christian doctrine. Once we dispose of reason, as the Pyrrhonists would have us do, how could we distinguish between true faith and absurd superstition? Both are equally abhorrent to right reason. This is precisely the philosophical rationalism that characterizes the approach to heresy in the clandestine manuscripts and that introduces the frontal attack on the authority of the Church. The clandestine philosophers were quick to draw the conclusion: "Man should not act except by reason: God himself only acts in that way" (*Examen,* ed. cit., p. 147). It is legitimate and necessary for each man to examine his faith in the light of reason and to follow his conscience:

> Finally, if you are mistaken, you have a good, legitimate excuse if you have followed the light that God has given you to conduct yourself, and renounced the persuasion of blind and interested men. (*Difficultés,* ed. cit., p. 56)
>
> Reason and conscience are perfectly sufficient for the conduct of men. (ibid, p. 141)

Thus, to my mind, by his rational—and not skeptical—analysis of the creed, Bayle reduces faith to two extremities: either it claims to be founded on reason and it is then subjected to infinite, insoluble objections and subtle nuances and ultimately, through the objection based on the existence of Evil, for example, reduced to silence, since God's nature remains fundamentally incompatible with reason according to Bayle's analysis; or faith is reduced to wishy-washy formulæ repeated parrot-fashion. At both extremities, orthodoxy is shown to be a sham (a political reality, of course, but a philosophical sham) and heresy to be inevitable because of the very nature of faith. Such was the lesson that Bayle passed on to the clandestine philosophers.

Notes

1. See, in particular, J. Solé, *Le Débat entre protestants et catholiques français de 1598 à 1685* (Paris: 1985), 4 vols.; Alan Charles Kors, *Atheism in France, 1650–1729,* vol. I: *The orthodox sources of disbelief* (Princeton: Princeton University Press, 1990).
2. *De la recherche de la vérité,* I.iii.§2: "En un mot, pour être Fidèle, il faut croire aveuglément, mais pour être Philosophe, il faut voir évidemment."
3. A. Fix, L. van Bunge
4. Pierre Bayle, *Commentaire philosophique,* éd. J.-M. Gros, Paris, 1992, pp. 86–87.
5. I analyze this formula in more detail and compare it to a text of Montaigne in ""Ils croient, ou ils croient croire": réflexions sur la foi chez Montaigne et chez Bayle," *Bulletin de l'Association des Amis de Montaigne,* VIII(11–12), 1998, pp. 61–71.

6. Ed. Alain Niderst, (Paris: Champion, 2000). The most recent inventory of clandestine philosophical manuscripts is that of Miguel Benitez, *La Face cachée des Lumières* (Oxford: The Voltaire Foundation, 1997).
7. On this theme, see also Voltaire, *Lettres philosophiques,* no. 3 on George Fox compared to Jesus Christ.
8. See the critical edition established by Gianluca Mori, (Oxford: The Voltaire Foundation, 1998).
9. See my edition of this text in "La controverse religieuse, une source de la philosophie clandestine: *De la conception et de la naissance de la Sainte Vierge et de sa généalogie,*" in *Materia actuosa. Antiquité, Renaissance, Lumières. Mélanges en l'honneur d'Olivier Bloch,* eds. Miguel Benitez, Antony McKenna, Gianni Paganini, and Jean Salem (Paris: Champion, 2000).
10. *NRL,* avril 1684, cat. iii; see also avril 1685, art. 3 *in fine,* and août 1686, art. 1, p.880.
11. *De l'examen de la religion,* ed. Sergio Landucci (Oxford: The Voltaire Foundation, 1996). The last chapter of this manuscript is identical to that of the text already mentioned: *Examen critique des apologistes de la Religion Chrétienne.*
12. "Vous dégoûtez un honnête homme d'avoir du zèle par le mauvais usage que vous faites du vôtre, supposé que vous en ayez" (ed. cit., p.65).
13. The influence of Paolo Sarpi's *Histoire du Concile de Trente* (London 1619; French translation by Jean Diodati, Geneva, 1621) seems here very probable.
14. See the edition of Charles Dellon established by Charles Amiel, Paris, 1997; that of Tyssot de Patot by Aubrey Rosenberg, Oxford, 1993, and my note in *La Lettre clandestine,* 7 (1998).
15. I present and edit this text in more detail in *Enlightenment Essays in memory of Robert Shackleton* (Oxford: 1988), pp. 163–173.
16. I present the complete text of this little-known work in *L'Intelligence du Passé, Mélanges . . . Jean Lafond* (Tours: 1988).

8

The Trinity as Heresy

Socinian Counter-Histories of Simon Magus, Orpheus, and Cerinthus

Martin Mulsow

Early modern histories of heresy took on a wide variety of forms. Even though most of them treated the same traditional material, diverse types of assertions were possible: accusations, defenses, apologies, or modifications. There were at least the following types:

1a. A person x, found traditionally to be a heretic by orthodoxy, is indeed a heretic because of reasons yz.

1b. A person x, found traditionally to be a heretic by orthodoxy, is indeed a heretic, especially in the light of current modifications of Christianity such as abc.

2a. A person x, found traditionally to be a heretic by orthodoxy, is indeed no heretic, because he or she did not advocate the condemned doctrine and was only denounced on grounds of historical circumstances yz.

2b. A person x, found traditionally to be a heretic by orthodoxy, is indeed no heretic, because although he or she advocated the condemned doctrine and thus erred, he or she was deep down a righteous and God-seeking person who was unjustly treated intolerantly.

2c. A person x, found traditionally to be a heretic by orthodoxy, is indeed no heretic, because the condemned doctrine he or she advocated was in fact the right doctrine.
3. A person x, found traditionally to be a heretic by orthodoxy, is indeed a pillar or a precursor of orthodoxy—an orthodoxy gone astray.
4. A person x, found traditionally to be a heretic by orthodoxy yet "saved" by heterodox positions for their purposes, cannot be usurped by either dogmatic position but must be understood in a historically unbiased manner from within the given context.

Traditional histories of heresy written by conservative theologians upheld positions of type 1. However, "conservative" in the sixteenth and seventeenth centuries already referred to many different orthodoxies and thus confessional differences were also reflected in historical assessments. To a Calvinist, for example, the Nestorian doctrine of Communion seemed by no means as heretical as to a Catholic theologian. Protestantism stood for a partial rehabilitation of heresies using statements of the type 2c. Already for this reason the significance of histories of heresies and dissidents for the development of tolerance and Enlightenment turns out to be a complex matter. One must take into consideration confessional debates within each orthodoxy, as well as deviations within particular orthodoxies or from a fundamental interdenominational consensus.

Type 2a could establish itself outside of any confessional historiography, as it did in Naudé's *Apology for all the Great Persons Who have been Falsely Accused of Magic* of 1625, which remained in Germany a prototype for a number of "acts of saving" (*Rettungen*) well into the eighteenth century.[1] Spiritualist and Pietist minorities of confessional churches, especially of Calvinists and Lutherans, and even spiritualists of independent sects developed a keen interest in bringing to light heresies with which they sympathized. Gottfried Arnold, for example, made statements of type 2b and 2c in his *Church and Heretic History* of 1699/1700, regardless of his claim for impartiality.[2]

Every time non-orthodox doctrines concerned not only particular doctrinal points but offered a systematic alternative to the official Church, a "counter-history" could be constructed or reconstructed. In these cases one not only encounters statements like type 2c, but it is also possible to find pointed and revealing assertions of type 3, which went beyond pure rehabilitation by offensively denouncing the official doctrine.

Type 4, which evolved only around the eighteenth century, was a reaction to insufficiencies in prior forms of histories of dissidence. I wish to emphasize that the road leading to a tolerant Enlightenment was not unequivocally deduced from only certain types of histories of heresy. Orthodox histories of heresy of type 1 could, for example, argue polemically against the Church Fathers' Platonism in the very same sense as did Socinians of type 2c or 3, and thereby unwittingly fortify the anti-essentialism of Enlightenment philosophy. Historiographers such as Gibbon were well-served by Church historians like Tillemont, Beausobre, or Mosheim as long as those historians presented their material in a philologically accurate manner. It was often precisely the synchronous effect of disparate types of a history of heresy, e.g., the ones Naudé and Arnold presented in Pietistic early Enlightenment Germany of the beginning of the eighteenth century, that brought off decisive consequences.[3]

Now I will focus on a compelling example of type 3 in the seventeenth century. As a prototype, let us review the story of Simon Magus, the "first heretic" according to Irenaeus's *Adversus haereses.* This example touches upon a delicate problem of the seventeenth century: the origins of gnosticism. Early modernity found gnosis in many ways precarious. It was associated with catch phrases like "enthusiasm," "esotericism," "the origin of evil." Treatment of Simon as a traditional example of type 1 can be found in the work of Daniel Ehregott Colberg, who followed Jakob Thomasius in drawing a straight line from Simon to the "enthusiastic" adherents of Paracelsus and Boehme.[4] But another key topic of the seventeenth century, the Trinity, comes into play in a Socinian counter-history of Simon of type 3. Radical proponents of the Enlightenment fought against the "irrationality" of Trinitarian dogma and for an essentially Unitarian understanding of Enlightenment: Christianity had to be "reasonable."

But was there something like a Unitarian counter-history of Simon Magus? Yes. It became possible at the very moment when Socinianism discovered its historical dimension beyond mere exegesis. This happened with the publication of the second volume of the Jesuit Denis Petau's *Opus de theologicis dogmaticus* in 1644, which for the first time presented Trinitarian dogma as historically evolved.[5] Unitarians such as Daniel Zwicker (and later Christoph Sand, Samuel Crell, Jacques Souverain, John Toland, or William Whiston) could now appeal to the "Arian" pre-Nicaean patristic doctrines, as well as interpret the successive development of Trinitarianism as a corruption of Christianity brought on especially through the reception of Platonism.[6] In order to study the tradition of a Christianity not yet corrupted, i.e., the early "Nazarethans" and "Ebionites" (Judeo-Christians or Jewish Christians[7]), one could reconstruct a line connecting the significance of the Ebionites to Islam and from Islam to the Socinians. Arabic studies in England since Edward Pococke had given radicals like Henry Stubbe material for such a "monotheistic" counter-history.[8]

Zwicker's assessment of the corruption of Christianity as not only caused by Platonism[9] but primarily by early gnosis introduced a further feature to the Unitarian construction of history. Now it was possible not only to reinterpret orthodoxy itself (e.g., Athanasius) as corruption, but furthermore—and even more subtly—expose precisely those who orthodoxy had condemned for heresy (like the gnostic Simon) to be orthodoxy's own predecessors. Hence Zwicker was able to construct a history of heresy of type 3.

Zwicker, an irenic theologian in Amsterdam whose eclectic arguments were influenced by both the Bohemian brothers and Socinianism,[10] established in 1657 the following hypothesis regarding dogmatic history. Simon Magus, who was mentioned in the Acts of the Apostles and since Irenaeus was regarded as the foremost arch-heretic Christianity ever knew, had indeed with his doctrine of the "great power" been the precursor of the Trinitarian creed that had transformed and corrupted Christianity into mere speculation. A pupil of Simon was alleged to have written texts that were pseudo-epigraphically attributed to Orpheus and that contained Trinitarian ideas. Further, Justin Martyr (mid second-century), apparently under the influence of both Simon and Pseudo-Orpheus, outlined his doctrine of the Logos, which eventually—through Tatian, Athenagoras, and others—led to the formulation of the Trinitarian dogma of Nicaea, just as described by Petau.[11]

Zwicker's hypothesis—although having no great influence, and certainly not widely accepted—is a significant one, worth a "thick description" analogous to the ethnology of Clifford Geertz. For just as the Balinese cockfight can help explain given social structures, Zwicker's hypothesis can help us understand the structures (and transformations) of theological-historical discourse around 1650. His hypothesis shows the way both backward and forward. It points forward to mark the importance of the ongoing search in the next one hundred years to find a "monotheistic," non-Trinitarian origin of Christianity (including its offspring all the way to Socinianism and Deism) and aiming at the formation of Enlightenment. It points backward in that Zwicker's dismantling of Orpheus annuls with one exemplary case the entire "prisca theologia." Zwicker's strategy, as his thesis concerning dogmatic history reveals, consists in manifold reversals and rereadings of traditional interpretative patterns, based on advanced historical criticism.

From the perspective of current research, Zwicker's hypothesis is certainly highly speculative, but not entirely erroneous. Whether the traditional accounts of Simonian gnosis can indeed be attributed to the historic Simon is highly controversial.[12] Both the Orphic "testament" and the Orphic "oath" that Zwicker refers to are today regarded as undoubtedly Jewish counterfeits (pseudo-epigraphs), albeit—as in the case of the "testament"—possibly dated earlier than Zwicker did (at the turn to the Christian era,[13] or even before Aristobulos' time, 250–150 B.C.).[14] In spite of his otherwise critical stance, Petau still cited the "oath" without any apprehension:

> I beseech Thee in Heaven, the wise
> Creation of the great
> God, and beseech Thee truly in the first
> Words of the Father;
> He was the Creator of the Universe, as was demanded
> by His Will[15]

Zwicker, however, seems to have been influenced by Gerard Joannes Vossius's sceptical attitude concerning the Orphica,[16] and thus rearranged Petau's tableau of dogmatic history.

The influence of pseudo-orphic testimony on Justin seems undeniable, since it is cited in the "Cohortatio ad Graecos." One must consider, however, that the "Cohortatio" mentioned by Zwicker is today regarded as only pseudo-Justinian, and has been placed in the second half of the third century or even to the beginning of the fourth century.[17] Neither has Zwicker's highly speculative association of Simonian "power" (as gnosis—criticized by Irenaeus) with Jewish-Hellenistic logos-theories found any confirmation. Yet his argument that Jewish-Hellenistic sources like Simon and pseudo-Orpheus had been more familiar to the theologian Justin from Palestine cannot wholly be dismissed by scholarship on Justin.

Zwicker's hypothesis has two points. First, as a statement of type 3 it reverses the ascription of orthodoxy: Simon and Orpheus did not deviate from orthodoxy but were indeed its precursors. This, again, shows that orthodoxy itself leads us astray. Second, it is also a reversal and a dismantling of the "prisca theologia," the syncretism of the Renaissance. For scholars such as Ficino, Pico, Steuco, and others, Orpheus was the exemplary model of an early pagan "theologian,"[18] whose work already entailed traces of later Christian Trinity. Thus, he was either a disciple of the truth manifested to Moses, or already a witness to a partial revelation known to pagans, which was to prepare the way for a Christian revelation.[19] The central proofs of this were the Jupiter-hymn, the "testament," and the "oath." In interpreting these texts, first, as late counterfeits, and second, as sources of corruption within Christianity, he doubly denies and deconstructs the efforts of the Renaissance.

Cerinthus, too, plays a part in Zwicker's hypothesis. Because of the *Epistula apostolorum* and Irenaeus, Zwicker regards him as closely associated with the Simonian doctrine, and possibly a friend or a pupil of Simon. Further, he does not consider it impossible that Cerinthus himself was the counterfeiter of the Orphic fragments. Here, however, he seems to succumb to an error already present in his own sources, namely that of a too early dating of Cerinthus (to the time of the Apostles), due to a confusion between characters of St. John. Actually, Cerinthus should probably be dated to around 100.[20] Yet Zwicker was right in realizing that Cerinthus represented a distinctive Christology. He made quite a rigorous distinction between Jesus and Christ, and thus is considered the first precursor of the two-nature theory.[21] Here we find connections between Judeo-Christianity and gnosis as well as the reason for the difference in ascribing such characters. Jews wanted to see Jesus as nothing but a human being; on the other hand, a "Docetist" separation according to which Christ was something wholly different than Jesus's human appearance signified a clear step toward gnosticism. Even though Socinians were inclined to sympathize with Jewish notions, they found both gnosis and the two-nature theory to be mere speculation, deviating from the simplicity and rationality of the Bible.

In the counter-histories of the Socinians, Cerinthus had two possible meanings. First, they could adopt Epiphanius's account of an assumption prevalent in antiquity, according to which it was not St. John who wrote the Gospel against Cerinthus but in fact Cerinthus himself who was the author of the Gospel. This assumption must be understood in the light of an impression that seems plausible even today, namely that the Gospel of St. John seems to be marked by early gnosis. Thus one would identify—as done by John Biddle—[22] Cerinthus with the Gospel of St. John and denounce both. Second, one could view Cerinthus as close to Simon as a precursor of Justin's logos-theology, as done by Zwicker. Here, too, one wished to escape both gnosis and the Trinitarian doctrine.

The centrality of such speculations around Cerinthus can also be seen from John Locke's comprehensive engagement with this very character in his studies on Biddle, the correspondence with Le Clerc, Limborch, and Beresford, as well as in his manuscript on *Some General Reflections on the Beginning of St. John's Gospel* from the late 1690s.[23] In it Locke reacts to John Edwards's accusation according to which Locke had not stated in his *The Reasonableness of Christianity* that the main point of the Gospel of St. John was the revelation of Trinity in John 1:1. Locke replies by clearing up the context: the situation should not be understood to be such that St. John had written the Gospel as a Trinitarian against a so-called anti-Trinitarian Cerinthus. Following Irenaeus and in unison with Zwicker, Locke tries to show how Cerinthus had in fact been the first to actually introduce the new doctrine of the divinity and preexistence of Christ. The difference between Cerinthus and St. John was thus only a matter of degree. For Cerinthus the divine person had entered Jesus with Baptism, whereas for

St. John it was present from birth.[24] Thus St. John did not actually reveal Trinity, only the fact that Jesus was Christ and thus the Son of God; this could, however, be understood in alignment with doctrines of adoption and subordination.

There are several reasons why one should study very closely seventeenth-century counter-histories. Today, categories such as "orthodoxy" and "heterdoxy" are justly differentiated and looked at from an altered perspective. One notices plurality, development,[25] or an unintended influence on the Enlightenment[26] within "orthodoxy"; the internal logic of "heterodoxy" is being questioned, and its contribution to tolerance probed. Yet even in patristic studies on the second century, traditional categories of orthodoxy and heterodoxy were hazy. There, one finds clearly the open plurality of the situation, comparable to a "laboratory of Christian theology."[27] "Gnosis" is no longer simply considered an absurd deviation from "true" Christianity; the significance of Jewish Christianity and Christian Jews has been upgraded considerably.[28]

From this point of view, the rediscovery of Jewish Christianity by Socinians and Deists of the seventeenth century versed in history can today gain new meaning. One discovers that here the pluralization of early modern Europe[29] meets the pluralization of early Christianity. From the perspective of current research, with its awareness of the uncertainties of source-material, the quite daring Socinian hypotheses concerning dogmatic history thus seem less offensive than to the then-prevailing orthodoxy. The interest and resistance of seventeenth-century theorists vis-à-vis an "enthusiastic" gnosis moves into a new and precarious light when looked at from the political perspective of an esoteric immediacy on the one hand, and of the preservation of sociability on the other. In a pluralized early modern age, strategies for maintaining and preserving identities from early Christian times were reproduced, yet became historico-philologically more and more questionable.

It must to a large extent have been the counter-histories of anti-Trinitarians, Spiritualists, and Sectarians that, through a form of meta-criticism, pushed intellectuals of the conservative enlightenment to develop "impartial" histories of heresies (type 4). Authors such as Mosheim and Walch distinguish dogmatics and history, obviously also in order to suppress the dogmatic interest of Socinians and Deists in history.[30] And the Deists' historic and pyrrhonistic attacks on the "fides" of early Christian sources could be countered with historical criticism of such anti-Christian authorities as Philostratus (as in the case of Blount) or the alleged Gospel of Barnabas (in the case of Toland).[31] Condemnations of heretics were only carried forward in a historically sublimated way, for example in Mosheim's claims that an irrational "philosophia orientalis" was a fertile soil for gnosticism.[32]

With this development, however, one also cut off one's own dogmatic opinions—which in the case of Mosheim were quite orthodox—from historiography.[33] The coast was thus clear for a complete marginalization of dogmatic questions, which we find in such thinkers as Voltaire and Gibbon. Their overall refusal to discuss dogmatics—which, additionally, gave them a more or less implicit air of tolerance—was encouraged by the fact that enlightened orthodoxy itself had, as a reaction to counter-histories by the non-orthodox, already withdrawn to this position. It is lastly because of this development that we can learn a great deal from "significant" hypotheses put forward by Zwicker all the way to Toland.

Notes

1. See Lorenzo Bianchi, *Tradizione libertina e critica storica. Da Naudé a Bayle* (Milano: Angeli, 1988), ch. 1.
2. Gottfried Arnold, *Unpartheiische Kirchen - und Ketzerhistorie* (Frankfurt: 1699–1700). See A. Schindler: "Häresie II," in *Theologische Realenzyklopädie* vol. 14 (Berlin: 1985), pp. 318–341.
3. See Martin Mulsow, "Appunti sulla fortuna di Gabriel Naudé nella Germania del primo illuminismo," *Studi filosofici* 14–15 (1991–92), pp. 145–146.
4. Daniel Ehregott Colberg, *De origine ac progressu haeresiae* and *Das Platonisch-Hermetische Christenthum* (Frankfurt and Leipzig: 1690–91). On Thomasius see Ralph Häfner, "Jakob Thomasius und die Geschichte der Häresien," in Friedrich Vollhardt, ed., *Christian Thomasius* (Tübingen: 1997), pp. 141–164.
5. Denis Petau (Dionysius Petavius), *Opus de theologicis dogmatibus,* Vol. II (Paris: 1644). On Petau, see Michael Hofmann, *Theologie, Dogma und Dogmenentwicklung im theologischen Werk Denis Petau's* (Bern: 1976).
6. See the orthodox work, George Bull, *Defensio fidei Nicaenae* (Oxford: 1685). For the context, see M. Mulsow, *Moderne aus dem Untergrund. Radikale Frühaufklärung in Deutschland 1680–1720* (forthcoming).
7. See J. E. Taylor, "The Phenomenon of Early Jewish-Christianity: Reality or Scholarly Invention?" in *Vigiliae Christianae* 44 (1990), pp. 313–334; G. Stanton and G. Stroumsa, eds., *Tolerance and Intolerance in Early Judaism and Christianity* (Cambridge: Cambridge University Press, 1998); R. Popkin and G. Weiner, eds., *Jewish Christians and Christian Jews from the Renaissance to the Enlightenment* (Dordrecht: Kluwer, 1994).
8. See Justin Champion, *The Pillars of Priestcraft Shaken. The Church of England and its Enemies 1660–1740* (Cambridge: Cambridge University Press, 1992), ch. 4.
9. For the Platonism thesis see Walter Glawe, *Die Hellenisierung des Christentums in der Geschichte der Theologie von Luther bis auf die Gegenwart* (Berlin: 1912).

10. See Peter G. Bietenholz, *Daniel Zwicker 1612–1678: Peace, Tolerance and God the One and Only* (Florence: Olschki, 1997).
11. Daniel Zwicker, *Irenicum irenicorum* (Amsterdam: 1658), pp. 14ff.
12. See Karlmann Beyschlag, *Simon Magus und die christliche Gnosis* (Tübingen: 1974).
13. Nikolaus Walter, *Der Thoraausleger Aristobulos* (Berlin: 1964); N. Walter and Ernst Vogt, eds., *Poetische Schriften* (Jüdische Schriften aus hellenistisch-römischer Zeit, Vol. IV) (Gütersloh: 1983), pp. 217ff. For advice concerning ancient religion and theology I am grateful to Roland Kany.
14. Christoph Riedwig, *Jüdisch-Hellenistische Imitation eines orphischen hieros logos* (Zürich: 1998).
15. (Pseudo Justin), *Cohortatio ad Graecos,* cited in Petau (note 5), I, V, 10.
16. Gerardus Ioannes Vossius, *De poetis graecis.*
17. See Christoph Riedwig, ed., Pseudo-Justin (Markell von Ankyra?), *Ad Graecos de vera religione (bisher: Cohortatio ad Graecos)* (Basel: 1994).
18. See Daniel P. Walker, "Orpheus the Theologian," in his *The Ancient Theology* (London: 1970), pp. 22–41.
19. See Edgar Wind, *Pagan Mysteries in the Renaissance* (London: 1958).
20. See Christoph Markschies, "Kerinth: wer war er und was lehrte er?" in *Jahrbuch für Antike und Christentum* (1998), pp. 48–76.
21. According to Simone Petrement, *A Separate God: The Christian Origins of Gnosticism* (London: 1990), Zwicker's thesis could be labeled "The Gnostic Origins of Christianity."
22. John Biddle, *Considerations on the Explications of Trinity;* see also his *The Apostolical and True Opinion of the Trinity* (1653). On Biddle see H. McLachlan, *Socinianism in Seventeenth Century England* (Oxford: 1951).
23. See John Marshall, *John Locke: Resistance, Religion and Responsibility* (Cambridge: Cambridge University Press, 1994), pp. 418ff.
24. Ms Locke e17, 175–186; Marshall (note 23), pp. 422f.
25. See John G. A. Pocock, "Within the margins: The definitions of orthodoxy," in R. D. Lund, ed., *The Margins of Orthodoxy: Heterodox Writing and Cultural Response, 1660–1750* (Cambridge: Cambridge University Press, 1995), pp. 33–53.
26. Alan Charles Kors, *Atheism in France 1649–1728,* vol. I, *The Orthodox Sources of Disbelief* (Princeton: Princeton University Press, 1990).
27. Markschies (note 20), p. 49.
28. See Taylor (note 7).
29. See Martin Mulsow, "Pluralisierung," in Anette Völker-Rasor, ed., *Oldenbourg Geschichte Lehrbuch* (Munich: 2000), pp. 303–307.
30. Johann Lorenz Mosheim, *Versuch einer unparteiischen und gründlichen Ketzergeschichte,* 2 vols. (Helmstedt: 1746–1748), reprint ed. Martin Mulsow (Hildesheim: 1998–99). See Martin Mulsow "Eine 'Rettung' des Servet und der Ophiten? Der junge Mosheim und die häretische Tradition," in M. Mulsow et al., eds., *Johann Lorenz Mosheim (1693–1755). Theologie im Spannungsfeld von Philosophie, Philologie und Geschichte* (Wiesbaden: 1997),

pp. 45–92. In this essay I assign Mosheim to a "conservative enlightenment" (Pocock) in Germany. For an association of conservative enlightenment and "eclectic philosophy" see Mulsow, *Moderne aus dem Untergrund* (note 6), introduction. See Ch. W. F. Walch, *Entwurf einer vollständigen Geschichte der Ketzereien, Spaltungen und Religionsstreitigkeiten, bis auf die Zeiten der Reformation,* 10 vols. (Leipzig: 1762–1782).

31. Mosheim, "De existimatione celeberrimi philosophi, Apollonii Tyanaei," in his *Observationum sacrarum,* vol. 1 (Amsterdam: 1721), pp. 260–382; Mosheim, *Vindiciae antiquae Christianorum disciplinae adversus Tolandi Nazarenum* (Kiel: 1720). See Mulsow, "Eine 'Rettung'" (note 30).
32. Mosheim, *Ketzergeschichte,* vol. 1.
33. Walter Sparn, "Von der fides historica zur historichen Religion. Die Zweideutigkeit des Geschichtsbewußsteins der theologischen Aufklärung," in *Berichte zur Wissenschaftsgeschichte* 8 (1985), pp. 147–160; Walter Sparn, "Verünftiges Christentum. Über die geschichtliche Aufgabe der theologischen Aufklärung im 18. Jahrhundert," in Rudolf Vierhaus, ed., *Wissenschaften im Zeitalter der Aufklärung,* (Göttingen: 1985), pp. 18–57.

9

Two Jewish Heresies

Spinozism and Sabbatianism

Richard Popkin

Unlike Christianity, Judaism does not have central authorities for determining what is a heretical belief. In earlier times, when there was more coherence and control, certain views were deemed heretical, and this has stuck through the ages. In the eighth century A.D., the Caraites refused to accept the Talmud as the Oral Law, or to accept the authority of the rabbis to interpret Jewish law.[1] They were expelled as heretics, and have remained so to this day, as far as Orthodox Judaism is concerned. Rabbis are supposed to refuse to marry a Jew to a Caraite. The Caraites persist in pockets in Lithuania, the Crimea, Egypt, and Turkey, and act as if they are a continuation of the Jews of ancient Palestine. Even Hitler accepted the Jewish view on this, and ranked the Caraites as "other religions." Those in Nazi-dominated territory survived the Holocaust without suffering the horrors of the Jewish world.

In the first centuries of the Common Era, Jews who accepted Jesus as the Messiah were regarded as heretics. This problem still persists in the form of whether one can be a Jew and a Christian simultaneously. The "Jews for Jesus" movement claims to be doing this, and is rejected by most Jewish groups. The Israeli Supreme Court has ruled that a Jew who is a Christian has accepted another religion, and hence is not a Jew. However, there still is a small "Jews for Jesus" movement, even in Israel.

This chapter is about two other heresies that developed in the 1660s and 1670s and are still going in the Jewish world: Sabbatianism and

Spinozism. These views were and are also accepted by many outside of Judaism. What made and makes them a heresy for Jews is different from what makes them a heresy for Christians or other non-Jews. No grand survey history of these two heresies was written in the early modern period, and the history of these heresies inside Judaism is still to be written. This chapter will review half a dozen early modern sources on each heresy, with some comments on contemporary manifestations.

Sabbatianism

Sabbatianism gained strength in and around Amsterdam, as well as in the Sephardic Diaspora in the Ottoman Empire. Its central tenet is that a Jew from Smyrna, Shabbatai Zevi, was the actual, expected Jewish Messiah. In the 1640s and 50s, Shabbatai attracted attention in the Ottoman world for his Kabbalistic views and mystical practices. A rabbi, Nathan of Gaza, convinced Shabbatai to reveal his messianic role. More and more excitement occurred until on Jewish New Year's Day 1665–66, he announced that he was the Messiah. He proceeded to change Jewish practices and prayers and declared that his birthday, which is a fast day, Tisha b'Av, would now be a feast day. He also changed rules about Jewish dietary practices, among other things.[2]

Tremendous excitement developed throughout the Jewish world. Almost all the members of the Amsterdam Jewish community became Sabbatians. The Jewish community has on display, even now, the letter he sent them at the end of 1665, showing him sitting on a throne, announcing the beginning of the Messianic Age and appointing the 12 kings of the Messianic World, 6 of whom were Sephardic relatives living in the Ottoman Empire. The excitement brought Jews from Eastern Europe and elsewhere together to greet the Messiah. The Turkish government began to worry that Sabbatianism might be a revolutionary political movement that would attempt to overturn the authority of the sultan, so the Turkish authorities arrested Shabbatai Zevi and put him in prison on an island in the Bosphorus where he held court to all visiting Jewish dignitaries.

One of the visitors told the Turkish authorities that he thought Shabbatai Zevi was a fake and a fraud. This led the sultan to bring the purported Messiah before him and to tell him that they would test his divinity by shooting arrows at him. If he was the Messiah, he could of course deflect the arrows and the sultan would recognize him. At this point we read that Shabbatai Zevi told the sultan he was just a little Jew who wanted to be a Muslim, and he put on a fez. His public Messiahship lasted a year. He lived another ten years as a Muslim political official in Albania and Montenegro. He died attending a Yom Kippur service in northern Albania and is supposed to be buried there.

Sabbatianism is the belief that Shabbatai Zevi was, is, and always will be the Messiah. His conversion to Islam is taken to be only an outward step and that he would return at some point in full glory as the Jewish Messiah. One of the leading theologians of his movement, Abraham Cardosa, originally a Spanish forced convert to Christianity, reverted to Judaism in the Ottoman Empire, and claimed that Shabbatai Zevi had made the greatest sacrifice a human could make by his conversion. He took on all the sins of mankind and would thus redeem everybody when he returned. A sort of latter-day Christian story was grafted onto the events, with the Second Coming to be the real deliverance and the real messianic period. In the meantime, his followers should keep the faith and await his return.

Contemporary non-Jewish sources on the movement range from those who thought of it as a fraud to those who believed it. A plate from a German book of the period shows Shabbatai facing the Quaker James Naylor as two examples of religious fakers.[3] Jean-Baptiste Rocoles, Jacques Basnage, and Ludvig Holberg listed Shabbatai among false messiahs.[4] So did an Italian manuscript that has recently been made available.[5] Paul Rycaut's *Turkish Spy* and his *History of the Turkish Empire,* along with other English versions of the story, spread the news of the fraud throughout Europe.[6] On the other hand, Protestants in the Netherlands such as Petrus Serrarius and Jean de Labadie wrote with enthusiasm that this must, indeed, be the Jewish Messiah.[7]

There were so many followers in Amsterdam and Poland and elsewhere that when the shock of his conversion reached the Jewish world, the rabbinical authorities slowly tried to reclaim control of various Jewish communities. Estimates are that at least 90 percent of the Jewish world accepted Shabbatai Zevi. In a novel, *Satan in Goray,* Isaac Bashevis Singer purports to depict what happened in a Polish village when it first became wildly enthusiastic that the Messianic Age had begun.[8] The enthusiasm led to orgiastic sexual activity. Then came a great disillusionment until the community was saved by a rabbi who had wandered away during the Saturnalia. In historical fact, the process of reclamation was very slow and there was a general consensus of rabbis in Europe and the Ottoman Empire to not discuss the case for 50 years and not pass judgement on people. This half-century of hiatus left a strong underground current of Sabbatian believers who turn up especially in new Jewish communities in London, Amsterdam, and the New World. The strongest and most persistent group was in the Ottoman Empire and it exists until this day. They are called the "Dönmeh," meaning "those who changed." They are officially Muslims but everyone knows that they are fake Muslims and that they have their own special rituals and prayers. It is only in the last half century that much has become known about what they believe and practice, mainly through the efforts of Professor Gershom Scholem and his followers. Scholem discovered some of the Dönmeh's prayer books. They outwardly practice Islam and dress accordingly.

In the Turkish world, the Dönmeh are an elite—professors, doctors, and lawyers. Everybody seems to know who they are. They only marry among themselves and other Turks would not want to marry with them because of their beliefs. As an indication of the state of affairs, I was in Istanbul in 1966 to give two lectures at the Philosophy Department of the University of Istanbul. They took me to lunch and we had pretty much exhausted conversation about the weather and the differences between Southern California and Istanbul. I brightly asked if anyone was doing research on the seventeenth-century figure, Shabbatai Zevi. At this point, even though we had not ordered our lunch, one person there looked at his watch and said he had to go and within a minute or two I was left alone as everybody trouped out. I took a taxi back to our hotel and told my wife of this amazing development. Later on, a professor called who had been at the lunch and he said that he and his wife would like to take us for a drive to the Black Sea. As soon as we started off, I was told I should never have asked that question. Why? Because everybody knew that the rector of the university, who was at the lunch, was a Dönmeh, and they would never discuss the matter when any of the Dönmeh were present. Then we were treated to a long exposition of the role of the Dönmeh in Turkish society and finally to the strong belief of our hosts that the late Kemal Ataturk, the father of modern Turkey, was a Dönmeh who destroyed Islam in Turkey. We were told that Ataturk came from the Dönmeh quarter in Salonica. A few days later, we were in Jerusalem at a birthday party for Gershom Scholem and I told him about this. He told me that when he gave his first lecture about the Dönmeh, two people came to him after the lecture and asked if they could talk to him in private. And then they said to him, "Why didn't you discuss Ataturk?"

Recently, a young member tried to escape the netherworld of the Dönmeh and wrote an exposé of what he was trained to believe and do, which included Jewish practices and beliefs, the belief that Shabbatai Zevi would return, and special Sabbatian practices including an annual orgy.[9] The position of the Dönmeh comes out clearly in his revelations. The Turkish Jews do not accept him because of his heretical beliefs and practices, and the Turks do not accept him because they know he's not one of them. The young man's desire is to get out of this ghettoized world, stuck between the Jews and the Turks, but he does not seem to have any option except changing his name and moving to America. I have been told that part of the fallout of his revelations is that a large group of the Dönmeh want to move to Israel to be accepted as Jews with their own rabbinical council so that they can keep up their own religious practices and beliefs.

This then is a heresy within Judaism that has gone on for 330-odd years. It is only beginning to be possible to describe it and its history. In Turkey, the ramifications for what happened elsewhere are still to be studied and examined. We know that in the seventeenth century Sabbatianism was very strong in Poland, Lithuania, and Amsterdam. We are learning about an agent of Shabbatai Zevi who was active in Paris in the 1670s and 80s.[10] There were rabbis who were trained in Sabbatian seminaries in Palestine who turn up in London, Amsterdam, and Newport, Rhode Island. And the movement did not just disappear outside of Turkey. A strong successor movement developed in Central Europe around a figure named Jacob Frank at the end of the eighteenth century who claimed to be the reincarnation of Shabbatai Zevi. He and his followers were driven out of the Jewish community and he and they became fake Catholics. Frankism is still a somewhat secret movement, especially in Poland. I have seen claims that Freud's and Louis Brandeis's mothers were both Frankists. I have no idea whether this is just a curious genealogical fact or whether some sort of heretical belief and practice was being carried on in upper-class Jewish life in the Hapsburg Empire and in the United States.

Jewish historians, by and large, have tried to write off Sabbatianism as a lunacy or as a dishonest way of gaining control of ignorant Jews. Even Scholem, who tried to flesh out the context and background of Sabbatianism in the seventeenth century, still insisted that its hero, Shabbatai Zevi, was a manic depressive and that this somehow explained the phenomenon. Elesheva Carlebach, Matt Goldish, Michael Heyd, and others are now studying what happened to the followers both in the Jewish and Christian worlds in the seventeenth and eighteenth centuries, and may reveal the ways in which some of these beliefs persisted and took root elsewhere.[11]

Scholem claimed in one of the more disputed parts of one of his books that Sabbatianism in the course of the eighteenth century got transformed into what was to be called "Reform Judaism."[12] If it could be shown that the Jewish thinkers who proposed revising Jewish ritual and belief in the early nineteenth century were influenced by surviving Sabbatians, that would be very interesting. It is usually claimed that Reform Judaism grew out of the German Jewish Enlightenment, but I think it comes out first from events in Amsterdam and in the Sephardic world. A decade before there was any Jewish reform movement in Germany, one existed in Amsterdam, led by people who accepted Napoleon's modernization of Judaism. The first American Reformed synagogue was not in the German diaspora of Cincinnati but was actually in the sephardic world of Charleston, South Carolina. So there were forces outside of those operating in Germany that led to a new view of Jewish belief and ritual.

Within various kinds of Jewish communities now existing, there seems to be a constant fear of Sabbatianism turning up in another form. When the followers of the Lubavitcher Rebbe Menachem Mendel Schneerson of Brooklyn began suggesting and later insisting that he was the expected Messiah, their movement was denounced by rabbis and theologians of various communities as reintroducing the Sabbatian heresy. His followers are waiting for some definite word from their leader that he accepted his role as the Messiah. They set up an elaborate telecommunications system so that the instant such news occurs, they will all know it. They built an exact reproduction of his house in Brooklyn in Israel with duplicate furnishings so that the moment he officially becomes the Messiah, he can set up his office in the Holy Land. Unfortunately, Rebbe Schneerson had a stroke and for the last couple of years of his earthly existence, was unable to speak or communicate in any way. His followers would gather around him every day, waiting for a blink, a finger movement, or something that would indicate he had accepted his messianic role. In secular terms, he passed away in June 1994. Since then, some of his followers have been dancing and praying around his tomb, waiting for him to return, while others refuse to discuss the matter and just propagate his ideas. The strong denunciations by others that the movement is a form of Sabbatianism indicate that the heresy is still feared.

Another indication of this is that a leading Israeli theologian, Yeshayahu Leibowitz, inveighed against those who asserted that geographical territory in Palestine had to be held by Jews before the Messiah would come.[13] Such people, deriving their view from Rabbi Kook, saw the Six Day War as the birthpangs of the Messiah and have seen the development of Israeli settlements throughout Palestine as a necessary prelude to the appearance of the Messiah. Leibowitz denounced this as a new form of Sabbatianism in which the physical land was becoming endowed with messianic properties.

Others have tried to defuse the force of the Messianic Idea in Jewish religion by suggesting it does not apply to a person or a place but to the Jewish people in general or to a spiritual state. These proposals are in part ways of avoiding any recurrence of Sabbatianism.

Spinozism

The second Jewish heresy discussed here, Spinozism, is just as important in accounting for the development of Reformed Judaism. Spinozism is, of course, the acceptance of two central views of Baruch de Spinoza, who was excommunicated from the Amsterdam synagogue in 1656. One of his views was that Jewish ceremonial law has lost any validity or meaning and should be dispensed with. And the other is that God or nature is in everything and causes everything. This means that there is no special providential history going on but only a divine natural history.

We do not know how much of the above beliefs Spinoza held at the time of his excommunication at age 23. We are told in the text of the excommunication that he had abominable beliefs and practices and that because of this he should be separated from the Jewish community and no Jew should read anything he wrote.[14] The usual picture that has come down to us is that poor Spinoza was just cast away into a cold cruel world and had no subsequent career within Jewish history. However, in spite of the blood- curdling excommunication statement, we've learned in the last decades that after the excommunication, Spinoza went to a theological discussion group in Amsterdam which was held in the home of a Jewish doctor, and that several Jews attended the sessions.[15] There, Spinoza proclaimed the view that God exists but only philosophically, which would seem to be well on the way to his final pantheistic metaphysics.

But Spinoza was never completely cut off from the Jewish community. Fifteen years later, when Spinoza was living outside the Jewish community in The Hague, we learn that he received some Jewish visitors, at least two of whom can be identified. One was the son-in-law of the rabbi of Copenhagen, and the other was Dr. Morelli, a Marrano from Padua who was a physician to upper-class people in France and England.[16] Spinoza made charcoal drawings of the people who visited him. We know that at the time of his death there were 200 of these, including some portraits of ladies of quality. These portraits existed at least 25 years later, in the possession of his landlord. If these could be located we would have a fuller picture of who came to visit him and how many were members of the Jewish community.

When Spinoza's major work came out posthumously, one of the most important refutations was by Isaac Orobio de Castro, published in 1684. Orobio was the leading intellectual of the Amsterdam Jewish community, and he read Spinoza's text most carefully. His refutation was taken very seriously at the time and was included in a collection of refutations published in 1731 that included one by the archbishop of Cambrai, Fénelon.[17] The fact that Orobio studied the text of Spinoza would indicate that he was not impressed by the ban. In fact, he might not have known about it since he only joined the congregation in 1659 or 1660. He was a theologian and medical doctor trained in Spain who had been a professor at Toulouse before moving to Amsterdam. He immediately became involved in a controversy with physician Juan de Prado, who had been expelled from the community along with Spinoza. Orobio had been in school with Prado in Spain and was surprised at how heretical his classmate had become. But there is nothing in his answers to Prado that mentions Spinoza.

He only seems to have become interested when a Dutch Protestant friend of his, Johannes Bredenbourg, wrote an answer to Spinoza in 1684 that consisted of saying that each step in Spinoza's argument was unanswerable, but by faith one knows that Christianity is the true religion. Orobio apparently decided that Spinoza deserved a better answer than that and actually worked long and hard at a text. A manuscript of his answer in French exists in Bordeaux, which suggests that it was disseminated to other Jewish communities.

It is also worth mentioning that there is nothing at all in the archives of the Amsterdam synagogue about Spinoza after the excommunication. The secretary treasurer duly recorded that he had been expelled and gave the statement that Jews should not talk to him, read him, or have anything to do with him. And this is the last time he appears in the records. The Portuguese Synagogue was extremely reluctant to allow anyone to explore their archives until 1961. It was usually said that this was because they were hiding their hatred of Spinoza. When the archives became available, it turned out that what they were hiding was the involvement of so many of their dignitaries in the Shabbatai Zevi movement.

A curious point is that at the end of 1665, Spinoza was asked by the secretary of the Royal Society of London, Henry Oldenburg, who was Spinoza's best friend outside of the country, whether it was true that the King of the Jews had arrived. Oldenburg asked him to find out what the Jews of Amsterdam thought about this. There is no evidence that Spinoza responded to the letter, even though Oldenburg told him that if the King of the Jews had arrived it would have world-shaking importance. Instead, Spinoza's patron, the Protestant chiliast Peter Serrarius, who was Spinoza's point of contact with the outside world, receiving and sending Spinoza's mail, assured Oldenburg that it was true. Serrarius later bombarded Oldenburg and Oldenburg's patron Sir Robert Boyle with news of what Shabbatai Zevi was doing.[18] When the conversion episode occurred, Serrarius and Oldenburg agreed that God works in mysterious ways. Serrarius gave up his business in Amsterdam and went to meet the new Messiah, but unfortunately died en route. Spinoza, for his part, never mentions the Sabbatian matter. In his bitterest criticisms of Jewish beliefs and practices, he never points out how crazy his former cohorts had become.

Spinoza died in 1677 and his major works were published later that year. Two of the first lengthy discussions of his work were by the French Protestant Jacques Basnage and J. G. Wachter, a German Lutheran Protestant.[19] They saw Spinoza's radical theories as emanating from Judaism. Wachter came to Amsterdam in the late seventeenth century to see an old friend, a former Jesuit who became a rabbi in Amsterdam, named Moses Germanus. From their discussions he came up with the view that Spinozism was not new, but rather the metaphysical core of Judaism going back to ancient times. He concluded that all Jewish theologians, if they were honest, would be Spinozists. He and Basnage both saw Spinoza as a Kabbalist who was working out the Kabbalistic metaphysics of the earlier Amsterdam thinker Abraham Cohen Herrera, whose book appeared in Latin in 1677 at the same time and from the same publisher as Spinoza's *Ethics*.

Basnage was writing the first *History of the Jews* since Josephus and was making a serious effort to get the facts straight. He reports that he went to rabbis to see what Spinoza's theory was all about. He apparently spoke with the very rabbi who had read the excommunication statement back in 1656, Isaac Aboab de Fonseca, who told him that Spinoza was just a Kabbalist plagiarizing from the Jewish thinkers of the time and stating his views in fancy Cartesian language so that he could look original. Given this clue, Basnage then traced Spinoza's view to Cohen Herrera and it is a point of historical curiosity that Herrera's masterpiece, *The Gate to Heaven,* was published in a Hebrew abridgment in 1655. There is a phrase from it that appears in Spinoza's *Ethics* and it is claimed that this appears nowhere else in philosophical literature.[20]

In the eighteenth century Spinoza became a poster boy for modern atheism, especially after Pierre Bayle's interpretation.[21] Jean Baptiste Stouppe denounced him in *The Religion of the Hollanders,* which was designed to show that they did not have any religion.[22] One of the most notorious anti-religious writings of the early eighteenth century was the pamphlet put together by Jan Vroesen and called *The Three Imposters: Moses, Jesus and Mohammed, or the Spirit of Monsieur Spinoza.*[23] This work contains the French translation of Spinoza's polemic against organized religion from the *Ethics.* It also often appears with a very saintly biography of Spinoza. These items appeared in hundreds of manuscripts of the time.[24] In the first complete history of the Jews, Ludvig Holberg discussed Spinoza as a misguided atheist.[25]

Some of the early interpretations by Basnage and others suggest that Spinoza's views were imbedded in Judaism. The research of Harry Wolfson and others on the Jewish sources of Spinoza's ideas show that he was deeply involved in late medieval Jewish theology and philosophy.[26] For our purposes here, the question is, Was there much Jewish interest in Spinoza after he was condemned? As we have seen, he had Jewish friends. Spinoza's ideas also turn up in later Jewish discussions in Western Europe. David Ruderman has found that a helpful member of the London Jewish community published translations of Spinoza in Hebrew for his fellow believers.[27] The leading Jewish theologian in London, David Levi, kept complaining that the Jews in London were studying Spinoza, Voltaire, and Hume.[28] A major controversy known as the Spinoza-Streit developed when F. H. Jacobi accused G. E. Lessing of being a Spinozist, and Moses Mendelssohn felt obliged to defend his friend.[29] Other German thinkers like Solomon Maimon examined Spinoza's ideas in contrast to those of Immanuel Kant.[30] As German Enlightenment thinkers discovered that Spinoza was not an atheist but a God-intoxicated man, they began to use his pantheistic metaphysics as a basis for their Romantic idealism.[31]

During the nineteenth century, one finds Jewish intellectual historians coming to grips with whether Spinoza was "one of us or one of them." He is a continuing problem and it is obvious that he was being read fairly widely. Protestant missionary statements about the beliefs of Jews in Eastern Europe are full of complaints about the baleful effects of Spinozism on Jews. In the struggle for emancipation and modernization, Spinoza obviously played a vital role in convincing people that they were not really obligated to keep all the rituals and laws. He provided a basis for rational consideration of the merits of various Jewish customs. He provided a way of seeing them as interesting as antiquity but irrelevant in the present.

One of the aspects of this that needs much study is how Spinoza affected Jews in Eastern Europe in an underground way. I have been told by various people who were schooled in Eastern Europe at the end of the nineteenth and the beginning of the twentieth century that they were liberated from their orthodox communities when someone passed around a page or two of Spinoza. Apparently, it circulated in Yiddish. Rabbi Aaron Nadler is now studying the Yiddish versions of Spinoza. People who had their religiosity cracked by Spinoza's rationalism went into Zionism, Marxism, or other forms of non-orthodox Judaism. Spinoza became a hero for many "enlightened" Jews. Each time there is an anniversary of his birth, circumcision, Bar Mitzvah, excommunication, or death, there is a petition by "liberated" Jews to have the excommunication ban rescinded. They petition the Portuguese Synagogue of Amsterdam and are always told that since the synagogue has no record of the case, they are in no position to overrule their seventeenth-century predecessors. The case has been taken to the Israeli Supreme Court, which insists that it has no jurisdiction.

In 1977, I was invited to two celebrations of the 300th anniversary of Spinoza's death: one at the conservative Jewish Theological Seminary in New York, and the other at the Reformed Hebrew Union College in Cincinnati. At the Jewish Theological Seminary, they listened to my paper, discussed it, and then one of the deans said there was no point in having the celebration, since Spinoza was not a Jew and had nothing to do with Judaism. At Hebrew Union College, Spinoza had found a happy home. They had one of the greatest collections of Spinoziana in their library. They were happy to have discussions about his merits and demerits and regarded him as if he were a founding member.

About the same time there was a celebration in Jerusalem in which I participated. A learned rabbi, after listening to a few of us, asked why we were bothering to study this figure. Spinoza was not a Jew, he didn't believe in the dietary laws, he didn't keep the Sabbath, he didn't believe in this and that, so why didn't we go on to something else? The rabbi got up and left.

As anyone who has visited Jerusalem knows, streets are named after all sorts of obscure figures. Spinoza is not one of them. At most he has half a block. In recent decades, there has been an effort, especially coming from the direction of the Labor Party, to make him an acceptable Israeli hero.[32] A Spinoza Institute has been established. It is next to the Reformed Hebrew Union Seminary near the King David Hotel, and now has the magnificent collection of Spinoziana that used to be in Cincinnati. There was an attempt to make the study of Spinoza one of the central projects of Israeli intellectual life. A ten-year program was announced, to end in the year 2000. I was listed as one of the speakers for the final episode. As a sign that Israel has not yet been captured by Spinoza, the last communication I received announcing the forthcoming meeting informed me that they no longer had funds to defray the cost of scholars' travels or lodging. The earlier meetings had been very lavish. This time they offered to supply me with a list of cheap hotels and restaurants and they assured me that there would be a guided tour of Jerusalem at their expense.

Notes

1. R. Popkin, "The Lost Tribes, the Caraites, and the English Millenarians," *Journal of Jewish Studies* 37, 1986, pp. 213–227; Yosef Kaplan, "'Karaites' in Early Eighteenth-Century Amsterdam" in David Katz and Jonathan Israel, eds., *Sceptics, Millenarians, and Jews* (Leiden: Brill, 1990), pp. 196–236; "Karaites," *Encyclopaedia Judaica* (Jerusalem: Macmillan, 1971), pp. 761–785.
2. See Gershom Scholem, *Sabbatai Sevi: the Mystical Messiah, 1626–1676* (Princeton: Princeton University Press, 1971).
3. [Johann Friedrich Corvinus,] *Anabaptisticum et enthusiasticum pantheon und geistliches Rüsthaus wider die alten Quacker . . .* (Kothen: 1702).
4. Jean-Baptiste Rocoles, *Les Imposteurs insignes* (Amsterdam, 1683); Jacques Basnage, *Histoire des Juifs* (The Hague: Scheurleer, 1716), vol. 9, part 2, pp. 779–789; Ludvig Holberg, *Jüdische Geschichte von der Erschaffung der Welt bis auf gegenwärtige Zeiten* (Altona: Korte, 1747), vol. 2, pp. 643–47. On Rocoles, see Richard Popkin, "Christian Interest and Concerns about Sabbatai Zevi" in M. Goldish and R. Popkin, eds., *Jewish Messianism in the Early Modern World* (Dordrecht: Kluwer, 2001). Gerald Cerny's *Theology Politics and Letters at the Crossroads of European Civilization: Jacques Basnage and the Baylean Huguenot Refugees in the Dutch Republic* (Dordrecht: Nijhoff, 1987)

does not mention Basnage's comments on either Shabbatai or Spinoza. On Holberg, see J. C. Laursen, "Ludvig Holberg's History of the Jews" in H. Méchoulan, R. Popkin, G. Ricuperati, and L. Simonutti, eds., *La formazione storica dell'alterità: Studi di storia della toleranza* (Florence: Olschki, 2001).

5. See Henri Méchoulan, "Au dossier du sabbataisme: une relatione italienne du Xviième siècle" in Katz and Israel, eds., *Sceptics, Millenarians, and Jews,* pp. 185–195.
6. R. Popkin, "Two Unused Sources about Sabbatai Zevi and his Effect on European Communities," *Dutch-Jewish History* 2, 1989, pp. 67–74; R. Popkin, "Three English Tellings of the Sabbatai Zevi Story," *Jewish History,* vol. 8, 1994, pp. 43–54. See also R. Popkin, *The Third Force in Seventeenth-Century Thought* (Leiden: Brill, 1992), p. 363; R. Popkin, "Forward: The Leiden Seminar," in S. Berti, F. Charles-Daubert, and R. Popkin, eds., *Heterodoxy, Spinozism, and Free Thought in Early Eighteenth-Century Europe: Studies on the Traité des trois imposteurs* (Dordrecht: Kluwer, 1996), p. xiv.
7. See R. Popkin, *The Third Force,* pp. 362–3. Other early sources can be found in the bibliography in Scholem, *Sabbatai Sevi.*
8. Isaac Bashevis Singer, *Satan in Goray* (New York: Noonday, 1955).
9. Reported in *The Jerusalem Report.*
10. R. Popkin, *Isaac de la Peyrère (1596–1676): His Life, Work and Influence* (Leiden: Brill, 1987), pp. 9, 87, 179, 183.
11. Elesheva Carlebach, "The Last Deception: Failed Messiahs and Jewish Conversion in German Lands" in Popkin and Goldish, eds., *Jewish Messianism;* Michael Heyd, "The 'Jewish Quaker': Christian Perceptions of Sabbatai Zevi as an Enthusiast."
12. Gershom Scholem, *Major Trends in Jewish Mysticism* (New York: Schocken, 1946), p. 304.
13. Yeshayahu Leibowitz, *Judaism, Human Values, and the Jewish State* (Cambridge: Harvard University Press, 1992).
14. See Asa Kasher and Shlomo Biderman, "Why was Spinoza Excommunicated?" in Katz and Israel, eds., *Sceptics, Millenarians, and Jews,* pp. 98–141.
15. See the documents in I. S. Révah, *Spinoza y Juan de Prado* (Paris: Mouton, 1959).
16. See Michael Petry, "Spinoza and the Military," *Studia Spinozana,* vol. 1, 1985, pp. 359–369; on Morelli, see R. Popkin, "Serendipity at the Clark," *The Clark Newsletter* no. 16, 1986, pp. 4–7 and R. Popkin, "The First Published Reaction to Spinoza's Tractatus: Col. J. B. Stouppe, the Condé Circle, and the Rev. Jean Lebrun" in P. Cristofolini, ed., *The Spinozistic Heresy: The Debate on the Tractatus Theologico-Politicus, 1670–1677* (Amsterdam: APA-Holland University Press, 1995), pp. 6–12.
17. Isaac Orobio de Castro, *Certamen Philosophicum* (Amsterdam, 1684); reprinted in Lenglet Dufresnoy, ed., *Réfutation des Erreurs de Benoit de Spinosa* (Brussels: 1731). On Orobio, see Yosef Kaplan, *From Christianity to Judaism: The Life and Work of Orobio de Castro* (Oxford: Oxford University Press, 1989).

18. See R. Popkin, *The Third Force,* p 273.
19. Basnage, *Histoire des Juifs;* Johann Georg Wachter, *Der Spinozismus in Judenthums* (Amsterdam: 1699). See R. Popkin, "Spinoza, Neoplatonic Kabbalist?" in L. Goodman, ed., *Neoplatonism and Jewish Thought* (Albany: SUNY Press, 1992), pp. 343–362.
20. Harry Austryn Wolfson, *The Philosophy of Spinoza* (New York: Schocken, 1934, 1969), pp. i.17, 314–315, 366–77, etc.
21. Pierre Bayle, "Spinoza" in Richard Popkin, ed. and tr., *Pierre Bayle: Historical and Critical Dictionary* (Indianapolis: Hackett, 1991), pp. 288–338.
22. See R. Popkin, "The First Published Reaction to Spinoza's Tractatus" in P. Cristofolini, ed., *The Spinozistic Heresy.*
23. See Berti, Charles-Daubert, and Popkin, eds., *Heterodoxy, Spinozism, and Free Thought.*
24. See also J. J. Schudt, *Jüdische Merckwürdigkeiten* (Frankfurt, 1714). More on the reception of Spinoza can be found in W. van Bunge and W. Klever, eds., *Disguised and Overt Spinozism around 1700* (Leiden: Brill, 1996); H. Delf, J. H. Schoeps, and M. Walther, eds., *Spinoza in der europäischen Geistesgeschichte* (Berlin: Hentrich, 1994); Rüdiger Otto, *Studien zur Spinozarezeption in Deutschland im 18. Jahrhundert* (Frankfurt: Lang, 1994); J. C. Laursen, "Spinoza in Denmark and the Fall of Struensee," *Journal of the History of Ideas* 61, 2000, 189–202.
25. See J. C. Laursen, "Ludvig Holberg's History of the Jews."
26. Wolfson, *The Philosophy of Spinoza.*
27. David Ruderman, *Jewish Enlightenment in an English Key* (Princeton: Princeton University Press, 2000).
28. R. Popkin, "David Levi, Anglo-Jewish Theologian," *Jewish Quarterly Review* vol. 87, 1996, pp. 79–101.
29. Alexander Altmann, *Moses Mendelssohn: A Biographical Study* (Tuscaloosa: University of Alabama Press, 1973), pp. 638–711.
30. See Achim Engstler, *Untersuchungen zum Idealismus Salomon Maimons* (Stuttgart: Fromann-Holzboog, 1990).
31. See, e.g., Hermann Timm, *Gott und die Freiheit, I, Die Spinozarenaissance* (Frankfurt: Klostermann, 1974).
32. See Yirmiahu Yovel, *Spinoza and Other Heretics* (Princeton: Princeton University Press, 1989) and R. Popkin, "Notes from Underground: The Marranos of Amsterdam," *The New Republic,* May 21, 1990, pp. 35–40.

Part Four

A Plague on Many Houses

Joining Hands Against Enthusiasm

Introduction

John Christian Laursen

Gabriel Naudé's *Apology for Great Men Suspected of Magic* continued to play a part in debates throughout the eighteenth century. A new German translation in 1787 signaled its purpose of intervention in the campaign against "enthusiasm" by including that term in its title: *On Belief in Magic and other Enthusiasms.*[1] By this time, many authors no longer treated heresy and heretics as much of a threat qua heresy. Rather, they were the targets of a wide-reaching campaign against enthusiasm, which has been well-labeled as "the anti-self of enlightenment."[2] Radical Bayleans, deists, and freethinkers at one end of the spectrum and both the latitudinarian and the conservative orthodox at the other end could see a threat in what they termed the "enthusiasm" and "fanaticism" of the prophetic, the magicians, the inspired.

The Pietistic enthusiasm of Gottfried Arnold and others also prompted Johann Lorenz Mosheim to issue two volumes of an anti-Pietist *Attempt at a Nonpartisan and Thorough History of Heretics* (1746–1748). It re-hereticized many of the figures that Arnold had labored to legitimize and its anti-Platonism influenced Brucker and Gibbon. Translated into English in 1764 by A. Maclaine, this work was frequently reprinted in England and America as part of the campaign against enthusiasm and superstition.[3] Johann Christoph Adelung's contribution to the campaign against the enthusiasms of alchemy, witchcraft, and the occult was entitled *A History of Human Foolishness* (1785).[4]

Some "enthusiasts" were celebrated public intellectuals. The famous poet and physiognomist J. C. Lavater, as an example, offered his friendship to the Jewish philosopher Moses Mendelssohn in a transparent attempt to convert him, tried to enlist the philosopher Immanuel Kant in support of faith and prayer, and influenced Goethe.[5] But Mendelssohn and Kant rebuffed him and Heinrich Corrodi defended the "enlightened" reasonableness and rational theology of Steinbart and his mentor, Semler, against the more enthusiastic Lavater. Corrodi's contribution to the campaign against enthusiasm, as Simone Zurbuchen's chapter reveals, focused on chiliasm, or the near expectation of Christ's reign on earth. His pessimistic conclusion was that most people will never be reasonable, and thus that fanaticism will be an ineradicable part of history forever.

By the latter half of the eighteenth century, the history of heresy was a widespread scholarly enterprise. Certainly the most thorough was Christian Wilhelm Franz Walch's *Attempt at a Complete History of Heresies, Divisions, and Religious Disputes up until the Times of the Reformation,* issued in ten volumes between 1762 and 1782.[6] Carl Friedrich Bahrdt typically turned Gottfried Arnold's title into a humorous spoof of his enemies in *Church and Heresy Almanac* of 1781–1787.[7]

Enthusiasm, fanaticism, and superstition became key words in the histories of heresy put out by writers such as Hume, Diderot, and Voltaire. They attacked religious orthodoxies, but rather than defending heretics against the bloodthirsty orthodox, they called down a plague upon both their houses. Heresy and orthodoxy were two sides of the same bad penny, and the best thing was to ridicule them both and hope they would fade away. John Pocock's chapter in this volume shows how Gibbon developed this strand of thought by showing that the history of heresy rests on the history of philosophy as a corrupting factor in religion. In this account, the early Christians were rightly persecuted because of their philosophically inspired intolerance.

In Pocock's account, Gibbon relied on the histories of heresy of Le Clerc, Mosheim, Beausobre, and Nathaniel Lardner in order to trace the corrupting influence of Egyptian philosophy and Platonism on Christianity, creating the categories of orthodoxy and heresy. To Gibbon's dismay, Joseph Priestley's *History of the Corruptions of Christianity* (1782) could rely on the same sort of history to show that the best answer to this corruption was enthusiastic millenarian apocalypticism. Gibbon's own preference, on the ground that it would make for stability and peace, was for a state church manned by a free-thinking episcopate.

Hume, Diderot, Voltaire, and the other radicals were often too facile and superficial in their dismissals of orthodoxy. An irony of history is that Gibbon grew to respect establishment figures such as Athanasius. Pocock tells us that therefore he was no Mercutio like the others. Specific contingencies of history—such as his Englishness—and of historiography—Athanasius makes a good story—pushed him to celebrate the orthodox. This means that rather than the militant unbelief of a Voltaire, Gibbon's history of heresy led him to an established but undemanding church as the best guarantee of civil liberty against the twin dangers of clericalism and fanaticism that both authors recognized.

Notes

1. *Über den Zauberglauben und andere Schwärmerei; oder Vertheidigung berühmter Männer, die von ihren Zeitgenossen für Zauberer gehalten worden* (Leipzig, 1787). On the late-eighteenth-century campaign against enthusiasm in Germany, see Anthony La Vopa, "The Philosopher and the *Schwärmer:* On the Career of a German Epithet from Luther to Kant," *Huntington Library Quarterly* vol. 60, 1998, pp. 85–115.
2. J. G. A. Pocock, "Enthusiasm: the Anti-Self of Enlightenment," *Huntington Library Quarterly* vol. 60, 1998, pp. 7–28.
3. Martin Mulsow, "Einleitung" to Johann Lorenz Mosheim, *Versuch einer unparteiischen und gründlichen Ketzergeschichte* (orig. 1746), ed. Mulsow, vol. 1 (Hildesheim: Georg Olms, 1998), p. 7.
4. Johan Christoph Adelung, *Geschichte des menschlichen Narrheit* (1785).
5. See Alexander Altmann, *Moses Mendelssohn: A Biographical Study* (Tuscaloosa: University of Alabama Press, 1973); Arnulf Zweig, ed., *Kant: Philosophical Correspondence* (Chicago: University of Chicago Press, 1967), pp. 14, 79–82.
6. *Entwurf einer vollständiger Geschichte der Ketzereien, Spaltungen und Religionsstreitigkeiten, bis auf die Zeiten der Reformation* (Leipzig).
7. Carl Friedrich Bahrdt, *Kirchen- und KetzerAlmanach* (Zullichau: Frommann, 1781; Berlin: Vieweg, 1787).

10

Heinrich Corrodi's *Critical History of Chiliasm* [1781-1783]

Simone Zurbuchen

Introduction

> We dreamt of nothing but enlightenment [. . .] and believed to have brightened the environment by the light of reason in such measure that enthusiasm (Schwärmerei) certainly would not show up again. However, as we see, from the other side of the horizon the night with all its ghosts already rises again. With horror one sees that the evil is so efficacious and penetrating that enthusiasm (Schwärmerei) always works, while reason only talks. Lord Shaftesbury was wrong to think that wit and humor were the strongest remedies against the progress of harmful superstition. . . . [1]

With this statement, Johann Georg Zimmermann expressed the anxieties of the late German Enlightenment that saw itself confronted by a new wave of enthusiasts. By that time the terms "*Schwärmerei,*" enthusiasm, and fanaticism, often conflated, were no longer restricted to denoting religious enthusiasts such as the mystics and Pietists who were of great concern to Protestant orthodoxy in the early Enlightenment. The terms were also being applied to a wider range of irrational phenomena. Enthusiasm can be found in religion, politics, philosophy, and the sciences.[2] Because of the Enlightenment's endeavor to combat emerging irrationalisms, "*Schwärmerei*" has been called the Enlightenment's battle cry (*Kampfbegriff*).[3]

In his study *Melancholy and Enlightenment,* Schings demonstrated how the rationalist critique of enthusiasm was linked both with anthropological and medical studies of the same phenomenon in the seventeenth and early eighteenth centuries and with the orthodox critique of the "heresy" of Pietism in the early Enlightenment.[4] In the late Enlightenment, literature and anthropology were the disciplines that took the lead in analyzing and criticizing enthusiasm. As anthropologists took the lives of enthusiasts as the object of their studies, they were especially interested in the experiences of radical Pietists and their predecessors. They saw the Pietists as paradigm cases of religious fanaticism. For this reason Schings concludes that histories of enthusiasts, such as J. Ch. Adelung's *History of Human Folly*[5] or Ch. F. Duttenhofer's *History of Religious Enthusiasms in the Christian Church,*[6] "replaced" the histories of true believers that the Pietists had earlier composed to advocate their cause against the established Churches. The best known among the Pietist histories is Gottfried Arnold's *Impartial History of Church and Heresy.*[7] Since Arnold tried to "revert" the concept of heresy as it was applied by the orthodox, his work provides the key to understanding the transformation of the notion of heresy during the Enlightenment.

Like other histories of enthusiasm, Heinrich Corrodi's *Critical History of Chiliasm* of 1781–83[8] must be evaluated in the context of the growing interest in anthropology. In his preface, Corrodi considers the study of man to be the great topic of the times, the topic to which all fields of research should contribute. In his view, important contributions to the knowledge of man's soul were expected to emerge from the history of the illnesses of the soul that are classified as "fanaticism": "A history in which the aberrations of human reason and the disorders and extravagancies of its imagination are presented in a faithful picture, in which their sources and causes are revealed, should be of great interest to the lover of the study of man."[9]

Like many of his contemporaries, Corrodi does not distinguish sharply among fanaticism, enthusiasm, and "*Schwärmerei.*" He uses the terms interchangeably, although he does acknowledge that there exist different kinds of fanaticism or enthusiasm. Besides religious enthusiasm, he views romantic and alchemist enthusiasm, akin to the imagination of ghosts and devils, as variations of the same illness. However, he restricts his own investigation to religious fanaticism. Because a complete history of religious fanaticism alone would still be too great a task, Corrodi concentrates on just one branch of religious fanaticism, namely chiliasm. Corrodi describes chiliasm as "a system of fanatical expectations of near pleasures and lusts that originate in equally enthusiastic ideas of the other objects of religion" (vol. 1, p. vi).

However, there is a second and more important reason why Corrodi is interested in the history of chiliasm. He remarks that the illnesses of the soul are not only of great interest to the philosopher, but also to the theologian. Whereas the philosopher discovers in the history of religious fanaticism the extravagancies of imagination, the theologian takes it as "the history of sensuous Christianity, or of the religion of those Christians who always remain children, and never wish to grow in knowledge, who always feel and experience, and never wish to investigate and think."[10] Based on a distinction between reasonable and "sensuous" religion, Corrodi attempts to uncover the sources of chiliasm. He concludes that the origins "of the grossest and most miserable lies, dreams, and fancies" go back to ancient times (vol. 1, p. 2), namely to the Jewish religion that was later conflated with "true" religion by the Jewish Christians of the first centuries. The first two parts of the *History of Chiliasm* consist of an extended proof of these insights. Here Corrodi is mainly concerned with questions of scriptural interpretation. The first part is devoted to Jewish expectations of a worldly empire to be erected by the Messiah; the second part deals with the continuity of these doctrines among the early Christians.

In the third part of the work, Corrodi deals with the history of chiliasm since the Reformation. He writes that in the fifth century the doctrine of the millennium was lost in "unknown darkness." As Christians were no longer subjugated to the Romans, they were no longer in need of this consolation. However, immediately after the Reformation an infinite number of prophets emerged within the Lutheran and the Reformed Churches. Their visions anticipated the near downfall of the Roman Antichrist and proclaimed the victory of their own Church (vol. 3, pp. 1–3). In the seventeenth and eighteenth centuries, expectations of the coming destruction of the papacy were superseded by prophecies of a different kind. The enthusiasts of the seventeenth century, who split off from the established churches, pretended to apprehend the "invisible" church that would replace the "visible" church after the "great separation" (vol. 3, pp. 8–9). Enthusiasts of a pietistic cast are the main targets of Corrodi's criticism. As their example demonstrates, the first of the main sources of chiliastic opinions consists in erroneous interpretation of the Holy Scriptures. The second source of chiliasm is to be found in mystic or fanatic theology. In Corrodi's view, mystic theology points a new way to wisdom and felicity. It is "an arcanum to become wise and happy that is in possession of only a few, and that makes those who possess it the wisest and happiest mortals, who are destined to become the rulers of the rest of men and the mediators between them and God."[11] The main doctrine of the mystics is contained in their idea of "illumination." Illumination by God is understood as an "immediate revelation of God in the soul and a union of the soul with the divinity" (vol. 3, p. 11). Corrodi is aware of the fact that mystic theology goes back to ancient times and has continued throughout the centuries. However, it was mainly in the seventeenth century that mystic theology was used to justify apocalyptic visions. The *History of Chiliasm* thus seems to fulfil the task that motivated Corrodi's investigation. He expected it "to reveal the sources and causes of fanaticism" that he in fact discovers in the interpretation of the Holy Scriptures "according to the way of the Jews" (*judenzende Schriftauslegung*) on the one hand, and in the conception of "immediate revelation" on the other hand (vol. 3, pp. 10–16).

The main concern of this chapter is to more closely examine the two basic assumptions on which Corrodi's history of chiliasm is built. As we have seen, the first assumption can be described in terms of the opposition between "reasonable" religion and "sensuous" religion. The second one, closely linked with Corrodi's idea of "fanatic" theology, is manifest in the distinction between "true" and "deceitful" revelation. As I will try to demonstrate in the next section, the important question is how Corrodi wished to relate the two basic assumptions to each other. Since he identifies chiliasm as a kind of "sensuous" religion that originates in "deceitful" revelation, it is obvious that "reasonable" religion and "true" revelation also belong together in some way. The exact nature of the relation between reason and revelation was at the core of the debates provoked by Lessing's publication of *Fragments of an Unknown Person*. Lessing had attempted to force the neologians—the leading rational theologians of the times—to reply to the provocation of "deism." The last section of this chapter is devoted to Corrodi's notion of "mediate," "natural" revelation that proves to be the key to his understanding of the history of religion. Based on these insights, I will return to the question of how the *History of Chiliasm* is related to histories of heresy.

Reason and Revelation

The main issue in religious thought in the German Enlightenment was to elucidate the relations between reason and revelation, and between philosophy and theology.[12] This problem was a legacy of Christian Wolff's philosophy. He taught that a purely rational theology was possible, with proofs of the existence of God, the immortality of the soul, the creation of the world, and so forth. At the same time he was convinced that the truths demonstrated by reason were in harmony with revealed theology: there could be no conflict between reason and revelation. Beginning with S. J. Baumgarten, Wolffian theologians attempted to resolve the tensions between reason and revelation. Of special interest here is the position of the "neologians," above all Johann Salomo Semler, professor of theology at the University of Halle, whose lectures introduced Corrodi to the most intriguing theological questions of the times.[13]

As early as his first anonymous writings, Corrodi proved to be a faithful disciple of Semler. Provoked by his influential compatriot Johann Caspar Lavater, he defended the idea of reasonable religion against the accusation of deism. Lavater—who was suspected of promoting enthusiasm because of his interest in supernatural religious experience—used his synodal speech of 1779[14] to launch an attack against some of the leading neologians of the times, whom he accused of reducing Christianity to deism. It must be recalled that only a few years earlier, Lessing had published the *Fragments of an Unknown Person* that he had selected from Hermann Samuel Reimarus's *Apology or Defense of the Rational Worshippers of God (Apologie oder Schutzschrift für die vernünftigen Verehrer Gottes).* It is most likely that Lessing edited this defense of deism with a view to forcing a debate on the questions of reason and revelation among neologians.[15] Lavater intensified the pressure on the rationalist theologians by his attempt to demonstrate that Gotthilf Samuel Steinbart's *System of Pure Philosophy or Christianity's Doctrine of Felicity* (1778)[16] consisted of "refined deism—with some Christian coloring."[17]

In 1779 Semler had already published a *Response to the Fragments of an Unknown Person.*[18] He attempted to refute Reimarus's claim that the doctrine of Jesus had to be separated from the apostolic doctrine of Christ. He denied that the apostles' message of Jesus's redemption was an invention of his followers and insisted that the apostles' doctrine of Christ belonged to the fundamental contents of Christianity. When Lavater publicly accused Steinbart of being a deist, Semler felt compelled to clarify his own position. Since he did not wish to defend Steinbart's *System* in his own name, he left the refutation of Lavater to his disciple Corrodi. Imitating Lessing's publishing strategy, he presented Corrodi's opinions about Steinbart's *System of Pure Philosophy* as those of an "unknown person," to which he added his own annotations.[19]

This publication is of special interest here because it reveals an important difference between Semler's and Corrodi's understandings of Christianity. Steinbart's own account of the Christian doctrine of felicity can best be understood as a response to one of the questions that was raised by Johann Joachim Spalding's *The Destination of Man,* a bestseller of eighteenth-century thought.[20] This work is important because it initiated an anthropological shift in the domain of theology that was significant for the movement of neology as a whole.[21] In an appendix to the third edition of 1749,[22] Spalding replied to those of his critics who concluded that his anthropological account of religion could easily be used to demonstrate the self-sufficiency of natural religion. The critics maintained that it was possible to refer to the excellence of natural religion that he had advocated as a basis for denying the truth of Christianity. Since nature introduces us to the concepts of God and of a future state of felicity and virtue, it seems no longer necessary to refer to revelation. In his response, Spalding claims that there would be no natural religion in the world if there were no revealed religion. Progress in the knowledge of man's nature is the most convincing proof that reason alone, without any assistance, will never be capable of transcending sensuality and attaining the truths of religion. Thus, the first "instruction" of man must have been divine revelation. The fact that revelation is antecedent to reason is best suited to explain how, after its general decay, natural religion was restored by divine revelation. For the same reason natural religion is best demonstrated and taught at places where the "light of the Gospel" has already illuminated people's minds. However, even if we suppose that some people are able to access the truths of natural religion without the assistance of revelation, we still have to conclude that natural religion could never be made "general" and "dominating." For the majority of people depend on the teaching of revelation.

The validity of Spalding's argument obviously depends on the assumption that there is no contradiction between natural and revealed religion. However, he does not confront the question of how to decide among competing interpretations of the Holy Scriptures. This task was left to Steinbart, who saw the "doctrine of felicity" as a means to concluding theological disputes about the fundamental truths of Christianity. In fact, Steinbart pretended to be the first to present and explain Christ's doctrine of felicity. In this way he claimed to bring clerical disputes about dogmas to a close. According to Steinbart, Greek and Roman philosophers were concerned with the question of human felicity. However, when Jesus's teaching was acknowledged as divine revelation about the way to felicity, the study of man was abandoned. Instead, Christian scholars tried to transform the "popular" doctrines of Jesus into "profound erudition" (*tiefsinnige Gelehrsamkeit*). In this way they perverted Jesus's doctrine, proclaimed with "sublime simplicity," into a "transcendent science," conflated with all kinds of philosophical theories. Steinbart sees the Reformation as the first serious effort to eliminate arbitrary theories from Christianity.[23] However, the reformers left their work incomplete, because they excluded the first five to six centuries of Church history from their critical examination. With the publication of his work, Steinbart hoped to complete the task of the reformers. The "doctrine of felicity" aimed at liberating thinking people from the tutelage of "arbitrary hypotheses" by showing them the way out of the "labyrinth of Church systems" and bringing them back to the truth (p. 4).

Since Steinbart derived the Christian "doctrine of felicity" from knowledge of man's destination, Lavater concluded that he in fact reduced Christianity to "refined deism." Lavater reproached Steinbart for missing the specifically Christian aspect of the doctrine of felicity, because he founds it on reason alone instead of relating it to the "document" ("*Urkunde*"), that is, the Holy Scriptures. According to Lavater, Christianity is singular among the religions of the world, and is thus only accessible through its "document." The fact that Steinbart is unable or unwilling to derive the doctrine of felicity from the "document" is the best proof that he is in fact a deist.[24]

In his reply, Corrodi argues that Lavater's accusations are based on a misunderstanding of the true nature of the Christian religion. According to Corrodi, Christianity is capable of an increase in perfection in different times and according to the mental capacities of different classes of people. The perfection (*Vervollkommnung*) of religion consists of its progress from sensuality to spiritual excellence (*Vollkommenheit*) (pp. 52–54). Religion reaches the height (*Gipfel*) of perfection when reason is fully developed and the motivation to fulfil moral duties is "pure" and "noble." The rational Christian does not obey moral laws because they are commanded by God, but rather the laws themselves are the motivation of his moral conduct (pp. 58–59). Religion is the more perfect the more it exercises and occupies the higher faculties of the soul. While "thinking" Christians do not need anything except reason, men "in the state of childhood" need assistance: they depend on the testimony of revelation. However, it was God's purpose that knowledge depending on testimony eventually be replaced by knowledge based on reasonable thinking and inquiry (pp. 60–64).

In his defense of Steinbart, Corrodi refers to some of the arguments that Semler had developed. Semler is usually considered the most important representative of the historical-critical theology of the Enlightenment. Against the tendency of Orthodoxy to make primitive Christianity the norm of true religion, Semler called for a historical-critical interpretation of primitive Christianity.[25] He argued that the Holy Scriptures contain eternal and fundamental truths on the one hand, and secondary and mediating truths on the other hand. Whereas the former are the testimony of revelation, the latter have to be explained as a result of their accommodation to the perceptions and opinions of the times. According to Semler this distinction is reflected in the "double way of teaching" (*Lehrart*) that we find in the Gospels. The first teaching method, mirroring the character of the times and places, is "sensuous," full of pictures and expressions of the Jewish realm. The second way of teaching represents the pure content of Jesus's spiritual doctrine.[26] The distinction between the two ways of teaching is not limited to the Gospels. Semler takes it to be a general category that he applies to the history of religion. Thus the difference between the "essence" of Christianity and its accommodation to the perceptions and opinions of the times is taken up in his distinction between "religion" and "theology." The former refers to the internal, spiritual apprehension of religious truth that is manifest in the moral life of an individual. The latter embodies the doctrines of the theologians that vary according to the state of opinions and research. The inevitable tensions between "religion" and "theology" provide the principle of historical change.[27] According to Semler, Christianity is capable of the perfection that is brought about by the conflict between "religion" and "theology."[28]

Semler's interpretation of perfectibility may have been the reason why he added his own annotations to Steinbart when he published Corrodi's reply. As we have seen, the latter identifies reasonable religion and true Christianity when he states that fully developed reason corresponds to the highest perfection of religion. It seems safe to conclude that he accepted Steinbart's doctrine of felicity as a normative standard establishing the rational core of Jesus's teaching. In contrast, Semler is more reluctant about presenting a system of the fundamental truths of Christianity. In his view Steinbart's expectation of the resolution of theological disputes is premature. Based on the distinction between "religion" and "theology," he argues that theologians since the times of Clemens of Alexandria and Origen tried to fulfil the task of describing and elucidating Christ's doctrine of felicity. Theologians developed theories at all times and at all places. In this perspective Steinbart's "doctrine of felicity" is just another imperfect attempt to explain Jesus's doctrine, directed at those of his contemporaries who are able to think. In Semler's view the perfection of Christianity is an infinite process aiming at the general realization of the religion of love.[29]

The Distinction between "Natural" and "Supernatural" Revelation

Although Corrodi was convinced that reasonable religion made revelation unnecessary for individuals who are able to "think," he was less optimistic about its progress in history. It seemed improbable to him that reasonable religion will ever replace revelation in whole nations, because access to it is restricted to philosophers. While reasonable religion may be best suited to bringing theological disputes to a close, it is not capable of replacing revelation. As the majority of people depend on testimony, the critique of "false" or "deceitful" revelation remains an essential task of the rationalist theologian. The critique of "deceitful" revelation ultimately leads to the question of how the testimony of any revelation can be justified by reason. Corrodi answered this question in *On Revelation, Judaism, and Christianity for Researchers of Truth* (1785).[30]

Corrodi distinguished between three different meanings of "revelation." First, revelation can be mere fancy, something invented by enthusiasts and impostors to deceive the people. This is the opinion of the "naturalists" or "deists." Second, revelation may be understood as an "extraordinary institution of providence in order to liberate a part of mankind for a certain time from errors and vices, without driving on any other than natural movements." This is the concept of natural, mediate revelation. Third, revelation can be explained as "illumination" brought about by forces "higher" than natural ones. This is the notion of immediate, supernatural revelation (pp. 5–7).

Corrodi's main concern is to criticize the third meaning of revelation. As he demonstrated in the third part of the *History of Chiliasm,* the supposition of an "immediate" revelation was one of the main causes of chiliasm. As a consequence, he wished to abandon the notion of "immediate" revelation and replace it with his own concept of "natural," "mediate" revelation. Having rejected the testimony of miracles, he concluded that we should conceive of revelation as an "exceptional event" (*ausserordentliche Begebenheit*) brought about by natural forces. In fact, it consists in the "enlightenment" (*Aufklärung*) of one part of mankind, brought about by men (p. 42). The "exceptional" character of the event has two different aspects. To those who are unable to access truth by reason, the source of revelation remains unknown. They take it to be God's "immediate" instruction, a testimony of eternal truths that man must believe in order to obtain God's grace. Since common people are unable to discover truth by reason, they must believe (pp. 32–35). On the other hand, those who are capable of reasoning perceive revelation as an "exceptional" event because they recognize it as an institution of divine providence, that is, a means by which God realizes his purposes. The purpose of revelation is to instruct "sensuous" men, who are unable to think, about their true destination. This explains why it consists of sensuous instruction about the most sublime truths concerning God's nature, His relation to man, etc. (pp. 30–31). Based on the idea of the perfectibility of Christianity, Corrodi links "sensuous" teaching with the childhood of mankind. However, despite the progress of reason brought about by God's providence, men still depend on memory of the "exceptional" event. They remain attached to "vivid memories" of the "helpful deceits" of childhood (pp. 35–37). As the notion of "helpful deceit" reveals, Corrodi's concept of "natural," "mediate" revelation ultimately depends on knowledge of reasonable religion. Thus he concludes that in the last instance reason has to approve of the "credibility" of testimony and confirm the truths that are related to it.

According to this understanding of revelation, the critical evaluation of revealed religion belongs to the main concerns of the rational theologian. Thanks to his knowledge of reasonable religion he is able to distinguish between the advantages and deficiencies of revelation. While the former are perceived as the effect of God's providence, the latter are due to the fact that revelation was brought about by natural forces. They must be explained as a result of the need to adapt them to the dominating perceptions and opinions of the times. This is exactly what Corrodi accomplished with his critical history of Judaism and Christianity in the first two parts of the *History of Chiliasm*.

Based on his interpretation of revelation, Corrodi could hardly escape suspicion of deism. Although he did not wish to question the testimony of revelation, he made it ultimately depend on the approval of reason. His understanding of natural revelation confirms that he does not uphold the distinction between "religion" and "theology," which for Semler provides the principle of Christianity's historical change. In Corrodi's view, Christianity reached perfection in the age of Enlightenment. Therefore, he accepts Steinbart's doctrine of felicity as the normative standard for establishing the rational core of Christianity. For him, the only reason not to question revelation as the deists do is the fact that enlightenment has not yet completed its task. In fact, it will never be able to do so, because the majority of people are unable to access reason and will always depend on revelation. While Semler considered the *perfection of Christianity* as an infinite process, Corrodi perceived *enlightenment* as a never-ending task. Therefore, he was ultimately unable to abandon the concept of heresy.

The History of Chiliasm and Histories of Heresy

As I noted at the beginning of this chapter, Schings concluded from his study of seventeenth- and eighteenth-century theories of melancholy that the same set of terms could be used either to defend or to denounce different forms of inward and spiritual piety. From this perspective, histories of enthusiasm of the late eighteenth century replaced the Pietists' histories of "true" believers of the early Enlightenment. Both types of histories were linked with the concept of "heresy" in a complicated way. Both the Pietists and the philosophers and theologians of the late Enlightenment were opposed to the established Churches and their claim to espouse the true notions of Christianity, so they denounced orthodoxy for "inventing" heresies (*Ketzermacherei*) in order to suppress them. In their view, the concept of heresy was closely linked with intolerance and persecution, so they wished to abolish it. However, as they themselves claimed to know what "true" religion consists of, they could not dispense with the distinction between "true" and "false" religion. They had to retain the idea of heresy, although they abolished the name.

This is clear in Gottfried Arnold's *Impartial History of Church and Heresy* of 1699/1700, well known for its influence on eighteenth-century church history.[31] Arnold introduced new standards into church history, both with regard to critical evaluation of the sources (*Quellenkritik*) and with regard to his methodological program, indicated by the term "impartiality."

Arnold did not wish to analyze church history from the perspective of any "religious party" (*Religionspartei*) or confession, but rather tried to adopt a standpoint above and beyond the established Churches. However, this did not mean that he applied "objective" criteria in the modern sense of the word. Instead, he used his own idea of "true" religion to evaluate the history of Christianity. To the established Churches marked by clerical hierarchy, fixed dogmas, and regulated worship he opposed the primitive Church of the early centuries that consisted of sincere piety of the heart and a general priesthood of all believers. Based on this model of true Christianity, Arnold perceived the history of Christianity as a process of decay that was further advanced by the union of church and state brought about by Constantine. Although he recognized the attempt to renew faith at the beginning of the Reformation, the process of decay continued up to his own time. Since Arnold identified true Christianity with the invisible church of the spirit, he denounced all established Churches and glorified those who lived the lives of saints, retired in mystical speculation. In this way he attempted to rehabilitate persons and groups that had been denounced as "heretics" by the established Churches. He recognized Baptists, mystics, and spiritualists as unjustly persecuted witnesses of truth.

In Arnold's view, the meaning of "true" religion was itself perverted. As a result, the notion of heresy was abused by those who claimed to be in possession of true religion. Arnold thus speaks of heresy to denote the doctrine of those who oppose orthodoxy. However, despite his efforts to transcend partiality, he retains the concept of heresy, albeit under a different name. Thus he denounces the established Churches as "sects." A generation later, Arnold's *Impartial History* was perceived as an exaggerated reaction to orthodoxy. In his *Essay on an Impartial and Thorough History of Heresy,* Johann Lorenz Mosheim observed that Arnold was less concerned with truth than with attacking those who oppressed the so-called heretics. In his view, Arnold made the orthodox into heretics, and the heretics into orthodox.[32] Against the radicalism of the Pietists in the early Enlightenment, Mosheim attempted to defend a more moderate position.[33] Adopting "impartiality" and "thoroughness" as guiding principles of his history of heresy, he used the terms "church" and "sect" as neutral descriptions.[34]

Compared with Mosheim's shift to neutrality, Corrodi's *History of Chiliasm* announces a return to a value-laden concept of heresy. In fact, it looks at first glance like a "reversion" of Arnold's *Impartial History.* In the third part of the work, devoted to the history of chiliasm since the Reformation, Corrodi deals with many of the persons and groups whom Arnold had described as "true" Christians (i.e., as heretics in the first sense of the term). He then pursued the influence of their doctrines up to his own time. It is thus of no surprise that the list of persons who figure in this part of the work overlaps to a large extent the list of "radical" Pietists and their predecessors that we find in contemporary manuals such as Brecht's *History of Pietism*. However, these parallels are also misleading, because they suggest that Corrodi simply argued along the lines that were conceived in the early Enlightenment. But as I have demonstrated, Corrodi did not adopt preconceived categories to apply them to the history of religion, but rather established his own criteria. The combined application of the two criteria of "reasonable" religion and "true" revelation leads him to his own, original concept of history.

Influenced by the teachings and writings of neologians such as Steinbart and Semler, Corrodi conceived of the history of religion in terms of perfection. Therefore, he rejected the attempt to uncover "true" religion in the Christian Church of the early centuries. Unlike Arnold and many of the deists, he did not view the history of religion as a process of decay. Based on the theory of "accommodation" that he adopted from Semler, he conceived of the history of religion in terms of progress from sensuality to reason. As the *History of Chiliasm* reveals, this did not imply the idea of continuous progress from ancient times up to the eighteenth century. Rather, Corrodi considered the transition from sensuality to reason an achievement of the Enlightenment. With the application of reason as a criterion to evaluate religion, Corrodi transcends the notion of heresy (as opposed to orthodoxy) that was still present in Arnold's account of Church history. If reason sets the terms for resolving theological disputes, the distinction between orthodoxy and heresy will cease to exist.

As we have seen, however, Corrodi does not wish to discredit revelation in the manner of the deists. Based on the notion of divine providence, he attempts to save the notion of "true" revelation within the limits of reason. On that ground he is ultimately unable to dispense with the concept of heresy. This is best evidenced by his own observation that revelation necessarily contains "deficiencies." The most important deficiency consists in the fact that it always needs to be adapted to the opinions of the times, which in turn leads to opposition between reasonable and revealed religion. Even in enlightened times most people are unable to access religious truths by reason, so they still depend on revelation. This exposes them to the risk of being deceived by those who keep the remains of "sensuous" religion alive and who pretend to have them confirmed by "immediate" revelation. In the terms of the late eighteenth century, the "deceivers" are the enthusiasts or fanatics. For that reason it is the main task of the rational theologian to uncover the causes of fanaticism.

Notes

1. J. G. Zimmermann, *Über die Einsamkeit,* vol. 3 (Frankfurt and Leipzig: 1785), pp. 54–5

2. See Anthony J. La Vopa, "The Philosopher and the *Schwärmer:* On the Career of a German Epithet from Luther to Kant," *Huntington Library Quarterly* 60, 1998, pp. 85–115.
3. Norbert Hinske, "Die Aufklärung und die Schwärmer—Sinn und Funktionen einer Kampfidee," *Aufklärung,* 3, 1988, pp. 3–6.
4. Hans-Jürgen Schings, *Melancholie und Aufklärung. Melancholiker und ihre Kritiker in Erfahrungsseelenkunde und Literatur des 18. Jahrhunderts* (Stuttgart: 1977).
5. Johann Christoph Adelung, *Geschichte der menschlichen Narrheit, oder Lebensbeschreibungen berühmter Schwarzkünstler, Goldmacher, Teufelsbanner, Zeichen- und Liniendeuter, Schwärmer, Wahrsager, und anderer philosophischer Unholden,* 7 parts (Leipzig: 1785–89).
6. Christian Friedrich Duttenhofer, *Geschichte der Religionsschwärmereyen in der christlichen Kirche,* 4 vols. (Heilbronn a. Neckar: 1796–1802).
7. Gottfried Arnold, *Unpartheyische Kirchen- und Ketzerhistorie. Vom Anfang des Neuen Testaments Biss auff das Jahr Christi 1688 . . .* (Frankfurt/Main: 1699–1700).
8. I refer to the second edition of 1794: *Kritische Geschichte des Chiliasmus. Oder der Meynungen über das Tausendjährige Reich Christi,* 4 vols. (Zürich: 1794).
9. Corrodi, *Kritische Geschichte,* vol. 1, p. v.
10. Corrodi, *Kritische Geschichte,* vol. 3, p. v.
11. Corrodi, *Kritische Geschichte,* vol. 3, part 1, introduction, pp. 9–10.
12. See Karl Aner, *Die Theologie der Lessingzeit* (Halle: 1929). Walter Sparn, "Vernünftiges Christentum. Über die geschichtliche Aufgabe der theologischen Aufklärung im 18. Jahrhundert in Deutschland," in *Wissenschaften im Zeitalter der Aufklärung,* ed. Rudolf Vierhaus (Göttingen: 1985), pp. 18–57.
13. Corrodi was born in Zurich, the son of an embittered Pietist who led his family into great poverty. Due to physical weakness and the fragility of his voice, Corrodi faced serious problems in obtaining his ordination, because he seemed unable to serve as a pastor. It was only with the support of one of the influential theologians of Zurich that he eventually received ordination and the financial support to continue his studies first with Ernst Platner in Leipzig and later with Semler in Halle. In 1786 he was appointed professor of moral philosophy ("Sittenlehre") and natural law at the "Carolinum," the "high school" of Zurich. Between 1780 and 1793 he published a number of theological and philosophical writings, mainly concerned with the problem of the relation between reason and revelation.
14. *"Synodalrede":* Speech before the Church Synod.
15. For Lessing's position in the controversy, see Thomas Saine, *The Problem of Being Modern or the German Pursuit of Enlightenment from Leibniz to the French Revolution* (Detroit: Wayne State Press, 1997), pp. 213–279.
16. Gotthilf Samuel Steinbart, *System der reinen Philosophie oder Glückseligkeitslehre des Christenthums für die Bedürfnisse seiner aufgeklärten Landsleute und anderer die nach Weisheit fragen eingerichtet* (Züllichau: 1778).

17. This critique first appeared in the *Christliche Magazin* and was repeated in the synodal speech. Reprinted in *Herrn Caspar Lavaters und eines Ungenannten Urtheile über Hrn. C. R. Steinbarts System des reinen Christentums. Mit vielen Zusätzen von D. Joh. Sal. Semler* (Halle: 1780), p. 8. For an analysis of the controversy between Lavater and Corrodi, see Simone Zurbuchen, "Die Zürcher Popularphilosophie. Heinrich Corrodi und Leonhard Meister," in: *Alte Löcher—neue Blicke. Zürich im 18. Jahrhundert,* ed. Helmut Holzhey und Simone Zurbuchen (Zurich: 1997), pp. 329–343.
18. Johann Salomo Semler, *Beantwortung der Fragmente eines Ungenannten insbesondere zum Zweck Jesu und seiner Jünger* (Halle: 1779). See Gottfried Hornig, *Johann Salomo Semler. Studien zu Leben und Werk eines Hallenser Aufklärungstheologen* (Tübingen: 1996), pp. 139–43.
19. See *Hrn Caspar Lavaters und eines Ungenannten Urtheile.*
20. Johann Joachim Spalding, *Die Bestimmung des Menschen* (1748). Critical edition in *Aufklärung,* 11, 1999, pp. 69–95.
21. As Clemens Schwaiger has recently demonstrated, Spalding's *The destination of man* can be considered a synthesis of elements of Leibniz-Wolffian philosophy and basic ideas of Shaftesbury. See C. Schwaiger, "Zur Frage nach den Quellen von Spaldings Bestimmung des Menschen. Ein ungelöstes Rätsel der Aufklärungsforschung," *Aufklärung,* 11, 1999, pp. 7–19. For the influence of Spalding's work on eighteenth-century thought, see G. Alessandro, "Die Wiederkehr eines Leitworts. Die 'Bestimmung des Menschen' als theologische, anthropologische und geschichtsphilosophische Frage der deutschen Spätaufklärung," ibid., pp. 21–47.
22. Reproduced in *Aufklärung,* 11, 1999, pp. 89–92.
23. Steinbart, *System der reinen Philosophie,* pp. 2–3.
24. *Hrn Caspar Lavaters und eines Ungenannten Urtheile,* pp. 15–20.
25. See Hornig, *J. S. Semler,* pp. 279–81.
26. Wolfgang Schmittner, *Kritik und Apologetik in der Theologie J. S. Semlers* (München: 1963), pp. 40–6. See also Hornig, *J. S. Semler,* pp. 197–200.
27. *Hrn Caspar Lavaters und eines Ungenannten Urtheile,* pp. 135–141.
28. See Peter H. Reill, *The German Enlightenment and the Rise of Historicism* (Berkeley: 1975), pp. 165–172.
29. See Hornig, *J. S. Semler,* pp. 195–209.
30. [H. Corrodi], *Über Offenbarung, Judenthum, und Christenthum für Weisheitsforscher* (Berlin: 1785), pp. 56–62.
31. See Wolfgang Bienert, "Ketzer oder Wahrheitszeuge. Zum Ketzerbegriff Gottfried Arnolds," *Zeitschrift für Kirchengeschichte,* 88, 1977, pp. 230–46. Hans Schneider, "VII. Der radikale Pietismus im 17. Jahrhundert," in *Geschichte des Pietismus,* ed. Martin Brecht et al., vol. 1 (Göttingen: 1993), pp. 397–437.
32. Johann Lorenz Mosheim, *Versuch einer unparteiischen und gründlichen Ketzergeschichte,* vol. 1, ed. Martin Mulsow (Reprint Hildesheim: 1998), p. 13.
33. See Martin Mulsow, "Introduction" to Mosheim's *Versuch,* pp. 7–8.

34. Ulrich Johannes Schneider, "Zum Sektenproblem der Kirchengeschichte," in *Johann Lorenz Mosheim. Theologie im Spannungsfeld von Philosophie, Philologie und Geschichte 1693–1755* (Wiesbaden: 1997), pp. 147–97.

11

Gibbon and the History of Heresy

John G. A. Pocock

This chapter is a preparatory essay toward detailed studies some day to be written[1]: first, of the famous fifteenth and sixteenth chapters of the first volume of Edward Gibbon's *History of the Decline and Fall of the Roman Empire;* subsequently, of Gibbon's accounts of theological disputes and general councils of the Church in the Empire between the reigns of Constantine and Heraclius (when the advent of Islam altered the historical scene). Here I identify the principal historians of heresy of whom Gibbon seems to have made use, but detailed study of his interactions with any of them must be deferred to further work. It is of some importance to the study of Gibbon to consider who these historians were and why he chose to rely on them, but this information must be enclosed within the settings furnished by two questions: first, What was Gibbon's attitude toward the phenomena of orthodoxy and heresy?; second, What part did these play in the construction of his historical narrative?

In the first place, it seems important to emphasize that Gibbon's attitude on these matters was not what we ordinarily think of as the liberal one. He did not, that is to say, single out as his primary target the decision to define one set of beliefs as orthodox and all departing from them as heretical; and he did not assign a primary value to the liberty to define one's religious beliefs without the interference of authority. He lived, indeed, to have contemporaries for whom the liberty to define belief was virtually synonymous with belief itself and the liberty to seek God with the worship of God; but these were the Rational Dissenters and Unitarians active in England and the American colonies in the years of the Revolutionary War,[2] and among their intellectual leaders was Joseph Priestley, with whom Gibbon conducted a quarrel so bitter that he recommended Priestley to the attention of the civil magistrate. The separation of church and state—to employ a phrase I do not think he ever used—seemed to him likely to expose the state to the tyranny of the private conscience. This being so, heresy was no better than orthodoxy, and the purpose of toleration was not the freedom of religion, but a means of neutralizing it and hoping that it would extinguish itself. The most that can be said for his liberalism in this respect is that he would certainly have been horrified by a state that set out to repress and destroy all religion whatsoever; but this is probably a condition he did not envisage during his writing life. He thought of religion as so far a given in human affairs that a state that intervened in it was bound to espouse one religious position or another, thus weakening itself by involvement in that which could never do it good. By withholding its hands from religion, the state might hope to reduce it to privacy and insignificance[3]; tolerance was indeed a form of repression.

Gibbon did not think of heresy as something defined, rightly or wrongly, by orthodoxy, as at least some of his historians of heresy probably did; to him orthodoxy and heresy were phenomena of the same order, produced by the same set of workings of the human mind and by the same historical processes. To follow his history of these matters, we have to consult not only his preferred historians of heresy—though these form a defined and distinctive group—but his historians of religion and his historians of philosophy. In the former category we find David Hume's *Natural History of Religion,* which Gibbon may have read as early as 1760[4] and cites with related essays in the *Decline and Fall*[5]; it should be taken in conjunction with Hume's essay *Of Superstition and Enthusiasm* and the use made of both in his *History of England*. The *Natural History* is what it claims to be: a study of how the human mind produces the phenomena of religion and the order in which it would do so if unmodified by particular and contingent circumstances. Thus superstition displays the mind identifying the godhead with material objects; enthusiasm displays it identifying the godhead with ideas and images formed within itself. There are circumstances in which the latter destroys the former and then itself, leaving only the opinions formed by the mind as the result of material and social experience. Here Hume has begun to talk of specific historical occurrences: the movement of Anabaptists and Quakers from prophetic fanaticism to sober sociability, the counter-religious utopia of Confucian China, which has never known any religion but the veneration of social practices.[6] Natural history insists on manifesting itself in specific civil history, which however distorts it and takes it out of its natural path. He is about to identify the greatest of these distortions.

Hume opens with a statement in civil history, highly questionable in terms of contemporary knowledge: that until the date of the advent of Christianity, all men but for inconsiderable exceptions were polytheists.[7] Polytheist statements about the gods were what the ancients termed poetic theology; they were myths and tales that laid no claim to truth status and did not contradict one another, and the ancient world was perfectly tolerant in the sense that the distinction between tolerance and intolerance had never been thought of. This portrayal of paganism was a commonplace among both Christian and skeptical writers, and Gibbon—who was a deeply Humean thinker—was to present it with nostalgia in an early chapter of the *Decline and Fall*.[8] Two forces, Hume continued, had brought this world to an end: the advent of monotheisms, which identified a sole true god and implied or asserted the falsity of all others; and the advent of philosophy, which imported the possibility of truth status into statements about the divine, and further made it possible not only to identify God but to explicate his nature. All such statements might be either true or false; and from this moment, this eating of the forbidden fruit of knowledge, orthodoxy and heresy, tolerance and intolerance, authority and liberty, became features of the life of religion and the mind.

Philosophy, in alliance with monotheism, was thus the root of most evils the Enlightened mind desired to eradicate; a paradox we must understand if we are to grasp the deeply contested meaning of "philosophy" in the eighteenth-century mind. It was after all a premise with Hume that philosophy must abstain at all costs from substantiating statements about the nature of God, and might go no further than pointing out the philosophical meaninglessness of such statements. If therefore there was a history of philosophy as a root cause of theology, orthodoxy, and intolerance, it must be a history of philosophic error; and such a history, while rooted in the nature of the human mind, was likely to be a civil history, recoverable from the narrated and recorded past. The history of heresy, dependent upon a concurrent history of orthodoxy, must rest upon a history of philosophy; and such histories existed in a profusion of which Hume's *Natural History of Religion* is a cavalier abridgement. Against the false philosophies found in history must be pitted the philosophy that was their solvent; though this philosophy might have a history of its own.

We now situate Hume and Gibbon in the history of a critique of ancient philosophy and Christian theology that arises from more sources than can be detailed here, but is at least as old as Hobbes's *Leviathan* of 1657. Here we encounter the "kingdom of the fairies," which is a source of the "kingdom of darkness," and is traceable to the primary error of Greek philosophy: the proposition that the mind can grasp the ideas or essences that are the true constituents of reality.[9] This error was Pythagorean, Platonist, and

Aristotelian, in that historic order; and the greatest pre-Enlightened history of philosophy, that of Johann Jakob Brucker,[10] traced it back further, to magian thinkers of an axial age who, surrounded by the myths of poetic theology, had supposed each to be an allegory or image of a primal reality that might defy representation but with which the intellect might nevertheless communicate. The thought systems of these sages or mages, said Brucker, were cosmogonies rather than philosophies;[11] they took reality to be the genesis of God. They had not usually reached the point where orthodoxy must be separated from heterodoxy; but in their complex role as philosophers and priests, legislators and mystagogues, these Sarastros of pre-Hellenic antiquity were the ancestors of those who would reach it. We detect here the seed of what Diderot and others saw as the tragedy of the legislator; driven to invent mystery as the sanction of his rational commands, he saw his work perverted by the priests who arose to exploit his inventions.[12] Priests, however, were the masters of superstition; the invention of orthodoxy was rather the achievement of philosophers, whose besetting sin was the enthusiasm of believing the mind to be divine, or directly linked with ultimate being.

The history of philosophy proper began with Socrates, an Enlightened figure who had sought to divert the intellect from the pursuit of metaphysics into that of morality, from which the existence of God could be inferred if His nature could not be apprehended. After Socrates was executed, however, the greatest of his disciples had come under Pythagorean and Egyptian influences and had betrayed his work by restoring the primacy of an ultimate reality, which could be apprehended as idea and had furnished all things with essence and substance.[13] In this history of how philosophy became theology, the principal actor is always Platonism, the heir of Pythagorean and magian antiquity, which invents the Logos and, in the person of the Fourth Evangelist, seizes on the Pauline portrayal of Christ to proclaim that the Logos has been made flesh and dwelt among us. The intellect dwells upon this incarnation, and is capable of distinguishing and combining the substances that compose its union, but is also capable of describing it erroneously, in ways that diminish or deny its perfection. At this point orthodoxy and heterodoxy appear, as terms of art in an ongoing dispute over Christ's nature. We have reached the culmination of a history of philosophy into theology, one exhaustively constructed by Christian divines and only deconstructed, never abolished, by philosophical skeptics such as Hume and Gibbon.

The second, third, and fourth volumes of the *Decline and Fall* contain a history of Christological dispute in the era of the Fathers and the Councils, from Arius and Athanasius to Nestorius and Cyril—a history, as these names indicate, rather Alexandrian than Roman or Byzantine. It is a history of disputation, and Gibbon follows a line of skeptics and historians, Protestant, libertine, deist, and Voltairean, who had condemned Christian history precisely because it was disputatious; the aim of Enlightenment was to diminish intellectual investment rather than extend it, to replace disputation by conversation. They objected to disputation on two grounds: first, because it was meaningless, focusing the intelligence on essential substance it could never apprehend and could not affirm to exist; second, because it enhanced ecclesiastical authority at the expense of civil, obliging citizens to venerate and obey those who had so far succeeded in this debate over the unreal as to define their positions as orthodox and others' as heretical. Theology placed philosophy at the service of priests. Once the concept of authority was introduced, history became civil once more and something other than a history of philosophy—to say nothing of a sacred history—was needed. Philosophy and theology, however, were indispensable to the civil history framed to contextualize them.

Gibbon, then, was constructing a history of orthodoxy no less than of heresy—he had no great preference for either of them—and of the conditions, including the history of philosophy, that made both possible. Given the subject of this volume, we wish to know what role in the process was played by the historians of heresy on whom he drew, and whether it was their own orthodoxy or heterodoxy that made them undertake the histories that they wrote. We must approach these questions, however, by way of a view of the stages by which the topic of theological disputation became part of the narrative of the *Decline and Fall* as a whole; and here I wish to propose that it did so belatedly and by stages Gibbon may not have planned or foreseen in detail. We have his word for it that he found the notorious fifteenth and sixteenth chapters difficult to write, and that as they stand they are an abridgement of what might have been a volume on its own.[14] These chapters deal with the history of the Church in the two centuries before Constantine, not a period in which it had been necessary for councils to distinguish between orthodoxy and heresy, or for the imperial authority to concern itself with the question. On the contrary, it had been an age of persecution, though Gibbon did not make this a feature of his central narrative, to which chapter 16 forms a sort of appendix designed to diminish the quantity of martyrological narrative that had grown up around it; a strategy I find hard to reconcile with the view that he saw Christianity as a cause of decline in the pre-Constantinean period. One of his sources here is William Warburton's *Divine Legation of Moses,* a work certainly of questionable orthodoxy, though it contains some vigorous assaults on the Hellenic philosophers—"knaves in practice and fools in theory"—whose attempts to reduce Christianity to an allegory of neo-Platonist theosophy were a fruitful source of subsequent heresies.[15] Warburton helped Gibbon understand that the Christians were persecuted for what might be termed their intolerance: their denial of what he called "intercommunity,"[16] meaning the network of mutually supportive cults that held the pagan ecumene together, but that they condemned as the worship of demons in whose malignant existence they appear to have believed. Neither Christian intolerance nor imperial persecution, however, was connected with the distinction between orthodoxy and heresy, and one might question how far the Church before Constantine was concerned to establish it.

There did exist orthodox histories of this period, in which heresies and heresiarchs were shown being overcome by the apostles and by nascent tradition after them, and those written by Tillemont and Fleury were well known to Gibbon, who made great use of both. He does not, however, reiterate those parts of their narrative; we hear nothing of Simon Magus and the Apostle Paul, or of Cerinthus and the Apostle John, and the reason is not simply that Gibbon thought these tales apocryphal, though it is possible that he did. The deeper reason is that he chose to begin his narrative in post-apostolic times, saying nothing about Jesus, Paul, or those who may have known them, and leaving room for the safer and more subversive strategy of constructing a philosophic history of the Church in the second and third centuries.[17] This is certainly not a history of orthodoxy, or perhaps of heresy either; it may be a history of doctrines still too unformed to allow of either term or the attempt to impose them. Two features may be noted. At the end of the thirteenth chapter, a point in the *Decline and Fall* where Christians have never yet been mentioned, notice is taken of the advent of the neo-Platonist philosophers; theosophists and mystagogues of whom it is observed that they were to play a part in the history of Christian doctrine.[18] Chapter 14 recounts how Constantine defeated his rivals; and when the history of the Church begins at last in chapter 15, the early (but post-Pauline) Christian communities are shown divided into those resident at Jerusalem until its destruction—orthodox Jews in everything except their belief that Jesus was the Messiah, and therefore a man divinely sent but in no way divine in nature—and those dispersed throughout the Greco-Syrian world, where they became exposed to various forms of gnosticism, presenting Him as in one way or another one more avatar of the divine energy. Here is a history with no thread of orthodoxy holding it together, and one may suspect that the proto-Socinian Ebionites for whom Jesus was all man, and the proto-Platonist gnostics for whom Christ was all spirit, are being put forward as the poles between which Nicene theology was to attempt bridging the gap. The gnostics, however, are hesitantly presented as neo-Platonists, and one may doubt whether they are Christian heretics at all; they may be a continuing sect of older Iranian magianism and dualism, who happen to have encountered the figure of Christ and set about enrolling Him.

Chapters fifteen and sixteen, then, may belong in the scheme of the *Decline and Fall* to the prehistory rather than the history of orthodoxy and heresy; but the questions of the incarnation and the divine nature are already present and even dominant. It is early in the fifteenth chapter that Gibbon begins introducing those we may consider the historians of heresy who are to guide him through the succeeding books. Three names stand out: Jean Le Clerc, the Genevan refugee settled in Remonstrant Amsterdam, initially for his *Historia ecclesiastica duorum primorum a Christo nato saeculorum* (1716), but a major figure for Gibbon through the series of *Bibliothèques,* which were the great review journals of the Huguenot Republic of Letters; Johann Lorenz von Mosheim, professor at Halle and later Gottingen, initially for his *De rebus Christianorum ante Constantinum Magnum* (1753), but further for his *Institutionum historiae ecclesiasticae libri quattuor* (1755); and Isaac de Beausobre, pastor to the Huguenot congregation in Berlin, for his *Histoire critique de Manichée et du manichéisme* (1734–39).[19] To these it would be fair to add one English scholar, Nathaniel Lardner, whose works had appeared between 1741 and 1762, though the *History of the Heretics of the First Two Centuries after Christ* appeared posthumously in 1780. All four were Protestants, though the presence of Tillemont in Gibbon's mind is never to be left out; and where they were not unorthodox—Le Clerc was Arminian to the verge of deism, Lardner, a Nonconformist, was Arian in his theology—they were involved in those debates over a sober and rational Christianity that subjected the Nicene tradition to close and perilous scrutiny. I ought to say at this point that I am avoiding any serious attempt to locate Mosheim in the context of Lutheran scholarship to which he belongs, but that he was the author of a work *De turbata per recentiores Platonicos ecclesia* (1727).

Of these scholars, Mosheim's importance for Gibbon probably lasts longest in the pages of the *Decline and Fall,* but Beausobre ranks close to him in this particular, and his sense of irony must have been peculiarly congenial to Gibbon. He tells us that he embarked on his study of Manicheanism out of curiosity regarding the medieval precursors of Protestantism;[20] it was a commonplace to list these as the Albigensians and Waldensians, and Beausobre found himself enquiring how far the former had been Manichean in the fully dualist sense, and then how far Mani himself and his followers had been practicing a Persian dualism independent of Christianity and how far pursuing a view of Christ's nature, which orthodoxy established itself by condemning. Gibbon engaged in the same enquiry as late as the fifty-fourth chapter of the *Decline and Fall,* where he traced the origins of the Reformation through the Albigensians to the eastern Bogomils and Paulicians, and was inclined to regard the latter's dualism as gnostic rather than Manichean. There are two important perceptions at work here, for both of which Gibbon is indebted to Beausobre and Mosheim, and for one at least, less immediately to Hume. This is the perception that it was dangerous for the sober and rational Christian, in antiquity or in modernity, to pursue an honest understanding of the plain sense of Scripture, if he took it to contain a true statement of Christ's nature and His Father's. He could all too easily be led into the enthusiasm of supposing some reading of his own to have been revealed to him in direct contact with the divine, and the most fearful consequences—apocalyptic, antinomian, dualist—might follow, and in history repeatedly had. Here the *philosophe* found himself paradoxically allied with the orthodox cleric; both wanted to discourage the private interpretation of scripture, for reasons whose essential opposition was easier to perceive than to penetrate, and both might find themselves treading very shaky ground.

Hume had brutally concluded that it was impossible to study nonsense for very long without coming to believe some of it; from which follows the question whether Gibbon saw critical erudition and historical narrative as means of evading this trap. That the problems confronting the theologians of the early Church were philosophically unreal went without saying (though Gibbon regularly gives us to understand it). The historian could not treat them as capable of solution without risking falling into fanaticism himself; but he could not construct a historical narrative by the mere repetition that the actors in it were foolish and deluded fanatics. It is hard to account for the fact that Gibbon wrote a long and detailed history of the disputes over orthodoxy and heresy in which he did not believe, without supposing that he thought the behavior of actors in it capable of being made intelligible and even respectable. Historical irony was his recipe for narrating actions at once respectable and absurd; he had learned it from the greatest of his predecessors, among whom he mentions Pascal and Bayle, while excluding Voltaire as too often a vulgar scoffer to have sustained historical insight.

Here again Beausobre looks like a mind congenial to Gibbon. In a display of irony the latter must have relished, he cites Augustine as thanking God for introducing him to the writings of Plato in time to save him from the errors of the Manichaeans. He did well to thank God, muses Beausobre, since the reading of Plato ought to have made matters much worse; short of divine intervention, we should think of Augustine as one of those animals who by instinct find the antidote growing next to the poisonous herb.[21] This observation must be connected with Beausobre's remark elsewhere that his studies taught him to cease demonizing the great heresiarchs and regard them as honest men laboring with problems of enormous difficulty.[22] There is more here than liberal benevolence; we encounter the question whether the benevolent liberal thinks belief possible at all, or must live in a universe of ironic historical narratives instead. Gibbon was in no doubt about this; but there remains the question whether the heresiarch is amenable to irony if—like Mani?—he ceases to be an exegete and becomes a prophet instead. Gibbon reserved the question for Muhammad, a prophet but not a heretic.

Mosheim, Beausobre, and Gibbon are enough—though there is very much more—to inform us that, after gnosticism, the chief cause in civil history of the advent of disputation, orthodoxy, and heresy, was the presence of Platonism. It could be and was debated how far the judgement should be confined to the neo-Platonists of Alexandria, or how far they were continuing and elaborating an error originally Plato's own: that of supposing reality to consist of substances whose essence could be apprehended by human intelligence. This problem, however, belonged to the history of philosophy, a precondition of the history of orthodoxy and heresy. Gibbon brought the neo-Platonists into his history, not as a means of accounting for the gnostics, but as incidental to the rise of Constantine; and I wish now to argue that the conjunction of Christian monarchy and theological dispute was the central thread of a history Gibbon came to intend, but could not begin telling within the framework of chapters 15 and 16. These chapters, long taken as the index to Gibbon's views on religion and its history, are in fact ancillary and preliminary in the structure of the *Decline and Fall.*

In chapter 21, five chapters and five years later in the history of composition and publication, the history of heresy proper at last begins. Arius and Athanasius enter into dispute over the equality of the Son with the Father, and the great debate over Trinity and Incarnation is under way. It is under way, that is, as a formal debate; Constantine is obliged to call the Council of Nicaea and preside over it, and there exist parameters of theology and philosophy within which debate must be conducted. Gibbon now goes back into the Athenian past as far as Plato himself in search of the philosophy whose conjunction with the Christian narrative makes theology possible and necessary. He does not explore the history of that narrative itself, in the Gospels, the Acts, or the Epistles, and we do not know what status he attached to it apart from philosophy and theology; if it was another poetic tale about a god, it was very unlike the poetic theology of the Gentiles, nor has he anything to say about the prophets of the Exile. Gibbon was interested only in Christianity as metaphysics, and thought the only other thing it could have been was morality; this is probably the point at which Enlightenment itself becomes heretical, or at least erroneous, in a Christian perspective.

Gibbon was writing a civil history, in which ecclesiastical authority took shape and intervened, to the point where it could almost be said to have captured history itself. A major theme of chapter 21, therefore, is how Constantine discovered the consequences of his decision—partial and delayed though it was—to make Christianity a religion of state. He had not anticipated a debate between Arius and his opponents, an attempt to determine what was orthodoxy and what heresy, but when it occurred he found that he could not neglect it or place himself above it; he might be the equal of the apostles, but could rise no higher. Gibbon indicates that it would have been better for the empire if Constantine could have remained indifferent to theological dispute—and that indifference implies contempt—but is left facing the question of why he could not. The degree of Constantine's own Christian conviction is one point at issue; another is the increasing authority of a clergy who were making doctrine a source of it; but chapter 21 is predominantly a history of philosophy, a history of ideas as we should call it, in which philosophy is permitted to determine the intellectual and therefore political climate in which all must act. Gibbon does not share the modern conviction that it is the duty of the intellectual to eliminate intellect from history as far as possible.

"The historian," he says, "may . . . be permitted . . . to deduce the progress of reason and faith, of error and passion, from the school of Plato to the decline and fall of the empire"[23]; language associating Plato with decline and fall neither less nor more explicitly than Gibbon ever associated Christianity. He goes back to Athens and to Plato himself, indicating the formation of the concept of Logos in language which implies the presence in Plato of a philosophic trinity or triune godhead, in which Christian writers were accustomed to detect anticipation of the Trinity they proclaimed as theirs.[24] Gibbon traces the acceptance of Platonic philosophy first by Alexandrian Jews and then by Alexandrian Christians, culminating in the revolutionary language of the Fourth Gospel, which declared that the Logos had been incarnate. This gospel, he says, appeared under the reign of Nerva,[25] and there is a Christian tradition that declares it intended to confute the opposed heresies of the Ebionites and the gnostics.[26] The authorities he cites on this matter are Cudworth's *Intellectual System of the Universe;* the Huguenot Jacques Basnage's *Histoire des Juifs;* Le Clerc, Brucker, Mosheim, Beausobre; the Jesuit Denis Petau and the Anglican George Bull.[27] Non-orthodox writers were perhaps more likely than orthodox to lend themselves to Gibbon's purpose of including Platonic philosophy in a theology reduced to history, but he was prepared to rely on and respect both. It is important, I believe, to emphasize that chapter 21 is remarkably free from irony, innuendo, and the underlying facetiousness that many found offensive in the chapters published five years earlier. Gibbon's unbelief here took the form of a serious history, exploring and expounding a logic in which all actors in a Christian empire found themselves involved and—he did not need to emphasize—ensnared. He stresses that the dispute over the Word made Flesh was not an academic or esoteric matter, but a public debate in which any citizen, and more importantly any believer, might be passionately engaged.[28]

Perhaps Gibbon wishes that this had not happened, but he does not write, as Voltaire might have written, to convey derision and horror at the absurdity of the dispute before him. Certainly he thinks the Platonic Logos a profound error, a typical product of the human mind's attempt to go beyond the limits of its powers; but Brucker is only the most comprehensive of the writers he cites who would have informed him that there was a huge and complex history to be studied, covering a minimum of 2,000 years, before Bacon, Grotius, and Locke had delivered philosophy from its magian and Platonic origins. Gibbon's history does not reach the point of recounting how this modern revolution happened. His concern in chapter 21 was to reveal the Platonic origins of the entire debate over Christ's nature, in which both orthodoxy and heresy took shape. It had been a commonplace with Christians since pre-Constantinean times that Platonists were unable to accept that the Logos had been made Flesh; Gibbon agreed, but was concerned to point out that only a Platonist could have made the assertion in the first place. The Christian critique of his position, in so far as I am competent to put it forward, would seem to be that a history of how philosophy generated theology becomes a history of philosophy and nothing else. Two figures are signally lacking from his narrative: he does not tell us what Jesus was or what Paul did.[29] A history of Christianity centered on the Fourth Gospel makes it a product of Platonism; Gibbon does not explore the history of the figure of Christ or ask—as perhaps a modern would—how far it was a Pauline and how far a gnostic creation. Chapter 21 contains great history of philosophy, but nothing behind that.

To the historian of civil society, what mattered about the history of philosophy and theology was their impact on civil manners and civil authority. Here a central fact for Gibbon was Constantine's decision to make himself a Christian (more or less) and Christianity the religion of state, and his discovery at and after Nicaea that he had made himself a member of a church he must try to control without the assurance of success; it was the debate about Christ's nature that forced this discovery upon him. Here we must avoid ascribing to Gibbon our liberal solution of the separation of church and state; it was Joseph Priestley, the most energetic proponent of this solution, with whom Gibbon had the bitterest of his civil quarrels. He had no objection to the ancient solution that there should be a political as well as a poetic and a philosophical theology, and even in the modern world did not deny that the state might establish a religion in the process of governing and policing it. The problem was that the ancient solution had ceased to be relevant with the disappearance of a polytheism in which religion made no claim to truth status. The magistrate could not regard all Christian sects as equally false and equally useful,[30] given the premise to which all must subscribe that one of them must be true and the others false; it was the advent of philosophical theology that had led to this premise, accepted by the literate and the illiterate alike. The only solution for the modern world was to break the link between philosophy and theology, and accept that the philosopher and the magistrate were indifferent to what became of God after that.

This solution was not possible in the fourth century. Constantine might have stayed further from the Arian controversy than he did,[31] but the very fact that he delayed his own baptism showed that he could not avoid membership in the church he had established, or escape putting his imperial authority behind one of the theological positions claiming orthodoxy. In Gibbon's own time a liberal theology, based on a philosophy rather Lockean than Platonist, was leaving the nature of Christ's divinity sufficiently open to allow the civil sovereign escape from the necessity of decision; and liberal theology tended to have the support of those Arians or Socinians who were critical of Nicene orthodoxy. Enlightenment and anti-trinitarianism thus found themselves in various patterns of alliance, from which they attacked Trinity, orthodoxy, intolerance and the alliance of church and state as more or less interchangeable terms, and there was a consequent tendency to equate heresy with liberty; not only was the heretic entitled to his liberty, but his theology as well as his interest made him the more likely to claim it. It was therefore—as it has remained—a temptation for historians, who are liberals by avocation, to write on the assumption that the heretic was the more reasonable being. Gibbon avoids this temptation, principally of course because he considers all theology, orthodox and heterodox, to be products of the same Platonic error. The heretic was not a liberal modern, but as Platonist as his opponent, if not more so, as was evident once we saw the whole debate as one about the incarnation of the Logos.

But Gibbon is no deist Mercutio, professing a hearty contempt for both houses. Though we know from chapter 21 that he considered Arius and Athanasius partakers in the same error, he is capable of eloquent and insightful exposition of both sets of deeply held beliefs. This is probably less an indication that he was moving beyond Enlightenment toward historicism than a product of the ancient skills of rhetoric, of which historiography had been a form. Nor does his engagement in the story he is narrating end there. Perhaps the most striking feature of this chapter is the fascination and admiration, undiminished by the irony with which it is mixed, that Gibbon expresses for none other than Athanasius,[32] who as the champion of the Trinity was treated as an enemy and held in abhorrence by all unorthodox moderns, from Arians like Isaac Newton, to deists, most of whom had come to their position by way of one or other of the anti-trinitarian critiques. Gibbon, it is worth suggesting, was (like Hume) never a deist but always a pure sceptic; the two are not the same. This may help us understand, but cannot altogether explain, his admiration of Athanasius. Perhaps part of the answer is that in the life of that father, Gibbon had access to a vivid and sustained narrative. He liked a good story, and more importantly it was still the historian's function to retail the narratives he had inherited—Gibbon felt obliged to retail those of the *Historia Augusta,* in which he had no faith at all.[33] But he admired Athanasius, and saw him as something more than an indomitable believer. After the Arians and their imperial supporters, Athanasius discovered a more dangerous adversary in Julian the Apostate, and Gibbon is as deeply attracted by this figure as he is by Athanasius—much more so, of course, than by any Christian heretic. But his sympathy for the incorrigibly likeable Julian (that graduate student in purple) is at tension with his discovery that the emperor was not a philosopher—no Stoic magistrate, Ciceronian sceptic, or Lucretian materialist—but a neo-Platonist magician striving to build a polytheist church on a theosophical foundation.[34] The theology of Logos and Flesh was coherent and philosophical by comparison, and Gibbon is further insistent that Athanasius was the more responsible statesman of the two; Julian reached the verge of launching a general persecution and Athanasius withstood him.[35] We have reached an interesting tension in Gibbon's thinking. On the one hand, it was a great evil that ecclesiastical authority, with its foundations in doctrine before they were in superstition, arose as a challenge to civil society and its authority. On the other, in a world lacking any constitutional limits to imperial despotism, it was as well that the Church should present an alternative, and an orthodox churchman might uphold the cause of liberty more responsibly than a heretical enthusiast. When Gibbon considered the Latin Middle Ages, he observed that it was the Guelphs in Germany and Italy who had upheld "the banner of liberty and the church" against the Hohenstaufen.[36] He wrote that Pope Gregory VII—a demon figure in the litany of both Protestantism and Enlightenment—was "a great man, a second Athanasius in a more fortunate age of the church," adding that of all the portraits in his earlier volumes, it was that of Athanasius that gave him least dissatisfaction.[37] There is more than dramaturgy to Gibbon's espousal of this figure.

Gibbon was an Englishman and (if only just) an Anglican. He had before him the possibility that an established and undemanding Church subordinate to the state was the best guarantee of civil liberty against the twin dangers of clericalism and fanaticism. While doubtless believing that all varieties of religious opinion should be free to express themselves in conversation, he was less sure of their right to public profession and association. Of all his clerical critics he was on best terms with Richard Watson, bishop of Llandaff, whose opinions were probably Socinian but who had remained within the Church of England, publicly supporting an orthodoxy he was prepared to discuss in private. In a trinitarian establishment ruled by a free-thinking episcopate, Gibbon may have seen the closest approach possible in the modern world to that government of a polytheist culture by a philosophic magistracy that Christianity and Platonism had combined to render irrecoverable. Here he was within a conservative tradition descending from the ambiguities of John Locke, but there was a more radical alternative descending from the same source. Of all Gibbon's clerical critics, he was most deeply opposed to Joseph Priestley, whose *History of the Corruptions of Christianity* (1782) is written on the assumption that every one of these corruptions—which include the Incarnation, the Atonement, the Trinity and the separate existence of the soul—is a Platonic accretion of Alexandrian and Egyptian origin. As a narrative history of theology, it is remarkably like that which can be culled without difficulty from the chapters of the *Decline and Fall,* and this must have increased Gibbon's anger on realizing that Priestley saw the formation of opinions about Christ, unimpeded by theology or philosophy, as the necessary road to apprehension of his person as that of a man sent by God who might at any moment return. All established religions of state were anti-Christian, Priestley wrote, and for their dissolution "we must await the fall of the civil powers. Calamitous no doubt will that time be."[38] In this apocalyptic, revolutionary, and democratic language, Gibbon recognized the unmistakable voice of the enthusiast, for whom his own once rational opinions had become the voice of God. Here Gibbon traveled back beyond anti-Platonism to the earlier lessons of Hume. The established churches of his day were not instances of priestcraft so flagrant as to need the fiery purgation of enthusiasm. When he heard of the revolution in France, Gibbon's response would be Burkean before he read Burke.

Notes

1. A handlist of those so far completed: J. G. A. Pocock, "Superstition and Enthusiasm in Gibbon's History of Religion," *Eighteenth-Century Life* 8, 1982, pp. 83–94; "Clergy and Commerce: the Conservative Enlightenment in England," in *L'Età dei lumi: studi storici sul settecento europeo in onore di Franco Venturi* (Naples: Bibliopolis, 1985), pp. 523–62; "Gibbon and the Idol Fo: Chinese and Christian History in the Enlightenment," in David S. Katz

and Jonathan Israel, eds., *Sceptics, Millenarians and Jews* (Leiden: Brill, 1990), pp. 15–34; "Edward Gibbon in History: aspects of the text in *The History of the Decline and Fall of the Roman Empire*," *Tanner Lectures on Human Values* 11, 1990, esp. pp. 338–61; "Thomas Hobbes: Atheist or Enthusiast? His place in a Restoration Debate," *History of Political Thought* 11, 1990, pp. 737–49; "Cudworth, Brucker and Hume: the Paradox of the *Natural History of Religion*," in Gordon J. Schochet, ed., *Politics, Politeness, Patriotism: Proceedings of the Folger Institute Center for the History of British Political Thought* (Washington, D.C.: Folger Library, 1993), pp. 59–72; "Within the margins: the definitions of orthodoxy," in Roger D. Lund, ed., *The Margins of Orthodoxy: heterodox writing and cultural response, 1660–1750* (Cambridge: Cambridge University Press, 1995), pp. 33–53; "Enthusiasm: the Antiself of Enlightenment," *Huntington Library Quarterly* 60, 1998, pp. 7–28; "Gibbon and the Primitive Church," in Stefan Collini, Richard Whatmore, and Brian Young, eds., *Essays in British Intellectual History, 1750–1950* (Cambridge: Cambridge University Press, 2001). See, in general, *Barbarism and Religion*, I: *The Enlightenments of Edward Gibbon*, II: *Narratives of Civil Government* (Cambridge: Cambridge University Press, 1999).

2. Knud Haakonssen, ed., *Enlightenment and Religion: Rational Dissent in Eighteenth-Century Britain* (Cambridge: Cambridge University Press, 1996).
3. For a classic statement of this position, see David Hume, *History of England*, ed. William B. Todd (Indianapolis: Liberty Press, 1983), III, p. 433. Cf p. 136, where it is argued that the clergy should be supported by public salaries rather than private contributions, since this will make them lazy rather than zealous in propounding their nonsensical doctrines.
4. It was published in Hume's *Four Dissertations* of 1757 (reprinted Bristol, 1995) and included in his *Essays and Treatises on Several Subjects* (1760), of which Gibbon owned a copy. For its possible presence in Gibbon's *Essai sur l'Etude de la littèrature* (1762), see Pocock, *The Enlightenments of Edward Gibbon*, pp. 235–36.
5. *Decline and Fall*, chs 2, no. 3; 8, n. 24; 15, n. 114 (essay on enthusiasm); 23, n. 12, 28, n. 85, (essay on theism and polytheism).
6. Pocock, *Narratives of Civil Government*, pp. 193–96. For Voltaire's similar treatment of Anabaptists, pp. 140–41.
7. Hume, *Four Dissertations*, p. 3.
8. *Decline and Fall*, ch. 2 (ed. David Womersley, London: Penguin, 1994, I, pp. 56–60).
9. *Leviathan*, chs. 46–7.
10. *Historia critica philosophiae, a mundi incunabulis ad nostram usque aetatem deducta*, 6 volumes (Leipzig, 1742).
11. Brucker, I, pp. 142, 363.
12. A.H. Anquetil-Duperron, *Zend-Avesta, ouvrage de Zoroastre . . . traduit en francais . . .* (Paris, 1771), II, pp. 65–66; Raynal, *Histoire philosophique et politique du commerce et des établissemens des Européens dans les deux Indes* (Geneva, 1780), I, pp. 61–64. This passage is attributed to Diderot.

13. Brucker, I, pp. 522–83, 627–68.
14. Georges A. Bonnard, ed., *Edward Gibbon: Memoirs of My Life* (New York: Funk and Wagnalls, 1969), p. 156.
15. *The Works of the Right Reverend William Warburton* (London, 1811), III, p. 201.
16. Vol. II, pp. 307, 311–12.
17. See further "Gibbon and the Primitive Church," (n. 1, above).
18. *Decline and Fall,* ed. Womersley, I, pp. 398–99.
19. *Decline and Fall,* ch. 15, notes 18, 19, 21, 22, 26, 32.
20. Beausobre, I, pp. iii–v.
21. Beausobre, I, p. 479.
22. Beausobre, I, p. xxi.
23. Chapter 21; Womersley ed., I, p. 771.
24. Ed. cit., p. 775.
25. P. 773, and n. 21.
26. P. 774.
27. P. 771, n. 12; nn. 19, 21, 22, 30, 39.
28. Pp. 776–80.
29. Again see "Gibbon and the Primitive Church."
30. The phrase famously used of the philosopher and the magistrate in polytheist culture; ch. 2, Womersley ed., I, p. 56.
31. Pp. 789–90.
32. This begins to appear later in ch. 21; see pp. 796ff.
33. This historiographical problem may be traced through chs. 4–7 and 12–14.
34. This is the theme of Gibbon's 23rd chapter.
35. Pp. 904–08.
36. *Decline and Fall,* ch. 49; Womersley ed., III, p. 144. Muratori is the source of this perception, which appears in Gibbon's earlier writings; see Pocock, *Narratives of Civil Government,* pp. 384–85.
37. *Decline and Fall,* ch. 56, n. 83; Womersley ed., III, p. 504.
38. *History of the Corruptions of Christianity,* II, p. 484. For Gibbon's response see ch. 54, n. 42 (Womersley ed., III, p. 439) and *Memoirs of My Life,* p. 172. Priestley's language was also singled out by Burke in *Reflections on the Revolution in France,* ed. J. G. A. Pocock (Indianapolis: Hackett, 1987), p. 50. His attention may have been drawn to it by Gibbon's strictures, published in 1788.

Part Five

Enlightened Orthodoxy

Introduction

John Christian Laursen

In spite of all the attacks on orthodoxy that we have seen above, the following chapters in this book show that the established orthodoxies never gave up. No history of the histories of heresy of the eighteenth century should rest content with the view that heresy won and orthodoxy lost. As might have been expected, the orthodox learned to change their rhetoric to a moderation more acceptable to late-eighteenth-century tastes.

The work of Patrick Coleman's abbé Pluquet is well-characterized as "enlightened orthodoxy" because, on the one hand, he believes in modern science and rationality, and on the other hand, he believes that many heresies are still heresies. He mined Diderot's *Encyclopedia* for information, but reversed its strategy. Where Diderot made reference in the more orthodox articles to more subversive articles, Pluquet referred the reader of the more subversive articles to orthodox refutations of them. He was gentle with the Hussites and Moravians, but conceded nothing to Luther and Calvin. Voltaire, in turn, mined Pluquet for his own more subversive purposes.

Intriguingly, Pluquet's book was translated into Italian in 1771 and Spanish in 1792. There must have been an ideological purpose and a perceived market in those countries for these volumes. The text could play a rather different role there than it did in France. It would be interesting to know exactly what roles it was expected to play and actually played, and whether the translations were faithful to the original or tailored for local purposes. Every community needed its own guides to the understanding of heresy. Clorinda Donato's chapter in this volume reviews the *Encyclopedia of Yverdon,* the Swiss Protestant answer to Diderot's irreligious *Encyclopedia.* The Protestants retained the critiques of Catholicism, but omitted the attacks on Protestantism and added critiques of irreligion. The next Catholic encyclopedia did it the other way around. Kay Doig's chapter on the fin de siècle *Methodical Encyclopedia* (1788–91) explains that the author of its articles on heresy, the enlightened abbé Bergier, never abandoned Catholic orthodoxy even as he attempted a moderate reconstruction of it. The market for this sort of work is indicated by the stunning statistic that it was reprinted 31 times in the nineteenth century.

12

The Enlightened Orthodoxy of the Abbé Pluquet

Patrick Coleman

I

The *Dictionnaire des hérésies*[1] published in 1762 by François André Adrien Pluquet (1716–1790) offers an interesting example, in both its content and its form, of what might be called enlightened orthodoxy in eighteenth-century France. It makes no concessions to critiques of Roman Catholic doctrine, but its tone is irenic rather than combative. Pluquet expresses admiration for the moral integrity of some heretical groups (most notably the Moravians); he acknowledges the cruelty of the Inquisition and the occasional high-handedness of the papacy; and, most notably, he recommends a measure of civil tolerance for Protestants in France.

At the end of the article "Calvinistes," Pluquet writes:

> It has been much debated recently whether we should grant them civil tolerance: zealous citizens have judged that we should; Bishops fear that the Faithful who have been entrusted to them will be seduced, and are opposed. It is not perhaps in the interests of the State to let the Protestants multiply in France, but by treating them with humanity, with charity, with gentleness, could we not hope to reunite them to the Church? This notion, it seems, has not entered sufficiently into the considerations of some Authors who have written on these matters. (I, 568)

Is Pluquet's recommendation purely tactical, or does it represent a real modification of the close linkage of doctrinal conformity, ecclesiastical community, and political unity in Pluquet's acknowledged master, Bossuet?

Pluquet's identification with *citoyens zélés* is, I believe, genuine (the very expression carries important Enlightenment connotations, evoking both Rousseau and, more importantly for Pluquet, Montesquieu), but it does not lead to any reconsideration of his views on doctrine or ecclesiology. Pluquet's Gallican perspective can accommodate some degree of religious tolerance, but only on the level of practice. On the theoretical level, tolerance and orthodoxy are juxtaposed rather than integrated. Whether we are supposed to take this juxtaposition as an invitation to read Pluquet "between the lines" in a liberal sense or as a sign of a conceptual impasse is a difficult question.

One of the reasons for the difficulty may be found in the form of the work. In overall conception, it somewhat resembles the philosophes' *Encyclopédie.* It begins with a 225-page "Discours préliminaire," whose title echoes that of d'Alembert in the earlier work and whose style is that of Enlightenment philosophical history, but whose focus is on the *égarements* rather than the progress of the human spirit. Yet, since Pluquet ends his history with the start of the Reformation in the sixteenth century, he forgoes the opportunity to attack heretics of recent times. The "Discourse" is followed by an alphabetical dictionary of heresies and heretics, with a number of cross-references tying them together. Here, too, an interesting comparison can be made with the *Encyclopédie.* Especially in its early volumes (the only ones Pluquet would have read before publishing his dictionary), the *Encyclopédie* often refers the reader from an article offering an orthodox treatment of a controversial subject to another article, often with an obscure title, which contains a bolder treatment of the same topic.[2] Pluquet uses the device in the opposite way. At the end of articles the reader is likely to look at first, such as "Calvinistes," Pluquet refers to other articles under more recondite headings, in which, he tells us, the heretical beliefs he has just played down in the name of practical tolerance are refuted. The reader is thus left with the responsibility of weighing the work's different statements against each other.

Pluquet was born in Bayeux, studied in Caen, and, from 1742, in Paris, where he became a *bachelier* of the Sorbonne in 1745, and a *licencié* in 1750. To support himself, he served as tutor to the abbé de Choiseul, younger brother of the famous duc de Choiseul who was Foreign Minister of France when Pluquet's dictionary was published. By the same date, the abbé had become Archbishop of Alby.[3] Choiseul gave Pluquet a pension of two thousand *livres,* which allowed his former tutor to pursue his studies independently. In 1757, Pluquet published a three-volume *Examen du fatalisme,* in which he traced all the various forms of determinist thought from antiquity down to La Mettrie, offered a careful analysis of Spinoza, and defended the notion of man's moral freedom.[4] In addition to the *Dic-*

tionnaire des hérésies, Pluquet wrote many other works, including a treatise on sociability (1767). His writing clearly found official favor. Bachaumont reports that in 1764, the king gave a pension of two thousand *livres* to "M. l'abbé Pluquet, doctor of the Sorbonne, well-known in the literary world for his *Examination of Fatalism.* He has also given the public a *Dictionary of Heresies;* although the matters he treats are not within everybody's reach, he has found the way to bring them closer, and to be read with pleasure and profit."[5] In 1766 Pluquet was named professor of moral theology in the Collège de France, where he taught until 1782. In his retirement, he translated a Latin collection of the classic works of China (1784–86) and wrote a long essay on luxury (1786).

The *Examen* was noticed by the *philosophes,* who according to rumor tried to enlist Pluquet among the contributors to the *Encyclopédie.* Pluquet apparently refused, and later comments on his work from the philosophic camp are a mixture of respect for his honesty and erudition, and mockery of what they saw as a retreat from the anti-religious (or at least anti-dogmatic) implications of his own arguments. In a letter to Damilaville, for example, Voltaire, commenting on the *Examen,* says of Pluquet: "This man has his metaphysics down cold (*est ferré à glace sur la métaphysique*); but I wonder whether he has not furnished a supper of which several dishes would be to the taste of the Spinozists. I would like to know what the d'Alemberts, the Diderots think of this book."[6]

The *Dictionnaire des hérésies* was greeted with sarcasm by Grimm in the *Correspondance littéraire:*

> M. l'abbé Pluquet, who has already fought the hydra of fatalism with a big, heavy treatise, has just opened a general battle against all possible heresies. . . . As you can imagine, it is only the Roman religion that is always right . . . divine wisdom almost never fails to procure good solid benefices for those who fight for its cause; this is what I wish with all my heart for M. l'abbé Pluquet.[7]

D'Alembert, however, took the work much more seriously. In a letter to Voltaire, he writes:

> A man of wit, who unfortunately needs to be a theologian, or to appear as one, has just given us a *Dictionary of Heresies,* in two large volumes, which deserves to be read: he has with much good faith included the objections on one side and the answers on the other, and one can say that in doing so *faith* does not get too much benefit from *good faith.* Upon my word, this is a tremendous book, in my opinion, against that *infâme* you hate so much. What the author says, among other things, to explain transubstantiation . . . is altogether comical. . . . [8]

Voltaire in fact immediately ordered the work and read it with care.[9] His remarks are generally sarcastic, but he also marked many articles simply for future reference, to be quarried for his own, more subversive *Dictionnaire philosophique.*[10]

Pluquet's dictionary received an equally respectful response from the Jesuit periodical, the *Mémoires de Trévoux.* According to the anonymous author of the *extrait* (the series of extensive quotations interspersed with summary and brief commentary that was the characteristic form of reviews in this period), the importance of the book was such as to deserve a lengthy account, spread over three issues of the journal.[11] While the reviewer thought some parts of the "Discours préliminaire" resembled a "roman philosophique, mais rempli d'art et d'adresse" (2529), and wonders why Pluquet did not carry the story forward to the eighteenth century (2532), the others were "plus solides," and the general tone is very positive. As with Pluquet himself, it is not easy to discover the reviewer's actual opinion on many of the passages he cites, for he often quotes without comment. However, the absence of negative reaction suggests that the reviewer tacitly endorses many of Pluquet's views as well as his strategy of avoiding controversy wherever possible. Such a conclusion is supported by the reviewer's relief at not finding any articles labeled "Jansénisme," "Molinisme," and "Quesnelisme" (even though some of the issues involved in these quarrels are tucked away under other headings),[12] for these are "odious titles that a wise prudence has excluded from this Dictionary" (2846). We shall return to some of the reviewer's other remarks.

II

While the *philosophes* may have considered Pluquet their unwitting, and perhaps also occasionally witting, ally, he certainly did not see himself in this light. It is true that in the *Examen* he praised freedom of thought as "so necessary to the progress of the mind" (iii), and he shared with the *philosophes* a belief in the role of the state as a religious peacemaker and a bulwark against fanaticism. For Pluquet, however, fanaticism is not the product of religion alone or as such. As he argues in the "Discours préliminaire," fanaticism derives from the sheer restlessness of the human spirit, which at every stage of historical development is subject to *égarements.* Itself, this restlessness is morally neutral: it is the motor of progress as well as of error—a conception that places Pluquet closer to Rousseau's view of perfectibility than to Pascal's notion of vain *divertissement.*[13] Society must, in fact, make room for restless curiosity rather than seek to stifle it entirely. Pluquet regrets, for example, the decline of scientific activity that followed

Constantine's establishment of Christianity as the official religion of his empire (I, 125), and criticizes the "absolutism" of the Byzantine state.

Yet, even though Pluquet adopts an evolutionary view of *l'esprit humain*, the organizing principle of the "Discourse" is quite different from that of comparable Enlightenment works. Whereas the latter focus on epistemological and political development, Pluquet's discourse is centered on the mediating and moderating role of authority, which is conceived as a permanent and essentially reactive, regulatory function. In the best of circumstances, authority ensures stability amidst progress, but sometimes the undisciplined actions of men provoke a reaction that may be excessively violent.

> The Church, which watches over the deposit of faith, condemns error, and prescribes the most appropriate means for stopping its progress; but the errant Christian is often indocile to its voice, and the defender of truth does not remain within the boundaries that religion and the Church prescribe to his zeal. In moral matters as in physical, the reaction is equal to the action, and one thinks one must employ in favor of truth all that error allows itself against it (I, viii).

Pluquet formulates his position impartially, and in his article "Fanatisme" he stresses that fanatics may be found in every sect and every religion, such that "the History of Fanaticism is not one of the least interesting portions of the History of the human mind" (II, 52). Still, the word "fanatic" is only used of dissidents, never of rulers or hierarchs.[14] As the product of an "unruly imagination," fanaticism seems by definition not to apply to anyone bearing the burden of instituted authority, even when that exercise is unjust.

Pluquet's attitude belongs to a long French political tradition of enlightened *dirigisme*. As far as religious authority is concerned, it is also grounded in another belief. Revelation does not belong to history, and is therefore exempt from historical judgment. For all his interest in the "history of the human mind," Pluquet has no hesitation in defending the belief, orthodox in his time,[15] that humanity's first religion was monotheistic, and that polytheism resulted from the corruption of this original revelation. Hume's arguments to the contrary (in *The Natural History of Religion* [1757]) are rejected as historically implausible, given the difficulty of moving from polytheism to monotheism through the work of human faculties alone (I, 9ff.). This is not to deny the historical development of belief in the pagan era, or the fact that Christ was incarnated at a time and place suitable for the reception of his message. But Jesus's mission belongs to a timeless revelation, and the Church, as guardian of the eternal deposit of faith, is also to that extent outside history, even though particular actions by popes and others are explainable by human motives.

The Church, however, is only one of two agencies of authority Pluquet invokes in relation to heresy. The other is the state, for so close is the relation between an orderly religion and the "happiness of civil society" (I, ix) that religious and secular authority must be discussed together. Pluquet does not explicitly analyze the foundations of secular authority (although we will find an important clue in some of his case studies), but in a sense he does not need to. His practical conception of authority as an essentially regulatory agency conceived in action/reaction terms allows him to move seamlessly from church to state, since whatever the basis of the political order, authority's role as the regulator of disorder is or should be self-evident.

It may be objected that many heresies involve only matters of belief, and pose no challenge to the state. Pluquet admits as much, and indeed at least twice distinguishes between heresies that do and those that do not have political implications. In the "Discourse," he concludes a chapter by pointing out that, unlike the Monophysite conflicts provoked by Eutyches and Nestorius in the Eastern empire, the Western Pelagian heresy of the fifth century "did not disturb the States" (I, 141). And in the article "Pélage," he devotes a whole section to the question why, unlike Donatism, "Pelagianism died out without disturbing the State" (II, 401–403). The explanation is not theological but historical, involving such factors as the threat of external war and the importance of other factions within the empire. Pluquet was surely aware of the irony involved in the second example: the reappearance of free will/predestinarian debates in the Reformation (and in Catholic France, in the conflict between Jansenists and Jesuits) did far more than merely "disturb the States." Perhaps he is suggesting that, were it not for political entanglements, the new Pelagian and predestinarian extremists would fade away, too.

Pluquet condemns all heresies, of course, but in practice there are important distinctions between those where the issues are primarily theological (for Pluquet the most important of these are the contemporary versions of Manicheanism and Arianism), and those in which the regulatory role of the state is as important as that of orthodox faith. When he discusses these heresies in his dictionary, Pluquet will introduce a mediating idea between the state and the universal church, one that had no place in the religious-historical overview but that plays a key, if not always theologically grounded, role in the assessment of reactions to heresy in specific cases.

This is the Gallican model of the national church, operating within the framework of the universal church but governed by the same criteria of prudent oversight as the political body with which it coincides. Pluquet's Gallicanism emerges most explicitly in the article "Grecs," in which Pluquet discusses the conflict over papal primacy that contributed to the schism between the Western and the Eastern church—a conflict that of course had taken on renewed importance in the wake of the Reformation. Pluquet reaffirms the primacy of the pope, but limits its extent. He cites approvingly a statement by Grotius claiming that if the nature of the obedience we owe the pope had been better understood, "we would now have a reformed and united Church" (II, 95).[16] "The Clergy of France," he goes on, "and all the universities of the Kingdom acknowledged the same truth, without however believing the pope was infallible, or that he had any power over the temporal of Kings." The primacy of the pope is divinely ordained, but it is one "of honor and jurisdiction; it is up to him to ensure the canons of the Church are observed by everyone, to convoke the Councils, and to excommunicate those who refuse to appear before them." Pluquet adds that the pope is subject to Ecumenical Councils, and that he cannot depose a bishop or absolve subjects of their oath of loyalty to the king.[17]

Note that nothing is said here about the reformulation of doctrine. No doubt this is because Pluquet does not believe anything new or controversial could emanate from Rome. When he speaks of papal decisions, he describes them in terms of legislation instead. "The Pope can make new general Laws, & propose them to the Church; but they have force of Law only by reception [*acceptation*]" (II, 95). The view here is similar to the position of the French *parlement* with respect to royal legislation. Indeed Church canons were also regarded as legislation to be received by the *parlement*—or rejected, as in the case of the Council of Trent: only the dogmatic declarations of the Council (as opposed to those concerning governance) were accepted by French law. In making these distinctions, Pluquet is echoing traditional Gallican claims. But he also wants to denounce papal mistakes in dealing with heresy. In his article "Albigeois," he remarks pointedly that the bloody repression of the Albigensians might have been avoided had not the indiscreet zeal of the mendicant orders fanned the flames of the conflict. Like his patron Choiseul, Pluquet was an energetic advocate for the secular clergy, rooted in parish life and controlled by the local bishop, against the encroachments of the religious orders controlled by Rome.

A national church closely linked with other institutions concerned with the well-regulated flourishing of the state is in fact the best defense against heresy. If in the late Middle Ages France was able to avoid the strife associated with Wycliff in England and Hus in Bohemia, it was because its relationship with the papacy was more equitable. "The kingdom and church of France, filled with enlightened men, with learned theologians, with famous universities, conserved its liberty without deviating from the attachment and respect due to the Holy See; one saw only a few lapses [*écarts*] resulting from an indiscreet zeal. They were condemned as soon as they were noticed, and had no defenders" (217). Setting aside for the moment the question of intellectual enlightenment, we should note that Pluquet's defense of the rights of national churches is not limited to France. It is an important argument in his history of religious and political disputes in the Netherlands. In "Hollande," he clearly opposes the bull of Pope Paul IV (in office 1555–1559) creating three new bishoprics in Holland. The new bishops were to serve as Inquisitors in their dioceses, and Pluquet is no friend of the Inquisition. The new bishops would weaken the legitimate authority of the magistrates of Antwerp, Louvain, and other cities, and usurp the revenues of the local abbeys (II, 110–111). As a result, he says, even zealous Catholics joined the opposition to the bull. Three parties thus emerged: "Catholics enemies of the Inquisition and defenders of the privileges of the nation; Catholics devoted to the Spanish Court, who wanted to sacrifice everything in order to ruin the Reformed; and lastly fanatical Protestants who wanted to maintain their position and extend their supposed Reform" (113).

It is easy to imagine which party Pluquet admires, and to draw parallels with the situation of sixteenth-century France, divided among *politiques,* partisans of the League, and hard-line Calvinists.[18] What should also be noticed is Pluquet's use of the term "nation" in this context. By the middle of the eighteenth century, the term's meaning in France had begun to shift from the organized polity of France to the public-spirited consciousness of French citizens (see the reference to *citoyens zélés* above), but its use remained ambiguous, since the integrity of the French state itself was not in doubt. Here, however, the term is used to designate a "nation" in the process of asserting itself as an autonomous force, politically as well as ecclesiastically. The nation is treated as a "fact on the ground" that may be in need of regulation, but that by its very existence possesses its own legitimacy, and so sets some boundaries to that regulation. The same realism drives Pluquet's approach to the religious identity, or rather identities, of the Dutch. They, too, are facts to be recognized. "Tolerate sects but contain them" is the formula he appears to endorse (II, 113).[19] He approves of the Peace of Ghent, and praises the prudent statesmanship, in both political and religious matters, of Oldenbarnevelt as opposed to the "tyranny" of Maurice of Nassau (II, 125). Yet, while these positions emerge clearly enough from the narrative, they are not articulated as declarations of principle. At the end of his article, Pluquet can therefore issue a half-hearted disavowal: "The tolerance of the United Provinces has been much censured; M. Basnage has sought to justify it" (126). The fact that Basnage (1653–1723) was a Protestant historian ostensibly disqualifies what he says, but Pluquet's endorsement is transparent enough, since it parallels his own statement at the end of "Calvinistes."

Pluquet's analysis of the English Reformation in "Anglicane (religion)" displays a similar attitude. This article, too, is more focused on political than on theological judgments, although Pluquet reminds us that he refutes Anglican positions on transubstantiation, clerical celibacy, etc. in other articles. Pluquet cites Bossuet's judgment on Henry VIII's cruelty and arrogance, but presents Mary as equally cruel. Worse, perhaps, she was unable to turn the conflicting advice of her advisers into coherent policy (I, 329). Elizabeth, on the other hand, is presented as well aware of her country's internal divisions and largely successful in resolving them. If her persistence in schism is regrettable, the pope's provocative declaration on her accession that England was a fief of Rome (I, 330) is set in implicit contrast with Elizabeth's efforts to reconcile her subjects as far as possible (I, 332), and to "unite all her subjects in the profession of a single Religion. The execution of this project in fact made her reign a glorious epoch, she ensured the tranquillity of her States, and made her power more feared by foreigners" (I, 331).[20]

Pluquet is much less indulgent of leaders who possess no legitimate political authority, among them Luther and Calvin. While their protests against the abuse of indulgences and the corrupted discipline of the church were legitimate, as disturbers of the peace they must be condemned. Luther "was a man of violent temper, got carried away and overstepped the bounds of moderation" (II, 179); he usurped the role of prophet with a direct link to God (II, 184). Similarly, Calvin did not imitate the "discretion" of the early French reformers; his "impetuous zeal burst forth" (I, 514). This lack of measure could only lead to serious doctrinal distortion and to political upheaval, for not only do such leaders lack legitimacy, they fail to appreciate the statesman's need for prudence. The same is true of Muntzer and other "conquering Anabaptists," who were criminal in seeking to establish a new monarchy in Germany (I, 284).

On the other hand, Pluquet is much more tolerant of humble heretics who lead quiet, productive lives and submit to legitimate authority. The "pacific Anabaptists," the Moravians, and particularly the Hutterites are presented as living an edifying pastoral life (I, 292–93). The peaceful followers of these leaders, whose only error is mistaken belief, are not to be punished. On the contrary. "There were no complaints to be made," Pluquet says, summarizing the attitude of the nobility that profited from the Hutterites working their lands, "about a society all of whose rules had no other goal, it seems, than public utility" (I, 293). The persecution and exile of the Moravians by the Holy Roman Emperor was a foolish mistake, the mark of an unenlightened ruler. Even if the goal is not just to contain heresy but to eradicate it—and as a churchman this must be Pluquet's goal—one should use persuasion rather than force. In "Hussites," he writes,

> Would more harm have been done to Bohemia and to the Church, if after the condemnation of John Hus and his doctrine, the Emperor, instead of sending troops against the Hussites . . . had brought into Bohemia some skillful and moderate Theologians, who would have instructed the people and fought the errors of the Hussites with the arms of Religion, charity, and reason? (II, 145).

III

This effort of enlightenment is not to be confined to apologetics. While the real solution to heresy is to raise the people's level of knowledge generally, one must also restore the proper relationship between religious and other forms of knowledge. Thus, after detailing the cruelties of the crusade against the Albigensians in the twelfth century, Pluquet exclaims: "How many disorders, crimes, and misfortunes does this century offer the thoughtful Christian! However, people were very ignorant; there is hardly a century in which more excommunications were issued, more heretics burned, and the Sciences and Arts less cultivated" (I, 277).

Pluquet develops this point in his analysis of the Reformation. The abuse of indulgences and the political ambition of popes provided the spark that ignited the conflict, but the split between theology and other forms of learning that became apparent in the preceding century was a major factor in the success of Protestantism.

> Men skilled in history, in criticism, in languages, spurned the study of theology and treated with contempt the oracles of the School; theologians, on their side, denigrated the study of letters and claimed it was dangerous for religion.
>
> It was not thus that Luther had treated the men of letters and the learned; he showered them with praise, he attracted some famous writers; thus the disciples of the Reformers found the literary and the learned men favorably disposed to them ("Discourse," I, 223; repeated almost verbatim in "Calvinisme," I, 539).

In his *Dictionary,* therefore, Pluquet is at pains to show that religious and secular forms of knowledge need not be incompatible, and may even be reconciled. He goes out of his way to be civil in his treatment of the heresies that most appeal to the men of letters of his day. A 60-page refutation of Arianism addresses modern English deists (I, 364–426), while a 40-page article on Manicheanism is mostly taken up with problems of theodicy and evil raised by Bayle (II, 247–86). Pluquet's arguments are presented serenely, and accompanied by a detailed summary of his opponents' views that, like the analysis of Spinoza in the *Examen,* strives to be fair and to acknowledge difficulties.

The conflict between orthodox theology and other forms of learning is not to be resolved solely by showing how the one need not contradict the other. Religious strife should also diminish if theology is freed from distortions brought about by its subordination to philosophy.[21] This second aspect of the relationship between different forms of knowledge emerges in Pluquet's analysis of the intellectual origins of heresy. The controversies of the early Christian era, he claims, arose from errors caused by the natural human tendency to use old categories to bridge the gaps in a new teaching.

> The discovery of a fundamental truth makes a strong impression on the mind, it suspends in a sense the activity of the mind; all the difficulties that hindered it disappear.
>
> When this first impression has weakened, curiosity is reborn, one wants to use the principles one has discovered to resolve the difficulties that had weighed one down; and if the principle one has adopted does not resolve them, the mind recurs to its old principles, which it mixes with its new opinions (I, 104).

Almost all the heresies of the first three centuries, Pluquet argues, derive from a forced "union" between Christian ideas and the "principles" of Egyptian, Jewish, Persian, and other early "systems."[22] Pluquet sees in such a union the expression of what the mathematician d'Alembert designated (in the "Preliminary Discourse" of the *Encyclopédie*) as the proper (and properly comprehensive) philosophical outlook. It is the "esprit systématique" of the religious philosophers that is to blame for early heresy (I, 33).[23] When he reaches the thirteenth century in his historical discourse, Pluquet uses a similar argument against scholasticism. While of course praising Thomas Aquinas and Bonaventure, Pluquet devotes most of the chapter to deploring the effect on theology of Arab and Greek philosophy, and especially Aristotle, when they came to be studied intensively in the West.

> There soon appeared a kind of fanaticism for the Greek philosophers, and above all for Aristotle; it was not enough to study his logic, they studied his physics, his metaphysics; they adopted his opinions; and one saw some theologians and philosophers who taught the dogma of the universal soul, the eternity of the world, and absolute determinism [*fatalisme*].
>
> Others sought to reconcile the opinions of this philosopher with religion; without their noticing, it was religion that they tried to accommodate to the principles they found in Aristotle. (205)[24]

This application of the Enlightenment's suspicion of metaphysics to the defense of religious doctrine should be distinguished from the skeptical fideism of the late sixteenth and early seventeenth century. While avoiding arguments that lead to sectarian divisions is a purpose common to both, Pluquet is clearly a firm believer in the new science and in the progress of (secular) knowledge. He praises Bacon, Gassendi, and Descartes for their "war against prejudices" and their research into "the truths of nature" (I, 224–25). And whatever reservations he has about casting theology in philosophical terms, Pluquet is equally convinced of the rational truth of Christianity. This religion successfully explains what the human mind had sought in vain: "the origin of the world, the cause of evil, the nature and destiny of man." Of course the truth of Christianity is also manifested by "the most dazzling miracles" (I, 104). But because Pluquet is confident in the rationality of Christianity, he does not use miracles to "prove" its truth, as Pascal did. On the contrary, he criticizes the "religious Pyrrhonism" that constrains his contemporaries to accept as true only what they can see clearly. Having affirmed the independent integrity of theology, however, Pluquet does not discuss the criteria by which particular beliefs are to be evaluated. Although it is clear from his arguments against heresies that rational discussion of belief is necessary and appropriate, the meta-critical issue simply does not arise, since the "deposit of faith" is unchanging.

In order to sustain this position in the face of doctrinal controversies, that deposit must include all the current teachings of the Roman Catholic Church, even where they have been colored by particular philosophical ideas. Thus, in supporting the notion of Christ's real presence in the Eucharist, Pluquet asserts that the idea of transubstantiation was part of that doctrine from the very beginning of Christianity ("Bérenger [Berengar]," I, 491). The fact that the word is not in Scripture is not an argument against it, since "Trinity" and "consubstantial" are not scriptural either, and yet the Protestants do not reject the teachings of the early Councils. Here we see the effect of viewing doctrinal discussion in essentially strategic terms, as a quest for a basis of agreement using familiar categories, rather than as an occasion for renewed thinking.[25] Strictures against the "esprit systématique" lose their methodological bite, since ultimately they are only directed at dissenters. Pluquet engages his Manichean and Arian opponents with more substantial arguments, but in these cases confessional divisions are not really at issue. Here, Pluquet is happy to use Protestant as well as Catholic authors to make his case, which is also a case for theology against the encroachment of philosophy.

A third aspect of the relationship between religion and the progress of knowledge can be seen in the many articles in which Pluquet cites statements that, although they come from heretics, he appears to endorse. These do not involve dogmatic issues, but rather touch on points of moral virtue or political tolerance. A striking example is "Quakers," which cites a key passage on tolerance from William Penn's constitution for Pennsylvania (II, 460), and gives a lengthy summary, without any critical comment, of Robert Barclay's defense of Quaker morality (II, 461–69). This kind of juxtaposition presents a problem of interpretation. If a general disposition to be tolerant may be inferred from the context and the overall tone of his work, is it possible to go further and to draw conclusions about actual policy? Should we follow d'Alembert's example, and take these statements as evidence that the tolerant Pluquet is really an Enlightenment encyclopedist who dared not say all he thought, or think through the consequences of all he dared? This would be to ignore the bulk of the work, which vigorously reaffirms Catholic orthodoxy. It would also amount to assimilating this third discursive strategy to the first rather than to the second of Pluquet's approaches to the relation between knowledge and religion. But to what extent can moral integrity really be considered separately from correctness of belief?

The answer, I think, is that for Pluquet this can be done only up to a point. First, on the ecclesiological level, Pluquet is genuinely convinced that schism is never justified. As he argues at length in his article on the Donatists, the Church will always contain the unjust as well as the just, and "one [must] tolerate in the church the abuses and the crimes one [can] not correct" (II, 13–14). The primary focus of tolerance should be inward rather than outward, for moral lapses within the true church rather than for virtuous people separated from it. That church doctrine or ecclesiology might have to be reconsidered *theologically* in light of the fact of broken communion is not something Pluquet can contemplate.[26] Second, like many of his Enlightenment contemporaries, although of course in a different sense, Pluquet believes that error is, ultimately, accidental. Pushing his argument only a little, one might say that for Pluquet heresy is either an intellectual *égarement* that will be reabsorbed into the mainstream with the passage of time[27]; or, if it becomes entangled with political division, the religious scandal of faith takes second place, behind the political issue, which is to be "managed" like any other, under the cover of the pious hope that the heretics will be reconciled some day. Thus the existence of the virtuous heretic (not to mention the virtuous atheist) is not a fundamental challenge to Pluquet's way of thinking. If this is the case, one may wonder why Pluquet believes it necessary for the state to regulate religious confessions at all. Could church and state not stand separately side by side, as theology and policy do in his own text? From the point of view of a Gallican Catholic such as Pluquet, the answer must be "no." In addition to the difficulty of conceiving of such a separation in eighteenth-century France, Pluquet hints at another reason, both political and theological: without the counterweight of moderate Gallicans, the church of Rome would be unprotected from its

Ultramontane extremists, unfettered by considerations of prudence, and hostile to the legitimate development of secular reason. In the short run, at least, Pluquet had a point.

Notes

1. The full title is *Mémoires pour servir à l'histoire des égaremens de l'esprit humain par rapport à la religion chrétienne: ou dictionnaire des hérésies, des erreurs et des schismes, précédé d'un discours dans lequel on recherche quelle a été la religion primitive des hommes; les changemens qu'elle a soufferts jusqu'à la naissance du Christianisme, les causes générales, les liaisons & les effets des hérésies qui ont divisé les Chrétiens.* The author's name does not appear on the title page, but the dedication to Archbishop Choiseul of Alby is signed. First published in 1762, the book was reprinted in 1764 and 1776 (and with additions and modifications by other hands in 1817 and 1844). All references are to the 1776 edition, in two volumes (602 and 580pp.). Translations from the French are my own. The biographical information on Pluquet comes largely from the *Dictionnaire de théologie catholique* entry by J. Carreyre.
2. Among other studies, see Stephen Werner, "Diderot's article 'Agnus Scythicus,'" *Studies on Voltaire and the Eighteenth Century* 79, 1971, pp. 79–92.
3. Léopold-Charles de Choiseul-Stainville (1724–1774) was five years younger than his brother. In 1764, he became Archbishop of Cambrai, a position once held by Fénelon. Choiseul supported the expulsion of the Jesuits from France and supported the secular clergy. He also introduced the Gallican liturgy into the diocese of Cambrai. See the entry in the *Dictionnaire de biographie française,* ed. M. Prévost and Roman d'Amat (Paris: Letouzey et Ané, 1933-).
4. *Examen du fatalisme, ou exposition & réfutation des différens Systèmes de Fatalisme qui ont partagé les Philosophes sur l'origine du Monde, sur la nature de l'Ame, & sur les Principes des Actions humaines.*
5. Bachaumont, *Mémoires secrets,* 26 April 1764 vol. 2, p. 48.
6. Voltaire to Damilaville, 20 Feb. 1765, Best. D. 12411.
7. *Correspondance littéraire,* ed. M. Tourneux, 5:189 (November 1762).
8. D'Alembert to Voltaire, 2 October 1762, Best. D. 10740.
9. A week after D'Alembert's letter, he asks Damilaville to send him the book (Best. D. 10755, 10 October 1762). For Voltaire's annotations on his copy, see Edith Phillips and Jean A. Perkins, "Some Voltaire Marginalia," *Studies on Voltaire and the Eighteenth Century* 114, 1973, pp. 60–72.
10. See Voltaire, *Dictionnaire philosophique,* ed. Christiane Mervaud, in the *Complete Works of Voltaire* 35 (Oxford: Voltaire Foundation, 1968-), s.v. "Arius."
11. *Mémoires de Trévoux,* October 1762, 2521 (the second part of the review appeared in the November issue, the third in January 1763). The pagination for each year's issues is continuous, and page numbers will henceforth be given parenthetically in the text.

12. For example, discussion of the Jansenist-Molinist quarrel over grace appears in the article on the lesser-known Baïus.
13. Crébillon plays with the different connotations of the term in his well-known libertine novel, *Les Egarements du cœur et de l'esprit* (1736–38).
14. Fanaticism is associated with Catholics only occasionally, as in references to the assassination of France's Henri IV.
15. See Frank E. Manuel, *The Eighteenth Century Confronts the Gods* (1959; rpt. New York: Atheneum, 1967).
16. For Grotius's reception in France, see Jacques Le Brun, "La réception de la théologie de Grotius chez les catholiques de la second moitié du XVIIe siècle," in *The World of Hugo Grotius (1583–1645)* (Amsterdam: APA-Holland University Press, 1984), pp. 195–214. Le Brun points out that interest in Grotius in eighteenth-century France was rekindled by Levesque de Burigny's *Vie de Grotius* (1754).
17. This passage is cited at length, by the *Mémoires de Trévoux* reviewer, whose only comment is that it is "clear and unequivocal" (Jan. 1763, p. 63n.)
18. Pluquet's article "Calvinistes" is much more circumspect on this topic. Although he describes at length the early stages of the Religious Wars, and deplores the violence on both sides, he does not mention the proclamation of the Edict of Nantes—only its later confirmation in 1629 (I, 566)! The events surrounding the Edict's revocation are also touched on only lightly.
19. "Tolérer les sectes en les réprimant": the phrase is part of Pluquet's summary of representations made by the Prince of Orange's friends to the King of Spain.
20. Pluquet also underscores the fact, often overlooked by Catholic polemicists, that Elizabeth did not wish to define herself as Head of her Church, "the authority seeming to her too extensive, & and too close to the power of J. C." (I, 334). Pluquet does add, however, that if Elizabeth had not lived so long, another prudent monarch could have brought the people back to Catholicism.
21. Pluquet encourages theology to refer more to the writings of the Church Fathers instead, a position that connects him with a part of the Port-Royal heritage that has only recently begun to attract scholarly attention.
22. For Pluquet, those *philosophes* who find in the writings of the early Church Fathers passages implying a belief in the materiality of the soul, are guilty of insensitivity to the historical context and the vocabularies available at the time.
23. For recent discussion of this notion, see Julie Candler Hayes, *Reading the French Enlightenment: System and Subversion* (Cambridge: Cambridge University Press, 1999), pp. 42–45.
24. A similar phenomenon occurred in purely theological investigations even earlier: in the Schools, "knowledge of the Fathers and of ecclesiastical authors was no longer considered necessary to form a theologian; for their study was substituted the art of making a syllogism, and with this art they

undertook to discuss dogmas and explain mysteries. By this method, the mind tended to bring the mysteries closer to the notions or ideas given by reason, and so to alter them" (196); an example is Berengar's theology of the Eucharist. One might speculate, however, that Pluquet also places his discussion of transubstantiation in this article because Berengar was not excommunicated for holding his (mistaken) views. The same is true of Baïus (note 12 above).

25. On the importance of this distinction in the broader context of the history of Roman formulations of doctrine, see Bruno Neveu, *L'Erreur et son juge: remarques sur la censure doctrinale à l'époque moderne* (Naples: Bibliopolis, 1993).
26. For a powerful analysis along these lines, see Ephraim Radner, *The End of the Church: a pneumatology of Christian division in the West* (Grand Rapids: Eerdmans, 1998).
27. This is the only claim in Pluquet's book explicitly rejected by the *Mémoires de Trévoux* reviewer, although in polite terms. "There is in the Discourse an expression . . . that often recurs: that *time* disposes of everything; *time* produces all revolutions; *time* destroys everything; *time* restores everything. This frequent repetition, which may easily have slipped from the Author's pen, astonishes the reader, who as a Christian Philosopher thinks that in what concerns Religion *time* does nothing, but that everything is directed by *the one who has created all times* . . . it is to be supposed that this is indeed at bottom the Author's thought, too" (Oct. 1762, 2538–2539).

13

Heresy in the *Encyclopédie d'Yverdon* [1770-1780]

Clorinda Donato

What could a reader of eighteenth-century encyclopedias learn about heresy from a perusal of their folio and in-quarto tomes spanning the period 1751–1780? This question brings us squarely into the center of the debate that raged throughout Europe in Protestant and Catholic quarters alike over the representation of religion in Diderot and D'Alembert's 25-folio volume *Encyclopédie* (1751–1765). While every stripe of anti-philosophe had denounced the *Encyclopédie*'s heretical premises, European perception of the editors' religiously subversive intentions escalated substantially in 1759 when Pope Clement XIII placed the *Encyclopédie* on the index. Access to the compilation became difficult in Austria, Italy, and especially Spain, where the iron-clad efficiency of Inquisitorial censorship and the fear it instilled in booksellers cut off all but a privileged few from any work published in France.[1] Italy's lack of central leadership had made it possible for two expurgated folio reprints to surface in Tuscany, where despite the purported autonomy of the Tuscan duchy under Leopold III, the Church was simply too close at hand for the Tuscan publishers Ottaviano Diodati in Lucca and Giuseppe Aubert in Livorno to produce thoroughly unaltered editions.[2] However, even less is known about the 58-volume, in-quarto "Protestant" response published in the Swiss town of Yverdon-les-Bains, in 1770–1780.[3] I have deliberately put Protestant in quotation marks to call attention to the somewhat problematic nature of this term as used by Robert Darnton to characterize the work in his *Business in the Enlightenment: A Publishing History of the 'Encyclopédie' 1775–1800.* Published in

1979, Darnton's book reflected the dominant tendency to read the majority of eighteenth-century literary production through the prism of France, which often resulted in the drawing of gross comparative generalizations and very little serious or thorough reading of other texts, especially articles in other encyclopedias.

Thanks to the growing interest in the religious aspects of the Enlightenment to which this volume of essays on heresy provides ample testimony, incentive has grown to read the articles of both encyclopedias comparatively to understand what was at stake for a remarkably heterogeneous audience. Readers delighted in the *Encyclopédie*'s blasphemies against the Church in numerous articles that are now legendary. Hidden surprises lurked in articles such as "Aigle," with its mocking satire of the Holy Spirit, or in cross-references such as "Capuchon," where the reader of the respectful article "Cordeliers" was sent for a heavy dose of *philosophie* polemicizing against the waste and corruption of religious orders that waged the war for Christ by sparring over the width of their hoods.[4] But these oft-quoted examples of Voltaire's agenda to "écraser l'Infâme" had begun to tire even the sage of Ferney himself; in 1770 Voltaire wrote to Gabriel Cramer, that the editor of the Yverdon edition, Fortunato Bartolomeo De Felice, and his collaborators, had produced a work that corrected the many errors plaguing the *Encyclopédie:* "As far as I'm concerned," he concluded dramatically "I would buy the Yverdon edition and not the other."[5] Indeed, many people did just that.[6] Readers now sought more than *philosophie;* they demanded reliable content, and had expectations from the encyclopedic genre that the *Encyclopédie* could no longer deliver, as the appeal of the biting satire and irony of the high French Enlightenment had waned. From this perspective, the religious articles of the *Encyclopédie* proved superficial, simplistic, and flawed, despite the fact that "There are . . . articles on religion favorable to Catholicism, Protestantism, deism, skepticism, or atheism."[7] While this analysis is technically correct, it does not promote a deeper understanding of the context of European religious debate as presented in the *Encyclopédie.* By adopting the comparative methodology of reading the articles of the *Encyclopédie* side by side with the articles of the Yverdon edition, the parameters of the religious issues can be seized and defined rather than left to some of the vague analyses of content that have plagued *Encyclopédie* scholarship. We have adopted this method of reading for the articles addressing heresy. The highly intricate interplay of Catholic religious orthodoxy, anti-Protestantism, and *philosophie* that characterizes the numerous *Encyclopédie* articles dealing with religion, and the encyclopedists who penned them, comes to light.[8]

Thanks to the work of Frank and Serena Kafker, we now know a great deal about the variety of encyclopedists who contributed to the *Ency-*

clopédie and the degree to which they were orthodox. However, the group's eclectic nature proved a stumbling block once knowledge rather than opposition to tradition became the primary quest of encyclopedia readers. In matters of religion, the encyclopedists who contributed to the *Encyclopédie d'Yverdon* were an extremely homogeneous group. They responded to the majority of religious articles taken from the *Encyclopédie* with revised versions; in many cases, they filled in gaps with articles that were completely new. Thanks to the database index of the *Encyclopédie d'Yverdon,* it is possible to accurately compute the number of entries in categories pertaining to religion that the Yverdon encyclopedists either rewrote, annotated, or provided as new articles.[9] Table 13.1 in this chapter has been compiled from that database and reflects most of the fields that are included, with the exception of article length, which is indicated in the database in increments of tenths of a column.

The statistics for articles in the categories of Ecclesiastical History, Sacred History, and Theology are indicative. The Yverdon encyclopedists designated 733 articles in the *Encyclopédie d'Yverdon* as pertaining to "Ecclesiastical History"; of those, 117 were rewritten and 179 were new, while 14 were corrected through the addition of long notes. Of the 422 articles that are unmarked, meaning that they have been reprised from the *Encyclopédie,* 95 percent extend over less than half a column, making them definitions rather than articles. Of the 592 articles under the rubric "Sacred History," 509 are new, 52 rewritten, and merely 31 taken over from the *Encyclopédie.* In the category "Theology," of the 620 articles that make up the total, 76 are new, 176 rewritten, and 10 heavily annotated. Of the remaining 368, the vast majority are less than half a column long. The electronic database makes it possible to express the differences between the two encyclopedias statistically, particularly in the cases where the Yverdon encyclopedists explicitly distanced themselves from the Paris edition by marking articles as new or rewritten. However, there is still a great deal of work to be done in comparing the articles in the two enyclopedias that appear to be the same. Fortunately, the articles regarding heresy that are marked "R" and "N" in Table 13.1 provide ample sources. An overview of the publishing history and cultural role of the *Encyclopédie d'Yverdon* is helpful in understanding the view of heresy in this Swiss compilation.

The *Encyclopédie d'Yverdon,* published in 58 quarto volumes between 1770 and 1780 in Yverdon-les-Bains, Switzerland, by the Italian, Fortunato Bartolomeo De Felice, was conceived of by a group of enlightened Protestant Vaudois pastors and Bernese patricians as a response to the irreligion and *philosophie* of Diderot and D'Alembert's *Encyclopédie,* and as a vehicle for disseminating a workable model of *Lumières* in which revelation, scientific discovery, and innovation would not only co-exist, but work together.

Previous work comparing different categories of these two encyclopedias has yielded tangible results regarding the significant differences between them and the divergent moments of eighteenth-century thought that both represent.[10] However, the articles concerning religion are among the most interesting to deal with, for they provide an enlightened, informed, intellectual response to the many points of orthodoxy and ecclesiastical history that were either overlooked by the authors of the *Encyclopédie,* or were given short shrift. This is not to say that there are no points of convergence—indeed, the two encyclopedias share a critique of Catholicism, which deserves attention. Concerning heresy, convergence can be found in the analysis of a number of ancient heretical movements such as the "Photiniens," the fourth-century sect that denied the divinity of Christ, or the "Originistes," who claimed that Christ was only the son of God by adoption and that demons would be delivered from the fires of hell. The articles describing this "ancient" group of heretical movements are identical in both encyclopedias. However, the bulk of the articles dealing with Protestantism or "proto-protestantism" has been rewritten to reflect the view that coincides with point 2c in Martin Mulsow's chapter, "Trinity as Heresy," in this collection. In delineating the variety of forms that early modern histories of heresy adopted, Mulsow defines position 2c as follows: "A person x, found traditionally to be a heretic by orthodoxy, is indeed no heretic, because the condemned doctrine he or she advocated was in fact the right doctrine." The rewritten articles "Hérésie" and "Hérétique" demonstrate a clearly defined position on heresy and its definition, culminating in an accusation of the *philosophes* as the true heretics.

Numerous factors contributed to the revision of articles on heresy in the *Encyclopédie d'Yverdon,* among them the autobiographical experiences of the enterprise's managing editor, F. B. De Felice. De Felice had had his share of personal experiences with heresy as a Franciscan in Italy in the 1750s who had been targeted as a Protestant sympathizer.[11] Disillusioned over the faltering Enlightement project in Italy following the death of Benedict XIV and the election of Clement XIII, De Felice sought any available means to leave Italy. He had been schooled in Brescia during the 1730s, when Protestant texts circulated freely in this small, intellectually curious northern Italian town, where religious debate was open.[12] He then went to Rome, where he took his orders, setting out a few years later for Naples to occupy a chair of experimental physics at Antonio Genovesi's behest. Ties between Naples and Berne were strong, and he had the opportunity to travel to the "pays hérétique" as the result of his oft-cited relationship with the Countess Panzutti. Forced to return to Italy, he began laying plans for a definitive escape to Switzerland, which he achieved in 1757.

From Switzerland, De Felice became Catholicism's unabashed critic in his work as journalist and encyclopedist. However, De Felice had already begun honing his anti-Catholic rhetorical skills before leaving Italy. In 1756 an eight-volume translation of Pascal's *Lettres Provinciales* with copious notes and commentary was published in Venice bearing his signature. Once De Felice fled Italy for Switzerland under a pseudonym, his reputation as a heretic in his native country was sealed. None of his former colleagues or collaborators dared correspond with him, making it impossible to enlist the services of his mentor, the Neapolitan Freemason Raimondo de Sangre, the Prince of San Severo, with whom, he wrote in his correspondence, he spent hours critiquing the Catholic Church.[13] His success in Swiss *Lumières* and the brisk publishing success of his encyclopedia began to rival that of Diderot and D'Alembert's *Encyclopédie*, and prompted Voltaire's infamous characterization of the Italian as an "apostate, moine défroqué," living in a "cloaque du Pays de Vaud."[14]

Nonetheless, the *Encyclopédie d'Yverdon* took the world of eighteenth-century letters by surprise. As much as the *philosophes* critiqued its humble origins, published far from the likes of Paris in the relatively insignificant Swiss town of Yverdon-les-Bains, the *Encyclopédie d'Yverdon* became one of the jewels in the crown of French-language publications produced in Romandie in the second half of the eighteenth century. Judging from the twice-yearly catalogues published to advertise book dealers' wares at the Leipzig book fair, the Yverdon edition was successfully marketed and sold there twice a year.[15] Book fair catalogue statistics clearly indicate a sudden and almost thorough shift in the publication of French-language books produced outside of France from Holland to Switzerland in the second half of the eighteenth century. De Felice's preface to the Yverdon encyclopedia explains why this small town in the Pays de Vaud under Bernese rule was the ideal place to publish an encyclopedia:

> The religion we profess, the government under which we have the privilege of living allow us to freely trace the tableau of human knowledge as it is, without following the authority of any laws other than those that require esteem for truth, love of goodness and guidance of enlightened reason, without the fear of being oppressed by a power as monstrous as it is fearful, of which ignorance, fanaticism, superstition, price, avarice and cruelty avail themselves to suffocate truth or to block its progress.[16]

Indeed, Berne and its French-speaking territories provided a haven of open expression, at least for the issues that were the most important to him, i.e., exposure of the Catholic Church *and* exposure of the fallacy of *philosophie*. De Felice's introduction can be read as a religious and political justification for *philosophie* as a strategy of necessity.

The preface also provides an overview of the criteria that he and his collaborators applied in their adaptation of the *Encyclopédie.* Acknowledging the French encyclopedia as their blueprint, De Felice explained that the relationships among articles belonging to the same branch of learning had been clarified so that the reader might better grasp the encyclopedic genre as a "système complet." He reported that the numerous gaps in the French edition had been filled, while articles that were no longer useful or interesting to "l'Europe éclairée" had been eliminated. Outdated or poorly written articles had been cast anew as well. Additionally, theology and church history figured prominently among the disciplines that had been completely rewritten.

Early testimony from Albrecht von Haller and Charles Bonnet bears witness to these changes. While harboring reservations about the enterprise in its early stages, Haller wrote in favor of De Felice's project to Charles Bonnet after reading the first few volumes in 1770: "The originality of De Felice's enterprise when compared to that of Paris cannot be contested. . . . In due course we will all be persuaded that the spirit is different, particularly concerning religion and philosophy . . ."[17]

In 1772 Charles Bonnet read 25 articles from the *Encyclopédie d'Yverdon* with a comparative mindset. Almost all of the articles that he annotated in a *Notice Raisonnée* dealt with religion and morality. Bonnet also critiqued numerous articles from the Paris encyclopedia as well as from Formey's *Dictionnaire portatif.*[18] Bonnet's judgement echoes contemporary Swiss thought about the work. On 13 November 1772, he wrote to Haller about his contrastive exercise: "I have just read a cross-section of 25 articles from the *Encyclopédie d'Yverdun.* If I judge the method of working from these articles, I prefer it to the one from de Paris."[19] Bonnet expressed particular interest in the articles written by Gabriel Mingard, the pastor in Lausanne who was De Felice's closest collaborator.[20] He was assigned the articles on theology and rational philosophy that Bonnet found particularly well done.[21] Bonnet complimented the reworked article "Certitude," which he found to be "good protection against Pyrrhonism."[22] Mingard's articles deserve close attention because they reveal the controversy concerning orthodoxy in the Pays de Vaud, which Berne controlled. Therefore Bernese censors read the Yverdon encyclopedia. In volume X of the Plates, the last volume of the *Encyclopedie d'Yverdon* to be published before the six volumes of the *Supplément* were produced, De Felice explained why Mingard signed his articles with two different sets of initials—either GM or MDB. Mingard had been accused of writing articles that were unorthodox, inviting close scrutiny of his signed work. For this reason, De Felice had him sign some articles MDB, a decision he explained in an *Avis de l'éditeur* inserted at the end of the tenth and last volume of plates. De Felice included this *Avis* in order to identify most of the authors whose initials had signed the majority of articles and to denounce those whose jealousy had prompted them to call Mingard unorthodox. He explained that he began to sign articles on religion MDB, while continuing to use GM for the rest.[23]

Table 13.1 shows articles on heresy signed both ways by Mingard. Further research is needed to pinpoint exactly why the articles signed MDB posed more of a threat to orthodoxy for the Bernese censors than the ones signed GM. The objections probably came from the censors themselves, who were pastors, as was Mingard.[24] According to Perret, the censorship policy established by Berne in 1768 stipulated that all manuscripts had to pass through the hands of the censors before the *Approbation* could be granted. However, as he also notes, application of the policy was lax, and many books were published, sold, and distributed in absence of the *Approbation*.[25]

In theogical matters, De Felice was also helped by one of Mingard's senior colleagues, Alexandre-Cesar Chavannes, who signed CC and wrote on the history of religions; Elie Bertrand (BC), a Vaudois pastor who carried on an intensive correspondence with Voltaire, also wrote about theology.[26] The list of *Encycopédie d'Yverdon* articles (Table 13.1) on heresy indicates articles that were either anonymous, signed by De Felice, Mingard (with both sets of initials), Chavannes, or Bertrand.

Table 13.1 provides a representative sampling for heresy; in order to do more thorough justice to the topic, the entire corpus of articles signed by each of the four authors on theological matters should be dealt with separately. For example, Mingard's articles (signed both GM and MDB) number over 1,371. Nonetheless, our work in this area allows us to establish the following general points of convergence and difference between the two encyclopedias. They establish an important working context for the topic of heresy:

1. The *Encyclopédie d'Yverdon* refuses and refutes all forms of "philosophical pride." The group of articles "Pyrrhonisme," "Conjecture," (not signed) and the articles "Comment" and "Evènement" signed by Mingard consistently attack atheism, materialism, and Spinozism, in particular D'Holbach's *Système de la Nature.* This refutation is found in Mingard's article "Nature," in which he cites the mistake of those who believe, as Spinoza did, that the universe "était la cause de lui-meme."
2. The *Encyclopédie d'Yverdon* repeatedly affirms the compatibility of revealed religion and natural religion. Mingard's article "Déiste" provides the best example, while De Felice peppers throughout his own articles entire passages of Rousseau's *Profession de foi du Vicaire Savoyard*.[27]

3. The Yverdon encyclopedists agree, however, with the *philosophes* that there is nothing worse than religious fanaticism. Catholics are taken to task for this in the articles "Jesuites," an unsigned, rewritten, 51-column article, and "Missionaire," also rewritten and signed by De Felice himself. However, the Yverdon encyclopedists consistently correct attributions of fanaticism to Protestant and proto-Protestant movements made in the Paris encyclopedia. This point will be addressed in detail in the discussion on the articles "Albigeois," "Hussites," and "Luther, Martin."
4. Strong attack against Catholicism, mainly coming from De Felice and Mingard. For example, De Felice adopted wholesale from the *Encyclopédie* articles that condemned the fanaticism of the Catholic Church, such as "Croisade"; rewritten articles such as Mingard's "Célibat" reinforce this position.
5. Critical of the persecution of heretics, even when the persecution was imposed by the Calvinists. For example, Elie Bertrand's articles on Calvinism and Jean Calvin criticize the "supplice de Servet" in which Jean Calvin put the Spaniard Michel Servet to death for his interpretation of Calvin's teachings; futhermore, Bertrand was bothered by the Calvinists' strict adherence to scripture, claiming in his articles that Calvin "n'est point du tout un docteur infaillible."[28]
6. By extension, the Yverdon encyclopedia is hostile to all forms of theological dogmatism, which is the position that informs the two articles "Hérésie" and "Hérétique." Gabriel Mingard's articles most consistently espouse anti-dogmatism.

The list of articles related to heresy in Table 13.1 is by no means exhaustive, yet it is thorough enough for the purposes of this chapter. At a glance it confirms the importance of a particular branch of religious inquiry in the Yverdon encyclopedia. It is helpful to keep in mind the autobiographical element of these encyclopedists, particularly in the case of Fortunato Bartolomeo De Felice. De Felice grew up and was ordained in a Rome whose Inquisition was:

> reconstituted in 1542 to combat the menace of Protestantism in the Italian peninsula, whereas the Spanish Inquisition had been created more than half a century earlier to deal with massive numbers of converted Jews. The nature of what was considered "heresy" in each system reflects these original concerns.[29]

Although almost 200 years separated the reconstitution of the Roman Inquisition to stamp out Protestant heresy and De Felice's decision in 1757 to leave Italy, it is easy to understand why De Felice identified with the Protestant mission to rectify the errors of the Catholic Church and to set a new course that was determined by the spiritual and intellectual engagement with Scripture that Luther had espoused at the time of the Reformation. Nowhere is this more clearly demonstrated than in the three articles "Albigeois," "Hussites," and "Luther, Martin." Although only the article "Hussites" is signed by De Felice, the other two are unsigned but bear his style and passion. These three articles completely rehabilitate the "heretics" described in the *Encyclopédie* articles of the same title by holding them up as shining examples of virtue and proponents of the right doctrine as defenders of reformed Christianity.

The Yverdon encyclopedia articles on Jean Hus, leader of the Hussites and Peter Waldus, founder of the Albigeois, read like hagiographies of reformation martyrs, stunning precursors in thought and deed to Martin Luther, whose biography constitutes the master narrative. The stories of these three men's critique of Catholicism follow similar patterns, with features that set them apart from accounts of Catholic martyrs in the lives of the saints genre. Hus, Waldus, and Luther shared an individual, intellectual engagement with religion that they discovered through the personal study of Scripture. The driving desire to allow each individual access to the word of God fueled the impetus to translate the Bible. Through the exempla of these three leaders, reformed religion is established as the domain of the reasonable man, in effect, the domain of the Yverdon encyclopedist's enlightened religion. There is no hierarchy whose authority in matters of religion reigns supreme. This point of view is summed up in Elie Bertrand's assessment of Jean Calvin as not being an infallible doctor.[30]

De Felice attacks the fanaticism of the Catholic Church, the "rage romaine" that persecuted Jean Hus and tried to block his teachings. More importantly though, De Felice described the ecclesiastical institutions in Prague, the Church and the University, as supporting Jean Hus in bucking the authority of Rome. Authority resides with truth, and that truth is to be found in Scripture, not in the structure of the Church or in the constructed infallibility of one man. The last lines of De Felice's article "Hussites" leave no doubt as to the Yverdon Encyclopedia's mission in defense of Protestantism, its precursors, and its proponents: "We have found it necessary to expose all of these things to the eyes of Christianity to prevent the faithful tricked by false imputations from making any attempts on the reputation of this just man or those who follow his doctrine."[31]

The article "Hussites" as it appears in the *Encyclopédie* serves up an entirely different interpretation of Jean Hus, the Hussites, and their aftermath. Less than one-quarter the length of the Yverdon article, Jaucourt focused on the religious wars that shook Bohemia in the wake of Hus's death. No information on Hus, the well-respected university professor, is provided. Instead, Jean Ziska, leader of Hus's followers in the wake of Hus's death, interests Jaucourt the most. He paints a violent image of religious war, characterizing the Hussites as savages.[32] Jaucourt's sarcasm is palpable, as is his scorn for anyone who would bring religious zeal to such a pitch. While De Felice emphasized the erudition and reason that guided Hus and his followers, Jaucourt focuses on Ziska, reminding us that the name means "borgne," or one-eyed (he had lost one of his eyes in battle), while the figurative meaning of dishonest and ill-reputed was lost on no one. Jaucourt also emphasizes the grotesque and the bizarre, ending his account with the image of Ziska's skin stretched over a drum that was used to lead the Hussites in battle after his death. Such a portrayal corresponds with the intent of the *philosophe* camp to point out the monsterous, superstitious, and fanatical repercussions of all religions.

The article "Albigeois" offers a similar comparative scenario. Though unsigned in the Yverdon edition, De Felice probably wrote it. As was the case for "Hussites," "Albigeois" has been greatly expanded with respect to the *Encyclopédie* article written by Mallet, a Catholic priest, as is evident from the first few lines of the article where Mallet cavalierly describes the Albigeois as "one of those sects made up of several heretics . . . whose main goal was to turn Christians away from receiving the sacraments, to overthrow the Church's hierarchy and upset the discipline of the Church."[33]

Mallet's style is noteworthy, as it constitutes a conflation of what we might call "philosophie catholique." Mallet employs the sarcastic, mocking tone used by the *philosophes* against the "Infâme," but his position purportedly represents that of the Catholic Church. This binary "us, Catholic" "them, heretics" opposition holds true throughout the article. Mallet plants the seed in the mind of the reader of a marauding band of various heretics who have come together to disrupt the institutional aspects of the Catholic Church. They are labeled as heretics from the first sentence of the article and are held in full contempt by Mallet for their "erreurs" and for having "infected" the provinces around Albi with their beliefs, which are disparagingly called opinions. Instead, the Yverdon version of the same article, asserts a different binary opposition: "us, Protestants and therefore heretics in the eyes of the Catholics" and "them, Catholics."[34]

The care taken in philological precision to accurately describe who they are and how they have evolved indicates the amount of respect the author accorded to this group. The author signals the central role the Albigeois played when he claims that the group constituted the heretical group whose beliefs and practices placed them at the origin of the illustrious genealogy of Christian reform that resulted in Protestantism: "Despite the rage of its persecutors, the doctrine of the Albigeois met with favor everywhere it spread. Born from it were the Viclesites, the Hussites of the brothers of Bohemia, and in the fifteenth century, the Protestants.[35]

This article also mentions the successful crusade Pope Innocent III started against them in 1208 and the promise of plenary indulgences to those who took up arms to stamp out heresy.[36] The article "Albigeois" contains a cross-reference to the article "Croisade," which has been reprised word for word from the same article in the *Encyclopédie.* This is not surprising, since "Croisade" takes a sober view of all crusades as prime examples of Catholic religious fanaticism.

The life of Peter Waldus is embedded in the article "Albigeois" to explain the use of the term Waldensiens, or followers of Waldus of Lyon, which is also used to refer to the Albigeois. A follower of Waldus, Olivier, brought Waldus' religious beliefs to the town of Albi, from whence they spread. Here we are given the real story of Waldus of Lyon, a rich merchant who gave generously to the poor while taking up a life of study and piety. His efforts produced a translation of the New Testament: "The people came to him in droves. He translated the New Testament for them in the vulgate, giving them a clearer and more judicious explanation than any they had ever heard publicly from the Doctors of the Church."

The parallels with Hus and Luther are evident. The Yverdon encyclopedists remind us in the article "Albigeois" that the Catholics themselves had made the case for viewing them as proto-Protestants since their beliefs did not differ from those of Luther.[37]

The 32.4-column article on Martin Luther's life, work, and teachings ranks the longest among the articles on heresy in the *Encyclopédie d'Yverdon.* Martin Luther's successful challenge to papal authority and the temporal sweep of the Catholic Church crown the culmination of a battle waged for reason and clarity by the Albigeois and the Hussites before him, and by the reformers who continued his work after him. Luther's translation of the Bible, indeed the very *act* of translating it, establishes the true moment of rupture with the Catholic Church. The author of this article underscores how the act of making the word of God known to the public constitutes the single most important cleavage between Catholicism and the Reformation, or what the Yverdon encyclopedists referred to as "notre réligion" in the passage describing Luther's transgressive act of translating *and* publishing his New Testament, which was read "by people of every class with extraordinary avidity and attention."[38]

Readers of the Bible were described as capable of understanding God's law for themselves once they had access to it in their own language. Moreover, the word of God contained everything a reasonable person needed to know. Through reading and reason everyone is capable of following and applying God's law and understanding where and when they might stray. This was the message of Luther, and of Waldo and Hus before him. The article also establishes the connection between Luther and the Renaissance. Faith, Renaissance erudition, and respect for man's faculties of reason join together in Luther, the embodiment of the Yverdon encyclopedists' beliefs. The article on Martin Luther in the *Encyclopédie d'Yverdon* elucidates how erudition and reform went hand in hand.

Martin Luther is styled as a man of scientific inquiry in pursuit of the truth contained in Scriptures, which reason dictate he tackle by means of the interpretive methods developed to extract the meaning of the Greek and Latin texts that had come to light during the Renaissance. Thus through reason, a human faculty, revelation is made known and is understandable.[39] Martin Luther fulfils the promise of Hus, Waldo, and Wyclif, whose efforts to make the Bible available in the vernacular are now crowned by the expertise of a Renaissance theologian whose erudition has made it possible to fully access the word of God. Renaissance, erudition, reason, "bel esprit," and even wit (*plaisanterie*) have been used to beat back the ignorance of the monks and priests who had persisted in promoting error that could no longer survive in the face of scientific truth. Once again, erudition and reform are undeniable partners. The individual who had studied the ancient texts immediately saw the fallacy of the Catholic Church and was drawn to the spirit of reform through reason: "The majority of intelligent men who applied themselves to the study of ancient literature at the end of the fifteenth century and at the beginning of the sixteenth saw the absurdity of many of the practices and opinions authorized by the Church without even having the intent or desire to overturn the system of established religion as a goal; they had seen the weakness of the arguments used by uneducated monks to spread them."[40]

The Yverdon encyclopedists saw their work as the continuation of Martin Luther's erudite reform. During the eighteenth century, when the *philosophes* were using reason and erudition to tear down religion, Bertrand, Chavannes, De Felice, and Mingard fought back with an entire encyclopedia whose working premise was the compatibility of reason and religion. The excerpts taken from the *Encyclopédie d'Yverdon* article on Martin Luther promoted the reformer as the embodiment of the "right doctrine" whose example continued to provide guidance. Instead, the brief *Encyclopédie* article "Luthéranisme" contained a portrait of the reformer that condemns him as the initiator of a new heresy in the first paragraph, disparagingly defining Lutheranism as Luther's "sentiments" about religion. He is described as a violent megalomaniac whose desire to see himself as "head of a party" was neither dampened by excommunication nor the condemnation of several universities.[41]

The article ends with the trivialization of Luther's ideas, and the damning evidence of the confusion and discord created by Luther in the spawning of 39 different sects, all listed for effect: "Il est sorti du luthéranisme trente-neuf sectes toutes différentes; savoir. . . ."[42]

Table 13.1 lists several articles, either rewritten or new, all of which serve to flesh out the position of the Yverdon encyclopedists on medieval heresies that were defended as prototype expressions of Protestantism. Articles such as "Cathares" or "Adamites ou Adamiens," both rewritten by Chavannes, present ideas that were close to those of the Albigeois, the Waldensiens, and the Hussites, but condemnable, along with the new article on Savonarola, whose assignment to the category "Fanatisme" leaves no doubt as to how his ideas of religious purity were seen. They see the thread of continuity running from the Albigenses, Waldensians, and Hussites through Luther, and indeed, both the Waldensians and the Hussites jumped immediately on the reformed bandwagon once exposed to the teachings of Luther.[43]

The story of reform is far from complete if we fail to press beyond Luther and examine a sampling of articles dealing with seventeenth- and eighteenth-century figures, not to mention the role they have assigned themselves as keepers first of the "right doctrine" and second as proponents of the tradition of Protestant erudition that Luther had established during the Renaissance. Indeed, one might read the entire *Encyclopédie d'Yverdon* as a monument to Protestant erudition, inclusive of Switzerland, to be sure, but also to the other two bastions of Protestant scholarship, namely, Holland and the German centers of Protestant erudition, i.e., Halle, Leipzig, Berlin, and Potsdam. Thus, the complementary set of articles we are about to examine focuses on seventeenth-century figures whose ideas and teachings set the stage for the controversy about heresy in the two encyclopedias; more importantly, they lay bare the opposing religious premises upon which each encyclopedia had built its intellectual edifice.

The most striking contrast among seventeenth-century scholars in the *Encyclopédie d'Yverdon* pits Pierre Bayle against the brothers Jacques Basnage and Henri Basnage de Beauval. The eighteenth-century battle over religion that plays itself out in the pages of the two encyclopedias finds a precedent in the comparison of their sources and the goals of their writings. New articles on all three appear in the *Encyclopédie d'Yverdon,* and source texts from all three are cited in other articles; for example, Bayle is quoted in the *Encyclopédie d'Yverdon* article "Hérésie," while Jacques Basnage's *Histoire Ecclesiastique* is cited in the article "Albigeois" as the authority who was used to correct the erroneous scholarship of Catholic Church historians like Bossuet.

The charges levied against Pierre Bayle are numerous. The tirade of accusations stems from the belief that Bayle willfully spread heresy through his writing; indeed, his *Dictionnaire Critique et Historique* paved the way for the *Encyclopédie,* as did Voltaire's glowing assessment of Bayle's method and erudition. The article tears down Bayle through a series of direct attacks that leave no stone unturned in their vehemence. However, the most serious of all his "crimes" is that of having opened the door to the *philosophes* through a method of argumentation that can be reduced "to the art of accumulating sophisms." Bayle is cited for his method of heaping one teaching on top of the other without imposing a teaching of "truth" or even seeking one. Bayle is the enemy of all religions and the precurser of the *philosophes* both in his procedures and in the goal of his erudition. For the Yverdon encyclopedists, Bayle placed reason and religion in a state of perpetual contradiction, resulting in their reciprocal destruction. Bayle is painted with the same brush that illustrated the worst of the heretics from the past (*manichéisme, pyrrhonisme*) to the present (*déisme, athéisme*).

The articles about the Basnage brothers are not remarkable in and of themselves, but when seen in contrast with the article on Bayle, the plan of the Yverdon encyclopedists emerges yet again. Their works are cited as reliable and solid, even though their mediocre style keeps them from attaining the pinnacle of erudition. The Yverdon encyclopedists also used these articles on Basnage to point out the frivolities of "La République des Lettres" that blocked the circulation of truth for reasons of personal gain and pettiness.

Further research continues to turn up figures who could be added to this picture of the seventeenth century, such as the theologian Philipp van Limborch, a Dutch Arminian whose work on the Inquisition is cited in the article "Hussites," or another famous Arminian, the jurist Hugo Grotius, whose work was greatly admired by the Yverdon encyclopedists; both Limborch and Grotius are the subject of new articles in the *Encyclopédie d'Yverdon.*

This assessment of the eighteenth-century representation of heresy brings us to the articles "Hérésiarque," "Hérésie," and "Hérétique," which are unsigned in the *Encyclopédie d'Yverdon.* The article "Hérésiarque" defines the heresiarch as the author of a particular heresy, or the head of a heretical sect, followed by a cross-reference to the article "Hérétique." The heretics listed are Cerinthe, Ebion, Basilides, Valentin, Marcion, Montan, Manès, Arius, Macedonius, Sabelius, Pélage, Nestorius, Eatyches, Socin, etc., but a comparison with the list in the *Encyclopédie* article reveals an important difference: seven heretics have been excised from the list that appears in the *Encyclopédie d'Yverdon* article. Their names appear in italics that we have added to the following citation from the *Encyclopédie:* "The principal heresiarchs were Cérinthe, Ebion, Basilides, Valentin, Marcion, Montan, Manes, Arius, Macedonius, Sabelius, Pélage, Nestorius, Eatyches, *Berenger, Wiclef, Jean Hus & Jerome de Prague, Luther, Calvin, Zuingle, Servet,* Socin, *Fox* etc."[44]

In the Yverdon encyclopedia, Berenger, Wyclif, Jean Hus, Jérome de Prague, Luther, Calvin, Zwingli, and Servet have been removed from the ranks of the heretics. Although we have not dwelt on these figures in this chapter, all of them or their followers are the subject of either new or rewritten articles in the *Encyclopédie d'Yverdon.*[45] The articles "Hérésie" and "Hérétique" need to be read as one, since they present a composite picture. Both encyclopedias define heresy as having changed in meaning from a positive term to a negative one. The Yverdon encyclopedists describe the heretic as a reasonable skeptic, akin to Descartes in his consideration and weighing of all assertions before committing to one.[46] Instead, the article claims, theologians have preferred to limit the meaning of heretic to those who express ideas relative to objects of faith and religion in a way that does not conform to the established and dominant system.

The author attributes all problems of the Church to its attempts to stamp out heresy. This position is not too far from that of Jaucourt's article on heresy in the *Encyclopédie.* Both encyclopedias reject the religious and political fanaticism that had given rise to strife and war throughout Europe. Jaucourt's article is specifically geared to the French monarchy and the wars of religion that tore France apart in the sixteenth century, something of a precursor (along with all of the articles related to the Catholic Church in the *Encyclopédie*) to Voltaire's *Traité sur la tolerance* in its political and consciousness-raising agenda.[47] Jaucourt is careful, however, of the powers-that-be, making it clear that he is not providing an apology for heresy. He wants French kings to be more concerned about maintaining peace in the kingdom than stamping out heresy. As Jean Haechler has pointed out in his discussion of the article "Hérésie," Jaucourt is challenging the Church's jurisdiction by juxtaposing dogma with reality.[48] The need for peace must prevail over the desire for a single religion.

Though Jaucourt's view of heresy converges substantially with that of the Yverdon encyclopedists, it is nonetheless the view of a philosophe. Jaucourt implies that all religion is based on opinion and that truth is only relative.[49] His simplistic call at the end of the article for peace, mildness, and tolerance with no real proposal of how to institute them, nor any real analysis of how, among contemporaries, heretics are formed, how they acquire a following, nor how they should be treated, constitutes an ideal example of what the Yverdon encyclopedists saw as empty *philosophie,* i.e., critique and commentary providing no real solutions. The *Encyclopédie d'Yverdon* instead asks the reader to consider the position of the purported heretic and the origin of his ideas, which were probably passed down through the milk suckled at his mother's breast.[50]

The author attempts to provide scenarios for how accusations of heresy should be treated in civil society that engages the reader in a consideration of individual rights within political structures, focusing on the boundaries between an individual's freedom of thought and the public good in a text that is reminiscent of Rousseau's *Social Contract* in its view of heresy as a phenomenon that society must take apart and address within the arena where individual rights and the common good spar. The author attacks Voltaire and the *philosophes,* criticizing them for recommending an empty tolerance in such cases. He accuses them for not understanding tolerance and for substituting it with reckless license and complete impunity, resulting in disorder. The author claims the *philosophes* have passed from one extreme to the other, the new extreme being the irreligion of *philosophie.* At one time, he tells us, superstition created heretics, sharpened knives, and lit inquisitorial fires; today irreligion threatens humanity with the same dangers by white-washing what is the blackest and most atrocious, and by conceiving of the human race as having "one single head so that it might be chopped off in a single blow."

This imputation against the *philosophes* as the new heretics whose ideas, methods, and practices are just as dangerous as those of Catholic orthodoxy in their establishment of a new orthodoxy *philosophique* constitutes a stunning rejoinder both to Jaucourt's entire article on heresy and his empty plea for toleration as well as to Voltaire, who is again criticized for his pronouncements on rights without defining what they are.

It is ultimately this position against the *philosophes* that makes the *Encyclopédie d'Yverdon* its own *machine de guerre,* one that is pointed against what has become the orthodoxy of *philosophie,* a claim not at all unlike the one levied against the French Enlightenment by Adorno and Horkheimer after World War II: the Enlightenment had failed to be self-critical.[51] We might ask ourselves, though, what doctrine or philosophy undergirded the Yverdon encyclopedists' position beyond their apology for enlightened Protestantism; moreover, we must ask what philosophy they were ultimately proposing to a late-eighteenth-century European audience. The answer is to be found in the teachings of Christian Wolff (1679–1754), who had argued for the "middle ground between the pure truths of revelation and those of reason, where reason can be used in support of revelation."[52] Fredrick the Great had been responsible for calling Wolff to Halle to establish religious tolerance throughout Prussia despite problems with the Pietists, who had kept enlightened thinkers from the professoriate. Among the handful of *Encyclopédie d'Yverdon* contributors who wrote anonymously from Potsdam was Jean Henri Samuel Formey, a transplanted

Table 13.1 New and Rewritten Articles Related to Heresy in the *Encyclopédie d'Yverdon*

Title of Article in Encyclopédie d'Yverdon	*Category*	*Relationship to* Encyclopédie[a]	*Country*	*Initials of Author*	*Length of Article in Columns (Quarto Edition)*
Acte de Foi	Hist Mod	R	Spain		0.6
Adamites ou Adamiens	Hist, Eccl	R		CC	1.8
Albigeois	Géod Mod	N	France	DG	0.1
Albigeois	Hist, Eccl	R			6.2
Apostasie	Théol, Morale	R		CC	4.6
Auto da Fé v. Acte de foi		S			
Auto de Fé ou Acte de foi	Hist Mod	R	Spain		5.2
Arianism	Hist, Eccl	R		CC	5.9 (4 renvois)
Berenger iens	Hist, Eccl	R		CC	1.5
Brescia	Géog	R	Italy		5.5
Calvin, Jean	Hist, Litt	N	Switz	BC	9.6
Calvinien, Calvienne	Théol	N	Switz	BC	0.2
Calvinisme	Théol	R	Switz	BC	5.8
Calviniste	Théol	R		BC	1.6
Campanella	Phil	S	Italy		5.1
Campanella, Thomas		R	Italy	GM	5.1
Cathares	Hist, Eccl	R		CC	1.3
Confessi on de Foi	Théol	N		CC	1.0
Déisme	Théol	R		GM	1.7

(continues)

Table 13.1 *(continued)*

Title of Article in Encyclopédie d'Yverdon	*Category*	*Relationship to* Encyclopédie[a]	*Country*	*Initials of Author*	*Length of Article in Columns (Quarto Edition)*
Déiste	Phil, Théol	R		GM	46.0
Démonomanie	Méd	S		S	0.9
Démonomanie	Médec	R		S	7.8
Dissidents, Dissidens	Hist, Eccl	R	Poland	BC	7.0
Erasme, Didier	Hist, Litt	N	Netherlands		4.1
Foi	Théol	R		GM	80.0
Foi,	Droit Nat & des Gens	N		DF	7.3
Foi, la	Myth	R			0.3
Exorcisme	Magie	R		GM	4.2
Hérésie	Théol	R		CC	4.5
Hussites	Hist, Eccl	R		DF	4.3
Inquisition	Hist	S			8.4
Jansenisme	Hist, Eccl	S			8.5
Jérome de Prague	Hist, Eccl	N			0.7
Jesuites, Ordre des	Hist, Eccl	R			51.9
Limborch, Philippe de	Hist, Litt	N	Holland		0.6
Lutherien	Théol	N			0.1
Luther, Martin	Hist, Eccl Mod	R			32.4

(continues)

Table 13.1 ***(continued)***

Title of Article in Encyclopédie d'Yverdon	*Category*	*Relationship to* Encyclopédie[a]	*Country*	*Initials of Author*	*Length of Article in Columns (Quarto Edition)*
Magie	Jurisprudence	N		MDB	22.4
Malabares, Philosophie	Hist	S			11.5
Manichéisme	Hist, Eccles	S			9.3
Matérialisme	Philosophie	N		GM	35.0
Menochius, Jacques	Hist, Litt	N	Italy		0.3
Nantes, Edit de	Hist, Mod	R		GM	9.0
Nestoriens	Théol	S			4.0
Ochino, Bernardino	Hist, Litt	N	Italy		2.1
Pape	Hist, Eccles	R		MDB	49.2
Photiniens	Hist, Eccl	S			0.6
Polythéisme	Histoire, Critique	R		MDB	10.8
Possession de démon (Renvoi d'Adjuraion, Théol	Théol	N			11.1
Primauté du Pape	Hist, Eccl	R		MDB	1.4
Protestans	Hist, Eccl	R		GM	7.4
Schisme	Hist, Eccl	R			4.8
Sociniens	Hist	R		GM	7.8
Sorcellerie	Hist des Superstitions	R			0.2
Sorciers et Sorcieres	Hist des Superstitions	R		MDB	6.0

(continues)

Table 13.1 ***(continued)***

Title of Article in Encyclopédie d'Yverdon	*Category*	*Relationship to* Encyclopédie[a]	*Country*	*Initials of Author*	*Length of Article in Columns (Quarto Edition)*
Sortilège	Magie, Jurisprudence	N		MDB	22.4
Unitaires	Hist, Eccles	R		MDB	3.6
Vampire	Hist des Superstitions	S			1.0
Vaudois	Hist, Eccl	S			3.5
Wiclef, Jean	Hist, Eccles & Littérature	R		BC	5.0
Wolf, Wolfius, Cristiern ou Chrétien de	Hist, Litt	N	Germany	DF	3.0
Zuingliens	Hist, Eccl	S			1.4

Note: [a]N = new, R = Rewritten, S = Same

Huguenot pastor who was the perpetual secretary to Fredrick the Great's Academy of Sciences and a strong proponent of Wolff. He penned many of the unsigned Yverdon articles pertaining to religion and was a proclaimed Wolffien.[53] Wolff's insistence on the tenets of reason and revelation provided guidance to the Yverdon encyclopedists in matters of religion and ecclesiastical history that explain their position on Hus, Luther, Waldo, and the rest, whose principal flaw in the eyes of orthodoxy was the application of reason to matters of religion. Although more work is needed to further uncover the presence of Wolff and his ideas in the *Encyclopédie d'Yverdon,* the Yverdon encyclopedists' position on heresy constitutes a solid foundation for future research. Moreover, this comparative sampling of articles provides us with concrete examples of the differences between the *Encyclopédie* and the *Encyclopédie d'Yverdon* that make it possible to articulate the views of both sets of encyclopedists in ways that bring us closer to a true understanding of different views of enlightenment over a highly contested issue.

Notes

1. See Robert Darnton, *The Business of Enlightenment: A Publishing History of the 'Encyclopédie' 1775–1800* (Cambridge: Harvard University Press, 1979), pp. 312–313 for an overview of the limited dissemination of the *Encyclopédie* in Spain.
2. See Madeleine F. Morris, "The Tuscan editions of the *Encyclopédie,*" in Frank Kafker, ed., *Notable Encyclopedias of the Late Eighteenth Century: Eleven Successors of the 'Encyclopédie'* (Oxford: Voltaire Foundation, 1994) pp. 51–84. Known as the Lucca (1758–1776) and Leghorn (1770–1778) folio editions, after the cities where they were published, both maintain the French language and are characterized by corrective notes to theological articles intended to keep the editions free from controversy, as well as by the "cartoni" or cardboard blinds that covered articles or passages thought to be controversial.
3. *Encyclopédie ou Dictionnaire universel raisonné des connoissances humaines,* ed. F. B. de Félice. Text: I-XLII; *Supplément:* I-VI; Tables: I-X Yverdon, 1770–1780. The earliest and still valuable study of De Felice and the Yverdon *Encyclopédie* remains Eugène Maccabez, *F. B. de Félice et son encyclopédie: Yverdon 1777–1780* (Basel: 1903). On De Felice's Yverdon publishing house and the publishing industry of Yverdon during the seventeenth and eighteenth centuries, see Jean-Pierre Perret, *Les Imprimeries d'Yverdon au XVIIe et au XVIIIe siècle* (Lausanne: 1945). A comparative index and discussion of the relationship between the *Encyclopédie* and the Yverdon edition can be found in Clorinda Donato, *Inventory of the 'Encyclopédie d'Yverdon': a Comparative Study with Diderot's 'Encyclopédie'*, dissertation, University of California, Los Angeles, 1987; and Clorinda Donato and

Robert M. Maniquis, eds., *The Encyclopédie and the Age of Revolution* (Boston: G. K. Hall, 1992). The best article-length treatment is Kathleen Doig, "The Yverdon *Encyclopédie,*" in Kafker, ed., *Notable Encyclopedias of the Late Eighteenth Century.*

4. *Encyclopédie, ou Dictionnaire raisonné des sciences, des arts, et des métiers (articles choisis)*, 2 vols. (Paris: Flammarion, 1986), "Aigle" I, 233–224 and "Capuchon," I, 287–288.
5. *Voltaire's Correspondence,* ed. Theodore Besterman (Geneve: 1962), LXXVII, p. 163.
6. See Darnton, p. 20: "the *Encyclopédie d'Yverdon* had a good reputation in the eighteenth-century, and not only in pietistic corners of Germany and Holland."
7. Frank A. Kafker and Serena L. Kafker, *The Encyclopedists as Individuals: a Biographical Dictionary of the Authors of the 'Encyclopédie'* (Oxford: Voltaire Foundation: 1988), p. xi.
8. See Jacques Proust, "Le protestantisme dans l'*Encyclopédie,*" *Dix-huitième siècle* 17, 1985, pp. 53–65.
9. The CD-Rom edition of this database will be available in 2002 from the Centre International d'Etude du XVIIIe Siècle, Ferney-Voltaire.
10. Charly Guyot, *Le Rayonnement de l'Encyclopédie en Suisse française* (Neuchâtel: 1955).
11. Information about De Felice's life from Giuseppe Gorani, "Aventures d'un homme célèbre," *Mémoires secrets et critiques des cours* (Paris: 1794).
12. There is a new, 5.5-column article on Brescia in *Encyclopédie d'Yverdon,* vol. VII, p. 335–338. Brescia was home to Arnaud de Brescia, a twelfth-century heretic involved with the sect known as the Patarines in Italy and the Publicains in France.
13. De Felice paid homage to his mentor in "Severo, Raimondo de Sangro, prince de San-," (N), Hist Litt., *Encyclopédie d'Yverdon,* vol. XXXVIII, pp. 482–485.
14. *Voltaire's Correspondence,* ed. Theodore Besterman, 107 vols. (Genève: 1953–65), n. 14699.
15. See my "From Switzerland to Europe through Leipzig: The Swiss Book Trade and the Leipziger Messe 1770–1780," *Leipziger Jahrbuch zur Buchgeschichte,* 4, 1994 on the marketing and sales history of the Yverdon encyclopedia based upon the Leipzig book fair catalogues and the archives of the *Société Typographique de Neuchâtel.*
16. *Encyclopédie d'Yverdon* vol. I, p. ix.
17. Ms. Bonnet, 50, fol. 60. Dec. 6, 1770.
18. Genève, Bibliothèque publique et universitaire, Ms. Bonnet 88/1, *Observations sur le Dictionnaire instructif de M. Formey; Notice raisonnée de divers Articles de l'Encyclopédie d'Yverdun; Notice raisonée de divers articles de l'Encyclopédie de Paris.*
19. *The Correspondence between Albrecht von Haller and Charles Bonnet,* ed. Otto Sontag (Berne: Huber, 1983), p. 1054.

20. Gabriel-Jean-Henri Mingard (1729–1786) pastor, theologian, and a founding member of the Société littéraire de Lausanne. For details on his collaboration with De Felice, see Etienne Hofmann, "Le Pasteur Gabriel Mingard, Collaborateur de l'*Encyclopédie* d'Yverdon," in *Le gout de l'histoire, des idées et des hommes,* eds. Alain Clarien and Bertrand Mueller, (Lausanne: 1996).
21. Bonnet to Haller, 26 May 1775 in Sonntag, p. 1164.
22. "Le nouvel article signé G.M. est bien d'une autre main, et porte l'empreinte de l'Esprit philosophique. . . . L'estimable auteur a posé les vrais fondemens de la Certitude, & a très bien caractérisé ses Espéces. On peut dire que ce Morceau est un bon Préservatif contre le Pyrrhonisme." *Notice raisonée de divers Articles de l'Encyclopédie d'Yverdun,"* "Certitude" 120.
23. *Encyclopédie, our Dictionnaire universel raisonné des connoissances humaines. Planches.* Tome X. TAR-VER. 1780. *L'Editeur,* p. 3. This table of contributors at the end of the volume is paginated separately. On page two of the *Avis,* De Felice also identifies Mingard as the author of articles signed GM: "(G. M.) M. Mingard de Lausanne."
24. Hofmann, p. 87.
25. Three censors signed the *Approbations* to the *Encyclopédie d'Yverdon* at the beginning of volume I, finding nothing objectionable. Subsequent volumes carry the *Approbations* at the end. The censors always signed the *Approbations.* Their names can be found in Perret, pp. 360–361.
26. Elie Bertrand (1713–1797): Naturalist and pastor who wrote 170 articles on theology, philosophy, and natural science; César-Alexandre Chavannes (1731–1800): Pastor and professor who wrote 276 articles on theology, sacred history, education, and anthropology. On *Encyclopédie d'Yverdon* collaborators, see Clorinda Donato and K. H. Doig, "Notices sur les auteurs des quarante-huit volumes de 'discours' de l'*Encyclopédie d'Yverdon,"* *Recherches sur Diderot et l'Encyclopédie* 11, 1991, pp. 133–141.
27. Guyot, p. 104.
28. This anti-Calvinist position is no different from that of the Genevan pastors such as Jacob Vernes, whom D'Alembert had painted as Socinians in the Paris *Encyclopédie* article "Genève." Genevan Calvinism of the 1750s and 1760s was far from the fire and brimstone of orthodox Calvinism, leaving many pastors in tacit agreement with Voltaire's denunciation of the torture and death of Servet at the hands of Calvin. See Graham Gargett, *Jacob Vernet, Geneva, and the philosophes* (Oxford: Voltaire Foundation, 1994), pp. 123–144.
29. John Davies, *A History of Europe* (Oxford: Oxford University Press, 1995), p. 496.
30. Ibid., "Calviniste," vol. VII, p. 128.
31. Ibid., pp. 590–591.
32. *Encyclopédie,* vol. VIII, p. 357.
33. *Encyclopédie,* vol. I, p. 245
34. *Encyclopédie d'Yverdon,* "Albigeois" (R) vol. II, p. 62. There are two articles with the title "Albigeois" on the same page: the first, only six lines long, is a new article (N), dealing with a geographical location; the second article

is the one we are referring to. It begins on page 62, follows the first article, and is rewritten (R).
35. Ibid., p. 65.
36. Ibid., p. 64.
37. See Harold O. J. Brown, *Heresies: The Image of Christ in the Mirror of Heresy and Orthodoxy from the Apostles to the Present* (New York: 1984), pp. 261–262.
38. Ibid., "Luther, Martin," vol. XXVI, pp. 769–770.
39. Ibid., pp. 769–770.
40. Ibid., pp. 769–770.
41. *Encyclopédie,* "Luthéranisme," vol. IX, p. 756.
42. Ibid., p. 757.
43. See Brown, p. 313.
44. *Encyclopédie,* vol. VIII, p. 158.
45. Berenger (or Berengar) of Tours challenged Scholastic authority in the eleventh century in the controversy over whether the Eucharist was literally flesh and blood, which did not seem reasonable, or whether this eating, which imparts to us the benefits of the Eucharist, is accomplished through belief in the word of God. Berenger's position called upon man's reason, understanding, and faith in the matter of the Eucharist. He was condemned and forced to sign a confession and a declaration of belief in transubstantiation in 1059. Jérome de Prague's classification as a heretic was also based on his view of transubstantiation, which he had learned about in Oxford, where he studied with Wyclif, who advocated the supreme authority of the Bible. Jean Hus learned of Wyclif's teachings through Jérome de Prague and made them his own. Michel Servet or Michael Servetus was put to death by Jean Calvin over issues of Church control. See Elie Bertrand's position on the controversy in the article "Jean Calvin" in the *Encyclopédie d'Yverdon.*
46. *Encyclopédie d'Yverdon,* vol. XXIII, pp. 142–143.
47. See Jean Haechler, *L'Encyclopédie de Diderot et de . . . Jaucourt: Essai biographique sur le chevalier Louis de Jaucourt* (Paris: Champion, 1995), p. 414. Jaucourt's attack on the Catholic Church is designed to convince his readers that tolerance is the only answer. Voltaire's *Traité sur la tolerance,* published in 1762, continues this discussion. Haechler pairs Jaucourt and Voltaire on these issues when commenting on Jaucourt's bold stance on matters of fanaticism and toleration in France in the 1760s (p. 443).
48. Ibid., p. 432.
49. *Encyclopédie,* vol. VIII, p. 158.
50. *Encyclopédie d'Yverdon,* vol. XXIII, p. 142.
51. Max Horkheimer and Theodor W. Adorno, *Dialectic of Enlightenment* (New York: Seabury, 1972).
52. Brown, p. 402.
53. See my "Jean Samuel Henri Formey and the *Encyclopédie d'Yverdon,*" *Die Schweizer in Berlin,* ed. Martin Fontius (Berlin: Akademie Verlag, 1995).

14

The Abbé Nicolas-Sylvestre Bergier and the History of Heresy

Kathleen Hardesty Doig

The history of heresy is often the history of the campaign for religious tolerance, which cannot be written without reference to the *philosophes* of the eighteenth century. Among their works the *Encyclopédie* must figure in the first rank. The number of pages on the subject is fairly limited; but the work as a whole acquired a reputation for fighting institutional authorities, and lent the metaphorical weight of its 28 folio volumes to the fight for reform. The epithet "encyclopédiste" became pejorative during the middle decades of the century, with the function of scholarship subsumed to that of criticism of political and ecclesiastical institutions in the minds of the reading public. By the 1770s a certain revisionism began to occur, illustrated most eloquently by the 58 quarto volumes of the revised *Encyclopédie* produced by Fortunato De Felice in Yverdon, Switzerland, the subject of the previous chapter in this collection. Just as that enterprise drew to an end in 1780, a new one, with intellectual and commercial aspirations on an unprecedentedly vast scale, was organized by the highly successful publisher, Charles Joseph Panckoucke.[1]

Panckoucke proposed yet another encyclopedia, the *Encyclopédie méthodique,* a corrected and greatly expanded edition organized into 26 subject dictionaries, each under the supervision of distinguished authorities in the field such as Jean François Marmontel in literature, Louis Jean Marie Daubenton in natural history, and Félix Vicq-d'Azyr in medicine.

The whole was to consist of 53 volumes and be completed by 1787. Neither of these promises was kept. The *Méthodique* ultimately numbered some 160 volumes, and it was not completed until 1832.[2] Not surprisingly, in view of this long production period, the *Méthodique* is characterized by repetitions among different series and by occasional contradictions, and the series that spanned several decades are not always intellectually coherent. It is not possible at present to compare the *Méthodique* in detail with the Diderot-d'Alembert *Encyclopédie,* but generally the later group of editors and contributors shared a common goal with their predecessors, advancing scientific knowledge and molding public opinion through strongly expressed views, within the limits of the encyclopedia genre.

Panckoucke's *Méthodique* included passages that fought the anti-religious spirit propounded in various notorious entries in the *Encyclopédie.*[3] This was especially true in the *Méthodique* series on theology. To make a new edition more palatable to censors, Panckoucke recruited as editor of the theology section perhaps the most formidable and respected apologist of the day, the abbé Nicolas-Sylvestre Bergier (1718–1790).[4] Lorrain by birth, Jesuit-trained, holder of a doctorate in theology from the university of Besançon, Bergier had spent 16 fruitful years ministering to the faithful in the obscure Vosges village of Flangebouche and adding to his stores of erudition. In 1765, after publishing *Le Déisme réfuté par lui-même,* a response to Rousseau's *Emile, Lettre à Mgr. de Beaumont,* and *Lettres de la Montagne,* Bergier was appointed principal of the collège at Besançon. In 1768 he refuted d'Holbach in *L'Apologie de la religion chrétienne,* and the following year was offered a canonry at Notre Dame in Paris. As an example of the less than hermetic seal between philosophic and Catholic circles, he frequented d'Holbach's salon for a time as an amiable observer,[5] but broke with the group after writing *Examen du matérialisme* in 1771, a response to d'Holbach's anonymously published *Système de la nature.* The same year Bergier moved to Versailles as confessor to the comtesse de Provence and Madame, the two royal sisters from Savoy.[6] The dictionary for the *Méthodique,* entitled simply *Théologie,* was undertaken in his sixties; it was Bergier's last major publication and the crowning achievement in a prolific career. In three quarto volumes totaling 2,375 pages, Bergier provided information and corrected factual errors in the *Encyclopédie,* and as an apologist he responded with brio to the Protestants and nonbelievers who stand as silent interlocutors in the penumbra of almost every article.

The apologetics practiced by Bergier, as Sylviane Albertan-Coppola has pointed out, is a genre that has been marginalized,[7] perhaps because what posterity has retained are the barbs of adversaries, Voltaire above all, rather than the apologetic texts themselves. Nigel Aston argues for the recovery of this work, given the current view of the Enlightenment as a "tendency of opinion" rather than a monolith, a tendency with ample room for clerical efforts to "reposition theology within the wider range of subjects competing for readers" and "to speak to men and women in the language of their time."[8]

There is no dearth of texts to recover. More than 900 works of Christian apologetics were published between 1715 and 1789,[9] and many of them do not deserve their reputation for obscurantism and absurdity. The most effective writers, lay and clerical, adapted the form of their presentations to the tenor of the times, hence the popularity of alphabetical works of apology[10] and the value their authors placed on good writing. They also accepted certain contemporary themes, stressing, for example, the concepts of usefulness and happiness, and making appeals to feeling. Most important, these apologists revered reason as much as any *philosophe.* As Guillaume de Maleville put it, "Each one is obliged to know why he is a Christian."[11]

The apologetic genre originated in complex refutations of early heresies (works that eighteenth-century apologists would tend to prefer to more recent productions of the seventeenth century), and the practitioners of Enlightenment followed suit. They used reason as a guide to truth, not as truth itself. According to the apologists, reason establishes the credibility of Tradition (i.e., the post-apostolic authoritative interpretation of Scripture), which must then be believed even if the events it recounts are not reasonable.[12] Likewise, the bases for accepting the claims of Scripture are explained in terms of reason rather than sentiment, faith, or some other appeal, and are held to be proven on the same bases; the content of Scripture must therefore be believed, even if certain points, such as the Trinity and transubstantiation, are non-scientific and illogical. The embrace of reason, in other words, could not accommodate those who reject Scripture. Virtually all apologists resisted such an unorthodoxy, no matter how tempting it may have been in terms of market appeal.

Among the most astute of these clerical apologists was Bergier. His willingness to meet the *philosophes* on their own terrain, both in the salons and in the encyclopedia, was lambasted by Jansenist contemporaries but condoned by archbishop Christophe de Beaumont of Paris. As various citations below suggest, Bergier also mastered both the rhetorical constructs of his adversaries and the art of succinct, clear summaries of complex theological points.[13] His skill as an apologist is best attested by Melchior Grimm's often-quoted acknowledgement[14] or by the lack of an answer to his works,[15] but recent appreciation of his skill is not lacking.[16]

Bergier wrote in the vulgarizing style of a *philosophe,* but devoted his professional career to opposing the philosophic movement. Enlightenment apologetics and *philosophie* may have agreed about method, but they were fundamentally incompatible in content. The chief area of discussion was the subject of this volume, the response to heresy, i.e., to the well-orchestrated demand for tolerance of varied religious beliefs (or of none at all) in a culture where possession of an established body of truths was taken as a given. Here, we shall distill the major elements of Bergier's defense of intolerance from articles distributed throughout *Théologie,* and then analyze several key entries where the impact of an extended development of these ideas on readers can be weighed.

Like most encyclopedic writing, *Théologie* is a study in intertextuality. In relation to the *Encyclopédie,* Bergier's stated position is that of corrector,[17] and in a certain sense the entire series is a response to the letter and the spirit of that work. Some articles respond to the *Encyclopédie* by name ("Priscillianisme"; "Vaudois," D. J.[18]), while the generalizing epithets that Bergier favors, "les incrédules," "les philosophes," and so forth, can be taken to include encyclopedists. But not all the latter, in a censored work like the *Encyclopédie,* were inimical to Bergier's purposes, so his stance is often that of borrower. The orthodox abbé Edme-François Mallet[19] occasionally provides an entire entry ("Bogarmiles"), more often a few sentences or a paragraph that Bergier then supplements ("Apollinaires," "Donatistes"). Borrowings from or positive references to other sources are made from opposing works that support his own views, especially in the case of historical accounts,[20] and Bergier traces advantageous textual genealogies.[21] Defense of the Fathers is crucial to Bergier's attack on heresy, and he undertakes philosophical and theological explanations of Augustine ("Prédestinatiens," "Donatistes"), Cyril and Origen ("Apologie," to counter the accusation of Platonism), and the Fathers as a group ("Matérialisme" and "Immatérialisme," to counter the accusation of materialism).

Bergier's chief use of Protestant sources is to refute them. In "Persécution," it is Jean Barbeyrac's *Traité de la Morale des Pères de l'Église* (Amsterdam, 1728); in "Albigeois" and many other articles, Jacques Basnage; in "Sabelliens," Isaac de Beausobre, "decided apologist of all heretics and errors" (3:398)[22]; in "Arnaldistes" and countless other articles, Johann Lorenz von Mosheim, likewise "declared apologist of all heretics" (1:137). The latter is the most frequently cited of the Protestants. He was also the most eminent and influential, having a pan-European reputation as a Lutheran minister, theologian, and professor.[23] Bergier had an encyclopedic knowledge of the details of Mosheim's works, and a shrewd grasp of its principal danger points, namely, opposition to the early Church Fathers (who validated Catholic belief), the pope (who represents its continuity), and ancient heretics (who prepared the way for the Reform) (see "Platonisme," "Papauté," "Jacobites"). Another recurring response to Protestant writers concerns Catholic missionary activity, which Bergier sees as validating the divine project of the Church. "Foreign missions" is particularly detailed in this respect, but many other articles vaunt the success of such activity ("Jacobites," "Eutychiens") and defend it against the attacks of Protestants ("Tartares").

Bergier's texts are often characterized by historical revisionism. In "Hussites," he stresses that it was the civil authorities, not members of the Council of Constance, who burned Jean Hus, a statement that is technically accurate. Bergier also holds that there were fewer bad popes than generally thought, and that there were often mitigating circumstances in cases of apparent misdeeds, e.g., popes did appoint emperors, but only in times of anarchy when no individual had a clear right to the throne ("Papauté"). In other articles, Bergier deals with factual errors published elsewhere. "Zèle," for example, corrects a mistake perpetrated by Jaucourt in the *Encyclopédie*:[24] it was not bishop Abdas, but one of his priests, who torched a pagan temple, setting off a long persecution.

Bergier met his intended audience on their terms by writing in a popular style. There is nothing of the theological treatise about *Théologie,* save perhaps the numbering of topics, which economizes on transitional expressions and reinforces the impression of intensity. The style is readable, no mean feat considering the subject matter, and characterized by affective terminology. Lexical groups include such terms as "calumny," "falsity," and "disguise" for the arguments of adversaries, and "violence" and "crimes" for their deeds. Terms of mental derangement as well as military metaphors ("combat the atheists," "take arms") are favored. Descriptive phrases, frequently structured around contrasts and often exclamatory or interrogative, are meant to inspire indignation: "What a difference between the tortuous march of heresy and the frank and firm conduct of the Catholic Church!" ("Arianisme," 1:130). Concluding sentences become pithy editorials (see, for example, "Donatistes" and "Martyrs"). But Bergier's style largely lacks wit and ironic distance. Had he possessed such Voltairean talents, his works would likely have drawn larger numbers of worldly readers to his side. As it was, Bergier chose to rely on the substance of his articles, rather than on their form. The heart of his argument, briefly, is that heresies are divergences from Church doctrine, that modern incredulity and heresies flow from ancient errors, and that the Church, as keeper of the truth, must not abide these aberrations.

The parameters of what constitutes heresy are set out in the two-page "Hérésie." Bergier holds that those who have never had the opportunity to know the truth through the Church are not in error, but those who choose to ignore the Church's magisterium are heretics. The specific errors of individual heresies vis-à-vis Catholic doctrine are spelled out in approximately 50 entries.[25] Bergier analyzes doctrinal divergences carefully, and insists that they are genuine and not just a matter of semantics (for example, "Consubstantiel" and "Arianisme," about the metaphysically difficult concept that Christ had both a fully human and a fully divine nature). He brings out qualities that characterize errant sects down to the present day. Perhaps the most noticeable is doctrinal disunity: "no sect of heretics was ever constant in its opinions; each doctor believes himself a master who can interpret and arrange them as he pleases" ("Albigeois," 1:37). This disagreement leads naturally to a proliferation of sects, which Bergier holds up as proof that these offshoots cannot claim filiation with a Church whose first claim is to unity. The early heresies, according to him, foreshadow what would happen during the Reformation. Bergier scorns the satisfaction that Protestants find in this evolution from the early Church, as simply a way to make error legitimate and to "give themselves ancestors" ("Henriciens," 2:169). He also disputes their counterclaims to fundamental unity in the face of uncontrollable multiplicity, since their principle of reliance on a personal interpretation of Scripture is divisive, not unifying (for example, "Église," 1:616). The emphasis on oneness and an implied corollary of central authority suggest the social and ecclesiastical model that Bergier endorses. The general lines can be summarized from "Autorité": humans are meant for society, social order requires subordination, anarchy is the greatest of evils (worse even than despotism), and the authority of sovereigns and pastors comes from God.

In religious matters, authority also derives from learning. In the late eighteenth century, such education was scarcely available to the lay masses, who are described as untrained to examine doctrine ("Raison"), and unreliable, their ignorance often coupled with fanaticism ("Anabaptistes"). Fanaticism is a major manifestation of another weakness that invalidates the claims of dissidents in Bergier's rationalist approach. Dissidents act in passion, which leads to a wide variety of disorders in Bergier's tour of heretical sects. The fourth-century Circoncellions, "enthusiastic brigands," cancelled debts and freed slaves ("Circoncelliens," 1:354). The Marcosians, like many heretics Bergier describes, were impostors guilty of conjuring "delusions" and favoring magic tricks and various other kinds of seduction ("Marcosiens," 2:563–424 [*sic* 564]). The Donatists spread their beliefs by "ruses, insinuations, captious writings, open violence, cruelties, and persecutions against Catholics" ("Donatistes," 1:566).[26]

This latter accusation, of violence perpetrated by various heretics, is a major point in Bergier's defense of Catholic reaction to heresy. He maintains, in "Albigeois," "Calvinistes," "Hussites," "Réformateur," and other articles, that heretics were guilty of bloody violence and had to be fought in self-defense. More frequently, they are accused of sedition. Bergier establishes the following principle in response. Heresy can be tolerated if its adherents do not disturb the peace, following the model of the early Christians ("Conscience [Liberté de]"). But any form of sedition, such as that detailed in "Donatistes," "Eutychiens," "Guerres de religion," "Vaudois," and many other articles, must, in the orderly society Bergier requires, be quelled.

He refuses to call this response a persecution: "one abuses the term when one calls *persecutors* princes who have used the penal laws to restrain seditious and turbulent heretics who want to become the masters, destroy the laws and the established religion" ("Persécuteur," 3:184). The victim of this repression, a "seditious fanatic," is not a martyr ("Persécuteur," 3:185), nor is he or she innocent, unless the errors are involuntary and invincible ("Persécuteur," "Conscience"), that is, unless the perpetrator had no opportunity to learn the truth. This rejection of the civil rights of the heretic goes, as Bergier rightly sees, against Bayle, Barbeyrac, and other Enlightenment theorists.[27] It also accounts for the more chilling passages in the *Dictionnaire.* Pierre de Bruys, for example, "merited the stake at which he died, he was a martyr, not for his opinions, but for the disorders and violence that he caused" ("Pétrobusiens," 3:194), and the 100 flagellants who were likewise burned to death in the early fifteenth century "to intimidate those who might be tempted to imitate them and to renew ancient disorders" ("Flagellans," 2:37) encounter no sympathy from him.[28]

Bergier is, however, occasionally willing to admit guilt on the Catholic side, although he does not stress on it. The structural weaknesses of the Church that inspired Calvinism are noted ("Calvinistes"), as are egregious cases of Ligue excesses during the sixteenth century ("Guerres de religion," 2:141–42). The apologist acknowledges guilt about Saint-Barthélemy, in a sentence that also hauls Calvinists to the bar: "They have tried to excuse their revolt on the grounds of the rightness of their intentions and for the public good, but that motive, always easy to feign, does not justify them any more than it excuses the cruelty of the Catholics" ("Barthélemy," 1:199). The cruelties during the Albigensian crusade are also admitted and attenuated by spreading the blame ("Albigeois," 1:38).

The tables are turned in further ways as Bergier develops his offensive. He accuses the weak of becoming intolerant when they become the masters in "Sociniens," and assumes the same pattern in "Cathares." Heretics are also guilty of obfuscation.[29] A recurring theme is that Protestants ask for concessions, such as tolerance, which they then refuse to grant to Catholics ("Luthéranisme," 2:503). And Protestants are accused of calumny against Catholic doctrine and clergy, not to mention the discrepancy between their touting of Scripture as the ultimate rule and their disputes about its meaning.[30] Among many counter-attacks on specific points, he claims that Protestants set up unreasonable conditions for Catholics (such as the requirement that they deny the papacy) for ecumenical meetings ("Syncrétistes").

Bergier elaborates a genealogical vision of heresy, where early errors were copied by medieval sectarians and then by the Reformers of the sixteenth century and the Jansenists, all of whom provided the deists and *philosophes* of the eighteenth century with ideas for attacking religion. The connection is constantly recalled in *Théologie*. The Jansenists are blamed for diminishing Church power, scorning the Fathers, and making all miracles suspect ("Jansénisme"). Contemporary nonbelievers repeat the "hideous portrait" that Protestants have limned of Catholicism ("Réformateur, Réformation, Réforme," 3:333), and Socinians and Protestants can be traced to the same origins ("Sociniens," 3:512).[31] Mosheim's criticism of Catholic missionaries ("Nord," 2:772)[32] and Beausobre's casting of doubt on the intellectual stature of early apologists is fodder for nonbelievers ("Manichéisme," 2:556). The slippery slope that Protestants opened by demanding the freedom to practice a religion other than the dominant one has led directly to the current demand by nonbelievers for the right to preach against religion; so why not a campaign for freedom of conscience for Turks, Jews, and pagans as well? ("Conscience [Liberté de]," 1:420).This claim for tolerance, ironic for Bergier but soon to be granted in various parts of Europe, is indicative of Bergier's attitude toward *philosophie* or incredulity. Incredulity is far more dangerous than Protestant sects. In Bergier's view, if the *philosophes* win, all religion, and not just the form he considers true, will disappear. To defend religion in general and Catholicism in particular against this new enemy, the apologist deploys the rhetorical arms described above, and engages the battle at every opportunity (the military metaphor is not an exaggeration), including entire articles devoted to rebutting philosophic positions, such as "Esprit," which attacks materialism.[33]

Bergier recognizes that limiting the authority allowed to reason is his main challenge, for reason alone will not accept revelation or the miracles that support it. So in carefully reasoned refutations—he is not contradicting himself, since he points out that reason must operate at this stage—Bergier sets up arguments against the *philosophes*. Reason is not our only guide to truth; instinct and feeling also play a role ("Raison," 3:317), an epistemological principle that he shares with Rousseau. Similarly, thanks to a "sentiment intérieur" and moral certitude, miracles are as certain as natural events. Modern *philosophes* confuse their dislike for believing a supernatural phenomenon with uncertainty that the phenomenon occurred ("Certitude," 1:305). Taking the offensive, Bergier mines the works of the *philosophes* for contradictions,[34] which he also finds between their message and their methods. *Philosophes* accuse defenders of the faith of mean-spiritedness and sharpness, yet are guilty of these faults themselves ("Zèle,"

3:827). The *philsophes* are intolerant (among many examples, "Anabaptistes," 1:69). They have brought the contentious spirit of philosophy into religion ("Hérésiarque"). They have tried to make the masses atheist; most are licentious and full of hatred, especially against the clergy; and they try to break the binds holding society together, from the paternal to the political ("Philosophe," 3:202).

The above overview conveys a sense of the structure and tone of Bergier's response to heresy. But most readers would not have appreciated this underlying thematic, since they probably did not read such a large corpus of articles on a single subject. To capture more accurately the experience of the average user of the dictionary, we shall now examine the content of several key entries with particular relevance to the subject of tolerance. While the conclusions do not differ from those outlined above, Bergier's arguments take on different contours in this more limited context.

Bergier goes far beyond simple definitions in two articles, "Apostasie, Apostat" and "Hérétique," that measure the guilt of those who abandon what he refers to simply, with the definite article, as "the true religion" ("Apostasie, Apostat," 1:109[35]). He brushes off secondary definitions of apostasy, saying that canonists can deal with them;[36] his interest is in answering the "detestable doctrine" of Hobbes, which allows a citizen to obey an infidel king in matters of religion (1:110). Hobbes sets up an hypothetical case of a Muslim ordered under pain of death to deny his faith. If we permit him to choose death, then we are accepting the principle of resistance to a sovereign in matters of religion. Bergier's answer is simplistic and oft-repeated in *Théologie*. The Muslim should allow himself to be instructed in the true religion; if he cannot then accept it, he should suffer death. This harsh pronouncement, implicitly giving the sovereign the authority to impose Christianity, is undercut by Bergier's insistence that the scenario itself is absurd, since Christ did not condone forced conversions or torture. But the pronouncement is firm and grounded on a fundamental principle that Bergier accepts through faith—that his religion is true. A related principle follows from this premise: "it is a sophism to compare the enlightened and upright conscience of a Christian with the erroneous and false conscience of a pagan or a Muslim" (1:110).

The notion that error is a crime if it is stubborn and voluntary is repeated in "Hérétique," in answer to Leibniz's protest that an erring conscience is innocent. Bergier presents the official Church view about the appropriate response to this kind of heresy: tolerate it if the heretics do not engage in sedition or violence. He devotes nearly half the article to demonstrating that the laws of Europe, both civil and ecclesiastical, respect that principle. The second question that the article answers is whether one has to honor one's word with heretics. The affirmative response requires Bergier to confront, once again, the execution of Jean Hus. The Bohemian dissident was guaranteed safe passage to the Council of Constance in 1414, but was burnt at the stake after being convicted of heresy. Bergier's defense of the punishment is based on the fact that Hus was granted only safe passage, not acquittal, and that he agreed to suffer punishment if convicted. According to this interpretation, he was executed legally because he was engaged in sedition. A third question, whether heretical books should be forbidden, is briefly discussed. Again, Bergier answers affirmatively, and he adds works by nonbelievers to the list. His reasoning is that the "simple faithful" are not capable of understanding such writings, and as a precedent he points to the apostles' refusal to allow early Christians to satisfy "frivolous curiosity" (2:177). Bergier would, of course, allow theologians to examine these works, but he makes no mention of an intermediary group, the educated faithful, and their capacity to evaluate *philosophie.*

The addition of the works of nonbelievers to the above list suggests that "Incrédules" will also express strong opinions. And so it does. Although only a little over ten pages in length, plus one and one-half pages for a supplementary "Incrédulité"—"we still have many reflections to make; this sad subject could furnish an infinity of them," explains Bergier (2:300)—the articles include many points that are developed elsewhere in *Théologie.* The apologist begins with a favorite tactic, a choice of texts from philosophic works that give a negative portrait of nonbelievers.[37] He follows with more proofs that "unbelievers are uneducated, not of good faith, not firm in their opinions, not happy, not good citizens, not excusable" (2:297). They copy each other, in a line stretching from contemporaries back to early-eighteenth-century writers, who in turn copied Bayle and the English deists, and so forth. They accuse Catholics of credulity, but are more credulous themselves. They are fragmented, united only in their hatred of Catholicism. They accuse the clergy of self-interest, and say that lovers of religion are fanatics; do they give less evidence themselves of these qualities?[38] They plead for tolerance, but are intolerant themselves. Bergier gives an abrupt quotation at this point from Helvétius's *De l'Esprit,* which says few men would avoid using violent means to establish their opinions if they could.[39] Their claim that those free of religious prejudices are happy is belied by their admission of the suffering of doubters and atheists.[40] They degrade humans to the status of brutes, decry the evils of society, and then show us their happiness by preaching suicide. They break the yoke of religion for the masses,[41] and claim that freedom of thought is a natural right.

This last contention elicits further commentary, including the following. Freedom of thought must not be confused with freedom of expression; enemies of public religion have always been punished; even if it were acceptable to attack dogma, morality must never be undermined. Bergier concludes that punishment of nonbelievers (he implies that banishment would not be excessive) may not be necessary; scorning them should suffice. In the companion "Incrédulité," he repeats the admonition about destroying religion among the masses and adds sociological observations about the connection between incredulity and materialistic, luxury-loving societies, as opposed to the simple faith of country folk, which Bergier would have known in his Vosges parish of Flangebouche. He concludes by pointing out that it was people acting on religious belief, not nonbelievers, who created charitable institutions. The cross-reference is to "Liberté de penser."

That article begins by repeating the maxim that freedom of thought does not equate with freedom to preach. Bergier revives the argument for religion as a bridle on the masses,[42] and then turns the tables on his adversaries in four additional attacks. Nonbelievers complain that religion inspires sects and disputes, but so do nonbelievers. Christian sovereigns should have as much right to reprove freedom of thought as infidel sovereigns have to proscribe Christianity. The Romans allowed complete freedom to write against religion, but at the same time fought all new religions. In this connection, Bergier inserts an opinion about the much-vaunted British freedom of the press: he surmises that an author across the Channel writing in favor of a return to Catholicism would be executed. And finally, he declares that the results of the freedom to publish in France during the last half-century have been a few apostacies among libertines, corrupted mœurs in all classes, suicides, and crimes.

Bergier adds the following entry, "Liberté politique," to his dictionary, even though it is not a theological subject, in order to respond to attacks by nonbelievers. His thesis is that Christianity is the best religion for restraining a sovereign. Bossuet's *Politique tirée des propres paroles de l'Écriture sainte* (Paris, 1709) provides his main argument, that kings are merely depositories of God-given power, and that there is a distinction between absolute and arbitrary power. Bergier adds additional arguments, e.g., that it is curious to see philosophers who deny natural liberty and free will insist on political freedom, and that a king without religion is more likely to act arbitrarily.

Bergier cannot avoid political discussion, since his form of apology so often involves the role of government. A key article in this respect is "Intolérance." Bergier's main argument, that non-violent, non-disruptive proselytizing should be tolerated, echoes the same entry in the *Encyclopédie.* The latter article, unsigned, consists of several pages of aphorisms, many in the same vein,[43] but the tone is entirely different, as the conclusion demonstrates: "Work for your salvation. Pray for mine, and believe that anything you do beyond that is an abominable injustice in the eyes of God and men" (8:844). Bergier would find abominable injustice elsewhere, for his apparent tolerance is significantly attenuated by the conviction that only one completely peaceful sect has ever existed—the early Christians. He holds that princes may protect religion with restrictive laws and by punishing turbulent infractors. Bergier undertakes some historical corrections, for example of Barbeyrac's claim[44] that the Fathers contradicted themselves in favoring the anti-pagan laws promulgated by Christian emperors. Bergier maintains that the laws covered blood sacrifices (presumably human) and other types of crimes. A good portion of the article is in fact devoted to refuting Barbeyrac. The latter argues that violence is not useful for effecting true conversions.[45] Bergier retorts that it *can* succeed, and cites the example of the Donatists. He hastens to add that violence should only be used against violent sects. Other points to which Bergier responds include Barbeyrac's rejection of infallibility and tradition, and his claim that intolerance depopulates the state.[46] Bergier answers this accusation in detail. He calls England and Holland worse off than a century ago, and France more populous, better cultivated, and more flourishing than in 1685. The conclusion to the article picks up a contradiction exploited earlier, that nonbelievers call priests self-interested for promoting religion, when they themselves act with the same self-interest.

The alphabet allows Bergier to return to the subject in "Tolérance, Intolérance," a significant article a bit over ten pages in length.[47] The essay first discusses four different meanings of the word "tolerance," then states two "truths" and asks several questions related to them, and finally refutes twelve claims made by enemies of the Church—again, Bayle and Barbeyrac are chief among the latter.

Of the meanings Bergier assigns to tolerance, we shall choose two as exemplars. Tolerance in general is said to be the fraternal charity that ought to always reign (3:646), but such charity does not preclude punishing the seditious, who disturb society's peace by disturbing the Church. To nonbelievers, Bergier adds, tolerance means indifference toward religion, and a claim to tolerance for themselves while they in turn are intolerant of Christianity. Aware that his tableau may appear exaggerated, Bergier refers readers to unspecified works by nonbelievers, and especially to the *Encyclopédie,* in the articles "Tolérance," "Intolérance," and "Persécution" (2:647).[48] This is one of the few occasions in *Théologie* where Bergier directly attacks the *Encyclopédie.*

Bergier then declares as truths the statements that religion is necessary in a society and that the Judaeo-Christian religion is the best one. He has little indulgence for any kind of cultural relativism, describing as "demented" the idea of "some unbelievers" that all religions are equally beneficial to society (3:648). Related questions concern specific moments in post-Reformation French history. They center around the rights or responsibilities of the king vis-à-vis proselytizers. Must the king, for example, respect agreements that were made in times of weakness, after he becomes stronger and the proselytizers more dangerous? This obvious reference to the revocation of the Edict of Nantes is a rhetorical question. Bergier calls it "a very strange jurisprudence" when Henri IV's word is more sacred than that of Charlemagne or St. Louis (3:648).

In the section comprising the 12 groups of responses to adversaries of Catholicism, Bergier reinforces arguments distributed throughout *Théologie* and cultivates a concentrated stridency. The distinction between freedom of thought and freedom of action is repeated. Bergier confines the latter, by force if need be, to a narrow space where the truth of the Catholic religion is the generating principle (3:649). Religion is deemed necessary to the temporal and spiritual well-being of society. The stubbornness of those who refuse to convert is a vice that does not deserve impunity. The early apologists did not call for an end to persecutions on the grounds of freedom of conscience, but because these Christians were peaceful. Allowing equal rights to erring consciences would justify crimes committed by fanatics. Bergier counter-attacks Barbeyrac, who says that intolerance fosters the proliferation of sects, by claiming that England and Holland prove the contrary. Barbeyrac points out that heresies seem to be inevitable in Christian sects, which should therefore bear with each other. Bergier comments that the Protestants indeed need to learn that lesson.

In other arguments, Bergier is on the defensive about the relative intolerance of Catholics versus Protestants, about the Catholic use of Scripture, and about several statements he calls calumnies, for example, that the Fathers influenced civil authorities to persecute heretics. In still others, the apologist offers a variant reading of history, as in his response to the claim that ancient peoples were more tolerant than Christians. He cites various examples to the contrary, complete with references to the *Histoire de l'Académie des Inscriptions* and other learned works.

"We would never finish if we had to refute in detail all of their false maxims," exclaims Bergier (3:655). But he does not resist one last argument: tolerance is a slippery slope, starting with deism and ending in atheism. Atheism is a harbinger of social unrest, therefore the masses must have a religion. The forces required to assure its maintenance are authority, law, and punishments. At precisely this moment in the article, when a philosophic reader might protest with the greatest energy, Bergier reveals the source from which he has copied the admonition—Bayle.[49]

In the last article of *Théologie,* "Zwingliens," Bergier repeats several attacks, one of which is that the sixteenth-century reformers were not the divinely inspired geniuses they claimed to be. They were simply the latest in a long line of sectarians, each copying the errors of his predecessors. For Bergier, this is mere plagiarism. He also stresses the doctrinal disunity among the many Protestant sects, between Luther and Zwingli about the Eucharist, for example. Bergier has the talent of cutting through dry and complex arguments to the essential, as when he discusses the ramifications of predestination. He concludes both the article and his dictionary with a rebuttal of the claim that Zwinglians were the most tolerant of all Protestants.

One might expect that Bergier's message that it is legitimate to exclude heretics from the community when any form of civil disobedience is involved would not weather a revolution that redefined the concept of civil disobedience. But in fact, probably because of the marshalling of intellectual forces in the Church's reaction to the Revolution, and certainly because of the wealth of information that *Théologie* contains on ecclesiastical history and systematic theology, the dictionary became one of the most widely diffused theological works in the nineteenth century. Reprinted in French 31 times down through 1882, it was by far the most durable portion of the *Encyclopédie méthodique* and must be counted among the most widely sold anti-heretical works of the nineteenth century. The exact extent to which *Théologie* contributed to the hardening of theological stances as the Church lashed back at heresy in the post-Revolutionary period is not known. It was only in the twentieth century that many theologians finally agreed that comprehensive freedom of religious practice could be defensible, i.e., that governments should not repress heresy. This leads back to the paradox behind so much intellectual history, and behind the truism that false logic can seem true. The justifications on which Bergier's arguments rest function perfectly only within a system where certain givens are accepted. Once these are altered, a whole new organization of knowledge, a whole new methodical encyclopedia, is necessary.

Notes

1. On Panckoucke, see George B. Watts, "'L'Atlas de la librairie française,'" *Studies on Voltaire and the Eighteenth Century* 68, 1969, pp. 67–205, and Suzanne Tucoo-Chala, *Charles-Joseph Panckoucke et la librairie française, 1736–1798* (Paris: Marrimpouey and Touzot, 1977).

2. See Tucoo-Chala, *Charles-Joseph Panckoucke,* pp. 323–44, and Christabel P. Braunrot and Kathleen Hardesty Doig, "The *Encyclopédie méthodique:* An Introduction," in *Studies on Voltaire and the Eighteenth Century* 327, 1995, pp. 1–152.
3. Some articles imbued by *philosophie:* "Existence," by Anne-Robert-Jacques Turgot; "Philosophe," by César Chesneau Dumarsais; "Causes finales," by Jean Le Rond d'Alembert.
4. As reported by [Louis Petit de] Bachaumont, *Mémoires secrets* 23 (London: John Adamson, 1783), pp. 135–36. See also Alfred J. Bingham, "The abbé Bergier: an Eigthteenth-Century Catholic Apologist," *Modern Language Review* 54, 1959, pp. 347–49.
5. On Bergier's contacts with the d'Holbach circle, see Alan Charles Kors, *D'Holbach's Coterie: An Enlightenment in Paris* (Princeton: Princeton University Press, 1976), pp. 113–17.
6. See "Notice historique sur M. Bergier," *Dictionnaire de Théologie,* new ed., 1 (Besançon: Gauthier, 1834), pp. v-xiv; E. Dublanchy, "Bergier," *Dictionnaire de Théologie catholique,* eds. A. Vacant and E. Manenot, 2 (Paris: Letouzey et Ané, 1905), pp. 742–45; Bingham, "The Abbé Bergier"; Guy Besse, "Une Lettre du Chanoine Bergier (1774)," *Dix-huitième siècle,* 28, 1996, pp. 259–66; and *Un Théologien au siècle des lumières: Bergier, Correspondance avec l'Abbé Trouillet, 1770–1790,* ed. Ambroise Jobert (Lyon: Centre André Latreille, 1988), pp. 15–44.
7. "L'Apologétique catholique française à l'Age des Lumières," *Revue de l'Histoire des Religions,* CCV, fasc. 2, 1988, p. 152. I thank Professor Albertan-Coppola for bringing this and other articles on Bergier to my attention.
8. "The Golden Autumn of Gallicanism? Religious History and Its Place in Current Writing on Eighteenth-Century France," *French History* 13, 1999, p. 209.
9. Sylviane Albertan-Coppola, "L'Apologétique catholique française," p. 152.
10. Sylviane Albertan-Coppola, "Les Réfutations catholiques du *Dictionnaire philosophique,*" in *Voltaire et ses combats,* eds. Ulla Kölving and Christiane Mervaud (Oxford: Voltaire Foundation, 1997), p. 786.
11. Sylviane Albertan-Coppola, "Bergier contre Voltaire, ou le combat de David contre Goliath," Colloque International d'Histoire Religieuse et de Théologie, Besançon et Flangebouche (1990), p. 6. Quoted from *La Religion naturelle et la révélée contre les Incrédules* (Paris et Nyon, 1756–1758).
12. R. R. Palmer, *Catholics and Unbelievers in Eighteenth Century France* (New York: Cooper Square, 1961 [1939]), pp. 77–102, analyzes the apologists.
13. Bergier speaks of the difficulty of making dry subjects attractive in a letter to Trouillet of 29 June 1770, *Bergier, Correspondance avec l'abbé Trouillet,* pp. 58–59.
14. Grimm, *Correspondance littéraire,* avril 1767, ed. M. Tourneux, 7 (Paris: Garnier Frères, 1879), pp. 294–95
15. They include *Le Déisme réfuté par lui-même: Examen des principes d'incrédulité répandus dans les divers ouvrages de M. Rousseau* (1765), particularly aimed at

the natural religion of *Émile; La certitude des preuves du christianisme* (1767), written to oppose the *Examen des apologistes,* compiled by d'Holbach from a manuscript by Jean Lévesque de Burigny; *Apologie de la religion chrétienne* (1768), against d'Holbach's *Christianisme dévoilé;* and *Examen du matérialisme* (1771), a refutation of d'Holbach's *Système de la nature.* The *Apologie* included *Remarques sur les principaux articles du Dictionnaire philosophique* of Voltaire. On the latter, see Sylviane Albertan-Coppola, "Les réfutations catholiques du *Dictionnaire philosophique,*" pp. 790–96.

16. Palmer ranks the *Examen du matérialisme* very high: "one of the best pieces of critical writing of the century" (*Catholics and Unbelievers*), p. 215. A. R. Desautels calls Bergier "the best apologist the Church in France produced during the second half of the 18th century," *New Catholic Encyclopaedia,* vol. 2 (New York: McGraw Hill, 1967, p. 323). See also Albert Monod, *De Pascal à Chateaubriand: Les Défenseurs français du christianisme de 1670 à 1802* (Geneva: Slatkine Repr., 1970, of 1916 ed.), p. 421.
17. See the section on theology in the prospectus of the *Encyclopédie méthodique,* reprinted in the series *Beaux Arts,* 1 (Paris, 1788), pp. xxxiii-xxxiv.
18. Louis, chevalier de Jaucourt, was the most prolific compiler of the *Encyclopédie.* See Frank A. Kafker and Serena L. Kafker, *The Encyclopedists as Individuals: A Biographical Dictionary of the Authors of the* Encyclopédie, SVEC 257 (Oxford: Voltaire Foundation, 1988), pp. 175–80; Madeleine F. Morris, *Le Chevalier de Jaucourt: un ami de la terre (1704–1780)* (Geneva: Droz, 1979); and on his views on toleration, Simone Zurbuchen, "Jaucourt, Republicanism, and Toleration," in *New Essays on the Political Thought of the Huguenots of the Refuge,* ed. John Christian Laursen (Leiden: E.J. Brill, 1995), pp. 155–69. References to *Théologie* (3 volumes, Paris, 1788–90[91]) will be given in the text where possible.
19. Bergier praises Mallet's work in the prospectus, and promises to retain his articles (*Beaux Arts,* I, pp. xxxiii-xxxiv). Bergier also praises Mallet in a letter to Trouillet, and dismisses the theological entries in the last ten volumes (chiefly produced after Mallet's death) as "mostly jumbled together by adventurers" (*Bergier, Correspondance avec l'Abbé Trouillet,* p. 208).
20. The background in "Hussites" is attributed to the Protestant Jacques Lenfant, *Histoire du concile de Constance* (Amsterdam, 1714). In "Guerre de Religion," Bayle, Hume, Rousseau, and Voltaire provide support for the claim that religion was only a pretext in these conflicts.
21. In "Calvinistes," he claims that the 1758 work defending the revocation of the Edict of Nantes (Jean Novi de Caveirac, *Apologie de Louis XIV*) repeats Bayle's *Avis important aux Réfugiés* and the *Réponse à la lettre d'un Réfugié* (1690), which accuse Calvinists of being more intolerant than Catholics (1:248).
22. His *Histoire critique de Manichée et du manichéïsme* (Amsterdam, 1734–1739) had, according to Bergier, been criticized by fellow Protestants J. J. Brucker and Mosheim ("Manichéisme").

23. In the summer of 1784 Bergier acquired two works by Mosheim: *Histoire ecclésiastique ancienne et moderne* (Yverdon, 1776) and *Dissertationem ad historiam ecclesiasticam* (Altona, 1767). Bergier says of his new acquisitions: "I found ample materials and an infinity of things to refute" (*Bergier, Correspondance avec l'abbé Trouillet,* p. 348).
24. "Zèle," in *Encyclopédie ou Dictionnaire raisonné des Sciences, des arts et des métiers,* ed. Denis Diderot, (Neufchatel [Paris], 1765) 17, p. 698. The database of American and French Research on the Treasury of the French Language (ARTFL), University of Chicago/Centre de Recherche national français, facilitated research on comparisons between the *Encyclopédie* and *Théologie.*
25. Bergier criticized the *Encyclopédie* for sloppy organization, including treating Arianism under 20 different headings (*Théologie,* "Avertissement," 1, p. v).
26. A phrase from Mallet's article in the *Encyclopédie,* 5, p. 50.
27. "Conscience" is an attempt to refute them.
28. The scholar Bergier laments the burning of several libraries during the Calvinist troubles ("Calvinistes").
29. "Pélagianisme":"the artifice of the heretics of disguising their doctrine and that of their adversaries in order to obscure the falsity of the one and the truth of the other" (3:160).
30. See "Réformateur, Réformation, Réforme" (3:330).
31. This five-page article takes a completely different tone from the thirteen-page "Unitaires" of the *Encyclopédie,* a famously sympathetic account by Jacques-André Naigeon.
32. Bergier refers to *Examen critique de la vie & des ouvrages de saint Paul* [Peter Annet, trans. d'Holbach, London, 1770], and promises to refute it in "Paul (Saint)." He does so without naming the author.
33. *Théologie* is nearly devoid of entries on a related word: religious spirituality. Bergier alludes to his preference for polemics in his correspondence: "[Dieu] a trouvé bon de me donner très peu d'inclination à prier et beaucoup à travailler" (*Bergier, Correspondance avec l'abbé Trouillet,* p. 211).
34. See, for example, "Liberté de penser," where he quotes an unidentified "déiste célèbre" who thought that blasphemers should be punished, and "Incrédules," which quotes selections from Dumarsais's *Essai sur les préjugés* (1770), Helvétius's *Le Vrai Sens du Système de la Nature* (1774), Jean Meslier's *Le Bon Sens* (1768), and P.-J. Gudin de La Brenellerie's *Aux Mânes de Louis XV* (1776).
35. This is also the phrase used in "Apostasie" in the *Encyclopédie* (1:535), signed "GH," probably a misprint for Mallet's "G."
36. The *Encyclopédie* entry covers several types: *à supererogatione, à mandatis Dei,* and *à fide.*
37. E.g., *Système de la nature,* (Paris: Fayard, 1990), 2, pp. 354–58, where d'Holbach concedes, as Bergier quotes, that many give up religion because of passion, moral corruption, and self-interest. But d'Holbach's stress is on the need to be disinterested in order to decide properly about the existence of God (which, of course, he denies in his conclusion).

38. Bergier gives references to Guillaume Raynal, *Histoire philosophique et politique des établissements et du commerce des Européens dans les deux Indes* (1770); an unidentified *Dialogue sur l'âme;* d'Holbach, *Essai sur les préjugés* (1770); *Œuvres de J.-J. Rousseau;* d'Holbach, *Système de la nature;* d'Holbach, *Le Christianisme dévoilé* (1767).
39. Claude Adrien Helvétius, *De l'esprit* (Paris: Fayard, 1988), p. 63 note e.
40. Bergier gives several references, including no. 22 of *Pensées philosophiques,* where Diderot pities true atheists (Geneva: Droz, 1965), p. 19.
41. Bergier blames Hume, Shaftesbury, Bolingbroke, Toland, and Bayle on this count.
42. On social utilitarianism, see William R. Everdell, *Christian Apologetics in France, 1730–1790: The Roots of Romantic Religion* (Lewiston: Edwin Mellen, 1987), pp. 109–61.
43. For example, "Instruction, persuasion & prayer, these are the sole legitimate means of spreading religion" ("Intolérance," 8:843).
44. In *Traité de la Morale des Pères de l'Église* (Amsterdam, 1728). Bergier refers the reader to chapter 12, which covers Gregory of Nazianzen. Barbeyrac implies the point that Bergier features, but is more interested in demonstrating that the Council of Constantinople was unfavorably weighted against the Arians. Bergier does not mention that Gregory opposed the "rage" of the emperor Theodosius against the Arians (p. 169).
45. *Traité de la Morale des Pères,* p. 169–70.
46. *Traité de la Morale,* p. 176–78, 180–81.
47. See Hervé Haquin, "L'Abbé Bergier et l'article 'Tolérance-Intolérance' de l'Encyclopédie méthodique (1790)," in *Michel Vovelle . . . Sur la Révolution, Approches plurielles,* eds. Jean-Paul Bertaud, Françoise Brunel, Catherine Duprat, and François Hincker (Paris: Société des robespierristes, 1997), pp. 53–57.
48. Bergier meant "Persécuteur." "Persécution" is a list of major offensives taken against the Church. "Tolérance," "Intolérance," and "Persécuteur" depict the Church as entirely at fault.
49. Bergier refers to the *Dictionnaire Historique et Critique,* article "Lubiéniezki," Remarks E and G, which include several possibly ironical statements.

Glossary of Heresies

For the convenience of the reader, many of the (alleged) heresies discussed in this volume are listed here by name of heresy or group of heretics, followed by information about the heresiarch (founder) or chief representatives, summary of doctrine, and cross-references to related heresies. Since there has always been much debate about the exact meaning of each heresy, this listing cannot claim to be complete or authoritative. The reader may find discussions of these heresies in this volume by using the index.

Albigenses (Albigensians)—named after town of Albi in southwest France. Antisacerdotal (denies powers of the clergy), ascetic theology first appeared in 1012–1020 in Limousin. Condemned by Council of Toulouse in 1119. Albigensian Crusade against them in thirteenth century; crushed by Inquisition in fourteenth century. Also called Cathars, and probably influenced by Bogomils and Manichaeans.

Anabaptists—Name for numerous groups that opposed infant baptism, also known as Baptists.

Anti-trinitarianism—Theology that contests the official Church doctrine concerning the Trinity (Father, Son, and Holy Ghost). Arians and Socinians were anti-trinitarian because they denied the equality of Father and Son.

Apollinarianism—Apollinarius the Younger, bishop of Laodicea (c.310–c.390). Denied the existence in Christ of a rational human soul, in order to combat Arianism.

Arianism—Arius, priest of Alexandria (c. 250–336). Emphasized the created, human nature of Christ, to be distinguished from divine nature of God, and thus God and his Son cannot be one. Condemned at Council of Nicaea (325 A.D.). See Anti-trinitarianism.

Bogomils—Bogomil, Balkan priest (tenth century A.D.). Morally austere neo-Manichaean dualists: the visible, material world created by the devil. Rejected baptism, the Eucharist, most Church institutions. Influenced Cathars, Albigenses.

Calvinism—Jean Calvin, founder of one major branch of Protestantism in Geneva (1509–1564). Stressed sinfulness of man, God's love. Rejected authority of Roman Catholic pope, criticized corruption of clergy. Differed with Luther about Eucharist, clergy, and role of good works.

Catharism—from Greek *katharos,* "pure." Morally austere neo-Manichaean dualists, rejected Roman Catholic Church as work of the devil. Known as Albigenses in France, also found in Italy, Germany, and Spain, influenced by Bogomil missionaries (eleventh to fifteenth centuries A.D.).

Donatism—Donatus, bishop of Carthage (d. ca. 355 A.D.). Moral ascetic, opposed state interference with church, rejected riches, inspired peasant rebels, sought maryrdom.

Epicureanism—Epicurus, Greek philosopher (341–270 B.C.); spread by Lucretius, Roman poet (ca. 95-ca. 50 B.C.). Atomist, materialist; soul dies with body; gods not concerned with the world.

Eutychianism—Eutyches, superior of monastery outside Constantinople (c.375–454). Emphasized the exclusive prevalence of divinity in Christ, against the Arians and Nestorians. Condemned at Council of Chalcedon in 451 A.D.

gnosticism—From Greek *gnosis,* "knowledge." Early figures include Apollos and Simon Magus (first century A.D.), Cerinthus, Marcion (100–160). Their "secret" knowledge included dualism: God and spirit are good; matter and body are evil.

Hobbism—Thomas Hobbes, English philosopher (1588–1679). Used materialist metaphysics to argue for subordination of church to princes.

Hussites—Followers of Jan (Jean) Hus, Moravian reformer (1373–1415). Proto-Protestant challenged corruption of Roman Catholics; emphasized Scripture, accepted some of Wycliffe's heresies. Burned at stake at Council of Constance; Pope Martin declared crusade against followers in 1520.

Lutheranism—Martin Luther, founder of one branch of Protestantism (1485–1546). Attacked corruption of Roman Catholic Church; translated Bible into vernacular. Salvation only by grace, not Calvinist works.

Magic—A heresy because it assumed supernatural powers beyond those accepted by other religions. Associated with devil worship and lewd behavior.

Manicheanism—Manes, or Mani, Iranian visionary and founder of religion (216–274(?) A.D.). Viewed the world as a fusion of spirit (good) and matter (evil). Ascetic. A type of gnosticism, related to Cathars, Albigensians.

Mennonites—Menno Simons, Dutch Anabaptist (1496–1561). Doubted Real Presence in Eucharist, rejected infant baptism.

Montanism—Montanus (second century A.D.). Ecstatic, prophetic, anti-materialist; rigorous morality. Tertulian (ca.155-ca.220 A.D.) was his most important follower.

Nestorianism—Nestorius, bishop of Constantinople (fifth century A.D.). Stressed double nature, divine and human, of Christ. Condemned at Council of Ephesus (431 A.D.).

Orthodox Catholic Church—also known as eastern orthodox, Byzantine Church. Church of Greek-speaking half of Roman Empire, split with Roman Catholics in 1054. Doctrine of Trinity different from Roman Catholic.

Pelagianism—Pelagius, British monk (c.354-after 418). Denied doctrine of original sin; stressed essential goodness of human nature and freedom of the will; believed human effort could lead to salvation. Condemned at Carthage in 418 and Ephesus in 431.

Priscillianism—Priscillian, bishop of Avila, first Christian heretic to receive capital punishment (c.340–385). Ascetic, similar to gnosticism and Manicheanism, dualistic doctrine that matter is evil and spirit good. Condemned at Synod of Bordeaux, 384.

Roman Catholicism—Christian church led by pope in Rome, claiming to derive from the Apostle Peter. Accused of heresy by orthodox, Protestants, Cathars, and returned charge.

Sabbatianism—Sabbatai Zevi (Shabbatai Sevi, Tzevi) (1626–1676), claimed to be the Jewish Messiah, but converted to Islam in 1666. Followers interpreted apostasy as a step toward full Messiahship, still to come.

Sabellianism—Sabellius (flourished 217–220 A.D.). Denied orthodox trinity. Excommunicated by Pope Calixtus. Followed by Miguel Servetus (1511–1553), who was executed by Calvin.

Socinianism—Laelius (1525–1562) and Faustus (1539–1604) Socinus. Italian reformers active in Switzerland, Poland, and Transylvania. Rationalist, anti-trinitarian theology.

Spinozism—Baruch (or Benedict) de Spinoza, philosopher and lens-grinder (1632–1677). Rejected Jewish law. Pantheistic, materialist, determinist philosophy was often understood as atheism. Biblical criticism was pioneering use of historical method on sacred text.

Waldensians (Valdenses)—Peter Waldo (Valdes), lay preacher at Lyons (flourished 1170–1184). Sect also known as "Poor of Lyons," banned by papal bull in 1184. Morally rigorous, emphasized poverty, also opposed Cathars.

Index

Printed in the United States
113690LV00002B/3/A

9 780312 294045